METHODS IN MOLECULAR BIOLOGY

Series Editor:
John M. Walker
School of Life and Medical Sciences
University of Hertfordshire
Hatfield, Hertfordshire, AL10 9AB, UK

For further volumes:
http://www.springer.com/series/7651

Mass Cytometry

Methods and Protocols

Edited by

Helen M. McGuire

Discipline of Pathology, The University of Sydney, Camperdown, NSW, Australia

Thomas M. Ashhurst

Sydney Cytometry Facility, The University of Sydney and Centenary Institute, Camperdown, NSW, Australia

Humana Press

Editors
Helen M. McGuire
Discipline of Pathology
The University of Sydney
Camperdown, NSW, Australia

Thomas M. Ashhurst
Sydney Cytometry Facility
The University of Sydney and Centenary Institute
Camperdown, NSW, Australia

ISSN 1064-3745 ISSN 1940-6029 (electronic)
Methods in Molecular Biology
ISBN 978-1-4939-9453-3 ISBN 978-1-4939-9454-0 (eBook)
https://doi.org/10.1007/978-1-4939-9454-0

This Humana Press imprint is published by the registered company Springer Science+Business Media, LLC, part of Springer Nature.
The registered company address is: 233 Spring Street, New York, NY 10013, U.S.A.

Preface

Mass cytometry is a state-of-the-art single-cell measurement technology with a multitude of applications. The use of heavy metal-conjugated antibodies allows for the detection of approximately 50 parameters per cell with minimal signal overlap. This has made mass cytometry the mostly highly multi-parametric method for the simultaneous measurement of protein at the single-cell level to date. The strength of this technology lies in its inherent flexibility, from the starting materials that can be examined to the types of markers that can be analyzed and methods of analysis that can be performed. The aim of this book was to assemble an up-to-date and cutting-edge compilation of protocols in mass cytometry, from the leading experts in the field. As a technology having only relatively recently emerged, these chapters capture the experience of those who pioneered the approach, as well as build on the collective knowledge that has seen it now actively embraced in the wider research community. As the application of mass cytometry has now reached a point of maturity in the field, these protocols represent both well-established methods and cutting-edge advances. The topics broadly cover the technology as a whole, considerations for implementation in a shared resource facility, experiment and panel design, as well as reagent preparation, specific examples of applications, and various tools and approaches used to analyze the resulting multi-parametric data. It is hoped that this book will have broad use for readers from a variety of backgrounds including from those that currently use the technology, as well as those who through the strength of examples illustrated throughout this book may now seek application of this technology in their own research in the future.

Camperdown, NSW, Australia

Helen M. McGuire
Thomas M. Ashhurst

Contents

Preface . . . . *v*
Contributors . . . . *ix*

Part I Setting Up a Facility

1 Setting Up Mass Cytometry in a Shared Resource Lab Environment 3
Sarah Warth and Désirée Kunkel

2 Acquisition, Processing, and Quality Control of Mass Cytometry Data 13
Brian H. Lee and Adeeb H. Rahman

Part II Panel Design and Reagent Preparation

3 Visualization of Mass Cytometry Signal Background to Enable Optimal Core Panel Customization and Signal Threshold Gating. 35
Amelia Au-Yeung, Chikara Takahashi, W. Rodney Mathews, and William E. O'Gorman

4 Method for Tagging Antibodies with Metals for Mass Cytometry Experiments. 47
Stephen Gregory Chang and Cynthia J. Guidos

5 Scalable Conjugation and Characterization of Immunoglobulins with Stable Mass Isotope Reporters for Single-Cell Mass Cytometry Analysis 55
Felix J. Hartmann, Erin F. Simonds, Nora Vivanco, Trevor Bruce, Luciene Borges, Garry P. Nolan, Matthew H. Spitzer, and Sean C. Bendall

6 Titration of Mass Cytometry Reagents 83
Caryn van Vreden, Paula Niewold, Helen M. McGuire, and Barbara Fazekas de St. Groth

7 Surface Barcoding of Live PBMC for Multiplexed Mass Cytometry 93
Axel Ronald Schulz and Henrik E. Mei

Part III Sample Preparation

8 Automated Cell Processing for Mass Cytometry Experiments. 111
Jaromir Mikes, Axel Olin, Tadepally Lakshmikanth, Yang Chen, and Petter Brodin

9 Live Cell Barcoding for Efficient Analysis of Small Samples by Mass Cytometry 125
Lisa E. Wagar

Part IV Specific Applications

10 Staining of Phosphorylated Signalling Markers Protocol for Mass Cytometry 139
Diana Shinko, Thomas M. Ashhurst, Helen M. McGuire, and Kellie A. Charles

11 Multiplex MHC Class I Tetramer Combined with Intranuclear Staining by Mass Cytometry 147
Yannick Simoni, Michael Fehlings, and Evan W. Newell

12 Analysis of the Murine Bone Marrow Hematopoietic System Using Mass and Flow Cytometry 159
Thomas M. Ashhurst, Darren A. Cox, Adrian L. Smith, and Nicholas J. C. King

13 Mass Cytometric Cell Cycle Analysis 193
Gregory K. Behbehani

14 Picturing Polarized Myeloid Phagocytes and Regulatory Cells by Mass Cytometry 217
Mikael Roussel, Todd Bartkowiak, and Jonathan M. Irish

15 Quantitative Measurement of Cell-Nanoparticle Interactions Using Mass Cytometry. 227
Andrew J. Mitchell, Angela Ivask, and Yi Ju

Part V Data Analysis

16 Data-Driven Flow Cytometry Analysis 245
Sherrie Wang and Ryan R. Brinkman

17 Analysis of Mass Cytometry Data 267
Christina B. Pedersen and Lars R. Olsen

18 Analysis of High-Dimensional Phenotype Data Generated by Mass Cytometry or High-Dimensional Flow Cytometry 281
Branko Cirovic, Natalie Katzmarski, and Andreas Schlitzer

19 Computational Analysis of High-Dimensional Mass Cytometry Data from Clinical Tissue Samples 295
Sam Norton and Roslyn Kemp

20 Supervised Machine Learning with CITRUS for Single Cell Biomarker Discovery. 309
Hannah G. Polikowsky and Katherine A. Drake

Correction to: Quantitative Measurement of Cell-Nanoparticle Interactions Using Mass Cytometry C1

Index *333*

Contributors

THOMAS M. ASHHURST • *Sydney Cytometry Facility, The University of Sydney and Centenary Institute, Camperdown, NSW, Australia; Viral Immunopathology Laboratory, Discipline of Pathology, Faculty of Medicine and Health, School of Medical Sciences, Charles Perkins Centre, The University of Sydney, Sydney, NSW, Australia; Ramaciotti Facility for Human Systems Biology (RFHSB), The University of Sydney and Centenary Institute, Sydney, NSW, Australia; Marie Bashir Institute for Emerging Infectious Disease (MBI), The University of Sydney, Sydney, NSW, Australia*

AMELIA AU-YEUNG • *OMNI Biomarkers, Development Sciences, Genentech, South San Francisco, CA, USA*

TODD BARTKOWIAK • *Department of Pathology, Microbiology and Immunology and Vanderbilt-Ingram Cancer Center, Vanderbilt University Medical Center, Nashville, TN, USA; Department of Cell and Developmental Biology, Vanderbilt University School of Medicine, Nashville, TN, USA*

GREGORY K. BEHBEHANI • *Division of Hematology, The Ohio State University and James Cancer Hospital, Columbus, OH, USA*

SEAN C. BENDALL • *Department of Pathology, Stanford University, Stanford, CA, USA; Immunology Program, Stanford University, Stanford, CA, USA*

LUCIENE BORGES • *Department of Pathology, Stanford University, Stanford, CA, USA*

RYAN R. BRINKMAN • *Terry Fox Laboratory, British Columbia Cancer Agency, Vancouver, BC, Canada*

PETTER BRODIN • *Science for Life Laboratory, Department of Women's and Children's Health, Karolinska Institutet, Stockholm, Sweden; Department of Newborn Medicine, Karolinska University Hospital, Stockholm, Sweden*

TREVOR BRUCE • *Department of Pathology, Stanford University, Stanford, CA, USA*

STEPHEN GREGORY CHANG • *The Hospital for Sick Children Research Institute, Toronto, ON, Canada*

KELLIE A. CHARLES • *Discipline of Pharmacology, School of Medical Sciences, Faculty of Medicine and Health, University of Sydney, Sydney, NSW, Australia*

YANG CHEN • *Science for Life Laboratory, Department of Women's and Children's Health, Karolinska Institutet, Stockholm, Sweden*

BRANKO CIROVIC • *Myeloid Cell Biology, Life and Medical Sciences Institute, University of Bonn, Bonn, Germany*

DARREN A. COX • *Viral Immunopathology Laboratory, Discipline of Pathology, Faculty of Medicine and Health, School of Medical Sciences, Charles Perkins Centre, The University of Sydney, Sydney, NSW, Australia; Marie Bashir Institute for Emerging Infectious Disease (MBI), The University of Sydney, Sydney, NSW, Australia*

KATHERINE A. DRAKE • *Cytobank, Inc., Santa Clara, CA, USA*

BARBARA FAZEKAS DE ST. GROTH • *Ramaciotti Facility for Human Systems Biology, The University of Sydney, Sydney, Australia; Charles Perkins Centre, The University of Sydney, Sydney, Australia; Discipline of Pathology, School of Medical Sciences, Faculty of Medicine and Health, The University of Sydney, Sydney, Australia*

MICHAEL FEHLINGS • *ImmunoSCAPE Pte. Ltd., Singapore, Singapore*
CYNTHIA J. GUIDOS • *The Hospital for Sick Children Research Institute, Toronto, ON, Canada; Department of Immunology, University of Toronto, Toronto, ON, Canada*
FELIX J. HARTMANN • *Department of Pathology, Stanford University, Stanford, CA, USA*
JONATHAN M. IRISH • *Department of Pathology, Microbiology and Immunology and Vanderbilt-Ingram Cancer Center, Vanderbilt University Medical Center, Nashville, TN, USA; Department of Cell and Developmental Biology, Vanderbilt University School of Medicine, Nashville, TN, USA*
ANGELA IVASK • *Laboratory of Chemical Physics and Biophysics, National Institute of Chemical Physics and Biophysics, Tallinn, Estonia*
YI JU • *Australian Research Council (ARC) Centre of Excellence in Convergent Bio-Nano Science and Technology and the Department of Chemical Engineering, The University of Melbourne, Parkville, VIC, Australia*
NATALIE KATZMARSKI • *Myeloid Cell Biology, Life and Medical Sciences Institute, University of Bonn, Bonn, Germany*
ROSLYN KEMP • *Department of Microbiology and Immunology, University of Otago, Dunedin, New Zealand*
NICHOLAS J. C. KING • *Sydney Cytometry Facility, Charles Perkins Centre, The University of Sydney, Sydney, NSW, Australia; Viral Immunopathology Laboratory, Discipline of Pathology, Faculty of Medicine and Health, School of Medical Sciences, Charles Perkins Centre, The University of Sydney, Sydney, NSW, Australia; Ramaciotti Facility for Human Systems Biology (RFHSB), The University of Sydney and Centenary Institute, Sydney, NSW, Australia; Marie Bashir Institute for Emerging Infectious Disease (MBI), The University of Sydney, Sydney, NSW, Australia; Australian Institute for Nanoscale Science and Technology, The University of Sydney, Sydney, NSW, Australia*
DÉSIRÉE KUNKEL • *BCRT Flow Cytometry Lab, Berlin-Brandenburg Center for Regenerative Therapies, Charité – Universitätsmedizin Berlin, Berlin, Germany*
TADEPALLY LAKSHMIKANTH • *Science for Life Laboratory, Department of Women's and Children's Health, Karolinska Institutet, Stockholm, Sweden*
BRIAN H. LEE • *Human Immune Monitoring Center, Icahn School of Medicine at Mt. Sinai, New York, NY, USA*
W. RODNEY MATHEWS • *OMNI Biomarkers, Development Sciences, Genentech, South San Francisco, CA, USA*
HELEN M. MCGUIRE • *Ramaciotti Facility for Human Systems Biology, The University of Sydney, Sydney, NSW, Australia; Charles Perkins Centre, The University of Sydney, Sydney, NSW, Australia; Discipline of Pathology, The University of Sydney, Camperdown, NSW, Australia*
HENRIK E. MEI • *Mass Cytometry Lab, German Rheumatism Research Center (DRFZ), A Leibniz Institute, Berlin, Germany*
JAROMIR MIKES • *Science for Life Laboratory, Department of Women's and Children's Health, Karolinska Institutet, Stockholm, Sweden*
ANDREW J. MITCHELL • *Materials Characterisation and Fabrication Platform, Department of Chemical Engineering, University of Melbourne, Melbourne, VIC, Australia*
EVAN W. NEWELL • *Agency for Science, Technology and Research (A*STAR), Singapore Immunology Network (SIgN), Singapore, Singapore; Fred Hutch Cancer Research Center, Vaccine and Infectious Disease Division, Seattle, WA, USA*

PAULA NIEWOLD • *Charles Perkins Centre, The University of Sydney, Sydney, Australia; Discipline of Pathology, School of Medical Sciences, Faculty of Medicine and Health, The University of Sydney, Sydney, Australia*

GARRY P. NOLAN • *Baxter Laboratory for Stem Cell Biology, Department of Microbiology and Immunology, Stanford University, Stanford, CA, USA*

SAM NORTON • *Department of Microbiology and Immunology, University of Otago, Dunedin, New Zealand*

WILLIAM E. O'GORMAN • *OMNI Biomarkers, Development Sciences, Genentech, South San Francisco, CA, USA*

AXEL OLIN • *Science for Life Laboratory, Department of Women's and Children's Health, Karolinska Institutet, Stockholm, Sweden*

LARS R. OLSEN • *Department of Health Technology, Technical University of Denmark, Lyngby, Denmark*

CHRISTINA B. PEDERSEN • *Department of Health Technology, Technical University of Denmark, Lyngby, Denmark*

HANNAH G. POLIKOWSKY • *Cytobank, Inc., Santa Clara, CA, USA*

ADEEB H. RAHMAN • *Human Immune Monitoring Center, Icahn School of Medicine at Mt. Sinai, New York, NY, USA; Department of Genetics and Genomic Sciences, Icahn School of Medicine at Mt. Sinai, New York, NY, USA*

MIKAEL ROUSSEL • *Univ Rennes, CHU Rennes, Inserm, MICMAC [(MIcroenvironment, Cell differentiation, iMmunology And Cancer)]—UMR_S 1236, EFS, Rennes, France*

ANDREAS SCHLITZER • *Myeloid Cell Biology, Life and Medical Sciences Institute, University of Bonn, Bonn, Germany*

AXEL RONALD SCHULZ • *Mass Cytometry Lab, German Rheumatism Research Center (DRFZ), A Leibniz Institute, Berlin, Germany*

DIANA SHINKO • *Discipline of Pharmacology, School of Medical Sciences, Faculty of Medicine and Health, University of Sydney, Sydney, NSW, Australia*

ERIN F. SIMONDS • *Department of Neurology, University of California San Francisco, San Francisco, CA, USA*

YANNICK SIMONI • *Agency for Science, Technology and Research (A*STAR), Singapore Immunology Network (SIgN), Singapore, Singapore; Fred Hutch Cancer Research Center, Vaccine and Infectious Disease Division, Seattle, WA, USA*

ADRIAN L. SMITH • *Sydney Cytometry Facility, Charles Perkins Centre, The University of Sydney and Centenary Institute, Sydney, NSW, Australia; Ramaciotti Facility for Human Systems Biology (RFHSB), The University of Sydney and Centenary Institute, Sydney, NSW, Australia*

MATTHEW H. SPITZER • *Department of Microbiology and Immunology, University of California San Francisco, San Francisco, CA, USA*

CHIKARA TAKAHASHI • *OMNI Biomarkers, Development Sciences, Genentech, South San Francisco, CA, USA*

CARYN VAN VREDEN • *Sydney Cytometry, The University of Sydney and Centenary Institute, Sydney, Australia; Ramaciotti Facility for Human Systems Biology, The University of Sydney, Sydney, Australia; Charles Perkins Centre, The University of Sydney, Sydney, Australia*

NORA VIVANCO • *Immunology Program, Stanford University, Stanford, CA, USA*

LISA E. WAGAR • *Stanford University School of Medicine, Department of Microbiology and Immunology, Stanford, CA, USA*

SHERRIE WANG • *Terry Fox Laboratory, British Columbia Cancer Agency, Vancouver, BC, Canada*

SARAH WARTH • *BCRT Flow Cytometry Lab, Berlin-Brandenburg Center for Regenerative Therapies, Charité – Universitätsmedizin Berlin, Berlin, Germany; Core Facility FACS/MC, Ulm University Medical Center, Ulm, Germany*

Part I

Setting Up a Facility

Chapter 1

Setting Up Mass Cytometry in a Shared Resource Lab Environment

Sarah Warth and Désirée Kunkel

Abstract

Core facilities or shared resource laboratories (SRL) ensure fast and direct access to high-end instrumentation that is often expensive and requires technical expertise to be operated and maintained. SRL have the responsibility and capability to ensure the possibility of generating reproducible quality data by implementing best practices, providing expert knowledge, and keeping the necessary instrumentation in good shape. Here we review critical steps involved in the integration of mass cytometry in a shared resource lab environment. Lab requirements and considerations for best practices are discussed, and critical aspects of mass cytometry project discussions as well as strategies for the management and documentation of samples are presented.

Key words Mass cytometry, CyTOF, Helios, Shared resource lab, Core facility, Best practices

1 Introduction

The integration of a new technology in a shared resource lab environment, which might also already contain other technologies, is challenging. One of the very first aspects to consider is the operation model. There are several options that later dictate accessibility to the lab, user guidelines, best practices, and staff availability. The mass cytometer (the latest version of the commercially available instrument is called *Helios*, earlier versions are *CyTOF* and *CyTOF2*; Fluidigm) can either be run completely staff operated or users can be trained to run the instrument by themselves. This can be restricted to frequent users or apply to all users. Depending on the operation model, procedures in education and training of users and staff need to be adapted. Moreover, the SRL needs to decide which services can be provided, considering lab space and staff availability. Best practices and standard operating procedures need to be developed and continuously assessed and adapted to the latest developments in the field. There are several ways to stay up to date with respect to these developments, including of course by

Helen M. McGuire and Thomas M. Ashhurst (eds.), *Mass Cytometry: Methods and Protocols*, Methods in Molecular Biology, vol. 1989, https://doi.org/10.1007/978-1-4939-9454-0_1, © Springer Science+Business Media, LLC, part of Springer Nature 2019

keeping in touch with other mass cytometry labs and by attending national and international conferences. There is also a very useful online mass cytometry user forum (cytoforum.stanford.edu).

2 Operation Model and Services

The mass cytometer can either be run completely staff operated or users can be trained to run the instrument by themselves. In the latter case, education and training programs for the users need to be developed.

Services that can be provided might include:

- Panel development.
- Conjugation of antibodies with metal isotopes (this can include or exclude validation of antibodies and titration).
- Establishment of in-house "core" panels that can be provided to the users to be filled up with experiment-specific markers (e.g., common leukocyte markers included in almost all immunological phenotyping panels).
- Full service for samples: isolation of cells, staining of cells, and measurement.
- Semi-full service: staining of cells and measurement.
- Fully prepared samples only: samples are delivered in the pellet after the last water washing step.
- Data pre-processing (this might include normalization of data with beads according to published methods [1] or the software provided by the manufacturer, de-barcoding, compensation [2]).
- Data analysis.

3 Requirements for the Laboratory Setup

While the preparation of cells for measurement on a mass cytometer is in many ways similar to the preparation required for fluorescent flow cytometry, the instrument itself has certain specifications that require room conditions that are very different to those for a flow cytometer. Before installation of the mass cytometer, an appropriate room setting needs to be ensured. A *site preparation checklist* is available from Fluidigm on how to prepare your lab for installation of the mass cytometer with respect to the technical aspects and it is strongly advised to carefully read this and make sure all conditions are met. Always ask for the latest version of this document.

The technical requirements include:

- Electricity (*see* **Note 1**).
- Argon supply (this can be cylinders of gas or liquid argon dewars, *see* **Note 2**).
- Ventilation (*see* **Notes 3** and **4**).
- Temperature regulation.
- Noise: the instrument and most of all the chiller produce a lot of noise, that is below the level that is physically harmful for the ears but can be stressful, especially if the mass cytometer is placed in a small room or operated with many other people working in the same room. The provision of hearing protection might be advisable (*see* **Note 5**).
- Space requirements: the instrument should be easily accessible by users and staff without getting in each other's way. The Helios has a very delicate sample capillary sticking out at the front of the instrument, so it is advisable to restrict the number of people passing it to a minimum.
- Interaction with other equipment: if the mass cytometer is placed in a room together with other equipment it needs to be verified that all instruments can run under the same conditions. The mass cytometer and the connected chiller produce a lot of heat. For equipment that needs tightly controlled temperature conditions, e.g., a cell sorter, the air conditioning system must be adapted to that (*see* **Note 6**).
- Data storage requirements (*see* **Note** 7).

4 Equipment and Materials that Should be in Place

The SRL can provide full service, including isolation of cells and staining of the samples, or accept fully prepared samples only. The equipment for staining the samples is similar to the equipment needed for flow cytometry and has been discussed elsewhere [3].

In addition, there are several things the SRL should have in place for preparation of instrument and samples before measurement:

- A torch flashlight for examination of the nebulizer spray.
- Spare parts for quick change of material if something breaks or clogs (*see* **Notes 8** and **9**).
- Small containers to keep nebulizers in water/Contrad/decon.
- Access to ultrapure (e.g., MilliQ) water for washes and final suspension of cells.
- Bench space for handling the reagents needed for washing and tuning of the mass cytometer and for cleaning its glassware and materials (*see* **Note 10**).

- A refrigerator to store the samples until they are run (*see* **Note 11**).
- Bench space for preparation of the samples directly before running them on the mass cytometer (resuspension in water, filtration, adding normalization beads, etc.).
- A device for cell counting (*see* **Note 12**).
- Cell strainers (*see* **Note 13**).
- Filtered pipette tips.

5 Best Practices in a Mass Cytometry Shared Resource Lab

There are multiple benefits to an SRL by implementing best practice standards in the lab. Leading to a more effective SRL, consistent production of quality data and provision of high-level services ultimately increase the probability of generating reproducible data for the users. Best practices for a mass cytometry SRL are essentially the same as for a flow cytometry SRL. A recent publication provides guidelines for setting up such best practices [4]. Here we present some possibilities for quality assurance in mass cytometry.

5.1 Instrument Quality Control and Performance Tracking

Before each use, a quality control of the instrument should be performed and documented for performance tracking. This quality control should at least include:

- Contamination check: before the instrument is tuned to the manufacturer's instructions, a short preview with ultrapure water will show remaining metal contamination (*see* **Note 14**).
- Tuning results show if the instrument is running according to specifications given by the manufacturer. Documenting these results each time the machine is tuned allows for comparison and checking against previous performance (*see* **Note 15**).
- Quality control: EQ four element beads from Fluidigm can be run for a defined period of time (e.g., 2 min) to control for yield and sensitivity, and can serve as a quality control before each experiment (*see* **Note 16**).
- Guidelines must be specified to help troubleshoot when any of the above quality assurance steps are not met.

5.2 Cleaning Procedures and Preventative Maintenance

Standard operating procedures should be in place for regular cleaning and preventative maintenance procedures. These should include:

- Washing steps between samples, taking into account the different sample types that might be run in the facility (*see* **Note 17**).
- Daily cleaning procedures (*see* **Note 18**).

- Weekly cleaning procedures (or more specifically, cleaning after a defined quantity of running time in hours or samples) (*see* **Note 19**).

5.3 Documentation

All instrument performance evaluations, quality control results, and maintenance interventions should be carefully documented. Especially if the instrument is solely staff operated, the user will rely on the operator to have this information ready to help troubleshooting if any anomalies are apparent when analyzing the data.

The SRL should also provide the user with the necessary information for the description of the mass cytometry experiment, based on the MiFlowCyt recommendations for flow cytometry experiments [5].

5.4 Project Discussions for Mass Cytometry Experiments

If the operation model of the SRL is to handle fully prepared samples only, the project discussion is one of the most important steps to ensure good sample quality and to reduce the risk of experimental errors. In our experience, if panel development is required, project discussions should be initiated at least 3 months before the first experimental samples are expected. If possible, the project discussion should occur even before samples are collected and stored, to be able to discuss optimal sample preparation and storage conditions beforehand.

The following points might be addressed during these meetings:

- Source of sample material (e.g., liquid biopsy, tissue, cell culture, particles other than cells) (*see* **Note 20**).
- State of sample material (fresh, already collected and stored, frozen or fixed) (*see* **Note 21**).
- Contamination of sample material with endogenous metals, e.g., from contrast agents, or with residual beads from pre-enrichment.
- Whether the antibodies will be surface stained only (for extracellular antigen detection) or the staining protocol will also incorporate intracellular/intranuclear steps (*see* **Note 22**).
- Whether the acquisition of samples will be done one by one or through means of sample barcoding, pooled and run together. Staining in barcoded batches, achieved by several different methods [6–8], offers the advantage of staining consistency.
- Whether "batch controls" or reference/"anchor" samples will be implemented for longitudinal experiments [9].
- The number of cells available per sample (*see* **Note 23**).
- The number of cells to be acquired per sample (*see* **Note 24**).
- Panel design, sources of spillover, validation, and titration of antibodies.

- Sample preparation protocols starting from the isolation of cells (*see* **Note 25**).
- Protocols for the different staining procedures.
- Stress the point of proper fixation of the cells to prevent lysis once they are resuspended in water (*see* **Note 26**).
- Data management and the different file types (*see* **Note 7**).
- Possibilities of data pre-processing (normalization [1], de-barcoding, compensation [2]).
- The options for data analysis.

5.5 Management of Samples and Documentation

Sample handover needs to be carefully documented. This starts with the announcement of a sample by the user and the information the SRL will need to prepare sample acquisition. The user should clearly define the file name and panel template for the sample to be acquired on, the expected cell number and events to be acquired, with clear notification of any deviations from what was agreed in the project planning discussion.

The SRL should document the actual cell count after resuspension of the sample in water directly before measurement, the adjusted cell concentration, and the resulting events/second during the acquisition. If a lot of background from residual free antibody is detected ("streaking"), a screenshot might be saved with the data for later reference. Any irregularities that happen during sample acquisition should be documented and the information should be passed on to the user if it might impact the experiment (*see* **Note 27**).

6 Notes

1. Check current local regulations, conditions may be different from those described for the USA.
2. A 50 L gas cylinder will last for approximately 8 h of operation. Having a system whereby two argon gas cylinders are utilized concurrently and that change automatically to a second set of two cylinders when empty, is advisable.
3. While the site preparation checklist only specifies exhaust values, be reminded that you will need equivalent specifications for air supply.
4. In our experience the most critical step is the venting system, and its installation often requires lengthy discussions with the engineers until it runs to the required specifications.
5. The use of individual in-ear hearing protection for musicians is best suited, as this will allow the operator to hear everything, but less loudly. That way communication is still possible.

6. In our experience running a cell sorter and a mass cytometer in the same room is perfectly possible; however sorting might need to be paused during the short period of time the air condition system needs to adjust the room temperature after starting the mass cytometer.
7. It will depend on your university's regulations as to which type of data needs to be kept for good laboratory practices. We advise the users to keep .imd files and .FCS files as raw data and to ask IT support for means to compress the .imd files.
8. Spare parts for Helios should include at least a set of two nebulizers, two sample capillaries, two sets of sample/probe lines, one glass torch body, one injector, and one set of cones.
9. The rather rigid sample and probe line may be replaced by a set of flexible sample lines from a BD FACSVerse instrument (Becton Dickinson). This way clogs are much easier to prevent and the line can be easily backflushed because the inner core has a wider diameter.
10. The reagents used (acidic/basic) might require special coating of the tabletops. This needs to be discussed with the local laboratory safety compliance authorities.
11. It is advisable to resuspend samples one by one and run samples with a large number of cells in aliquots of no more than 1 h runs (approx. one million cells), as the cells tend to disintegrate over time if they are kept in water, even if fixed well.
12. It is important to adjust the cell concentration directly before running the sample. A sample that is too dense might immediately block sample capillary and nebulizer, resulting in time-consuming efforts to clean and de-clog or the requirement to replace items that are very expensive. Optimal sample concentration might differ depending on the source of the cells or cell type. Isolated cells from liquid biopsies (e.g., blood, liquor, urine) are less prone to form clumps than cells isolated from tissue and can be run at higher concentrations.
13. All samples must be filtered through at least a 35 μm cell strainer, better even through a 20 μm cell strainer directly before measurement. This will minimize the risk of clump formation and clogging.
14. The xenon signal as an impurity of the argon will also indicate if a signal will be seen at all before running the acidic tuning solution. If there is no xenon signal, troubleshooting might include verification of the sample capillary and nebulizer spray for irregularities. This is also a good quality check for impurity of the argon gas.
15. Values to include in the documentation could be Tb count and %rCV (as one reference isotope), resolution, gases (nebulizer/

make-up), and oxides. A screenshot of the tuning result might be saved for documentation.

16. This is also a control for the beads themselves, as these are the same beads that will be run with the sample for later normalization. A low doublet count should be present.
17. This is dependent on signal intensity, background, and overall quality of the samples. Time for washing with washing solution and water needs to be validated by controlling for residues of metals and cells before running the next sample. When running a titration, begin with the control, then samples with increasing antibody concentrations. This way, washing between samples can be reduced to a minimum as a little carry-over of cells and metals does not interfere with the outcome.
18. Daily cleaning includes running washing solution and water until the instrument is clean, cleaning of the nebulizer, and back-flushing the sample capillary and sample/probe line.
19. This includes cleaning of cones, torch, injector, and spray chamber. To speed up this process a second set of torch and injector can be used. A second set of cones should at least have been tested to facilitate the change once the cones are worn out.
20. This information gives an idea about the probability of clogging, maximum cell concentration, and thus length of sample acquisition. If particles other than cells (e.g., compensation beads) are measured, the minimum event duration might need to be adapted.
21. Not all antigens will be detected after fixation of cells with paraformaldehyde. There might be loss of sensitive cell populations after freezing.
22. The cell number that is needed to start the staining should be adapted to this, as a lot more cells are lost during intracellular and intranuclear staining.
23. Below a certain number of starting cells, barcoded "filler" cells should be added to the sample to reduce the loss of cells during the staining procedure, which is much more pronounced than in flow cytometry due to the extensive number of washes. Make sure to accurately calculate the ratio of target cells and filler cells to later know the number of total cells that need to be in the data file to contain the desired quantity of target cells.
24. This dictates the maximum number of samples that can be run per day, especially if the instrument is staff operated. There is a maximum number of events/second that is around 300–500 events/s. Also keep in mind that the instrument will need cleaning after a certain number of cells have passed through.

25. Centrifugation speed for living cells above 300 × *g* might lead to a compromised cell membrane and later to clump formation and clogging of the instrument.
26. We recommend an overnight step in 2–4% paraformaldehyde for all samples regardless of the type of staining procedure. The volume needs to be adapted to the total cell number. Samples with very high cell numbers (>10 million cells) should be placed on a rotator to ensure proper fixation.
27. Documentation of some parameters during sample acquisition especially in the beginning is also helpful to get to know the "normal" state of the instrument and to detect reappearing issues that might be linked to certain cell types or sources of samples. Values to document include the sample acquisition pressure during the run and any variations in pressure and sample flow rate.

Acknowledgments

The CyTOF2/Helios in the BCRT Flow Cytometry Lab was funded by DFG grant INST 335/453-1 FUGG.

References

1. Finck R, Simonds EF, Jager A et al (2013) Normalization of mass cytometry data with bead standards. Cytometry A 83(5):483–494. https://doi.org/10.1002/cyto.a.22271
2. Chevrier S, Crowell HL, Zanotelli VRT et al (2018) Compensation of signal spillover in suspension and imaging mass cytometry. Cell Syst 6(5):612–620 e615. https://doi.org/10.1016/j.cels.2018.02.010
3. Leipold MD, Newell EW, Maecker HT (2015) Multiparameter phenotyping of human PBMCs using mass cytometry. Methods Mol Biol 1343:81–95. https://doi.org/10.1007/978-1-4939-2963-4_7
4. Barsky LW, Black M, Cochran M et al (2016) International Society for Advancement of Cytometry (ISAC) flow cytometry shared resource laboratory (SRL) best practices. Cytometry A 89(11):1017–1030. https://doi.org/10.1002/cyto.a.23016
5. Lee JA, Spidlen J, Boyce K et al (2008) MIFlowCyt: the minimum information about a flow cytometry experiment. Cytometry A 73(10):926–930. https://doi.org/10.1002/cyto.a.20623
6. Mei HE, Leipold MD, Schulz AR et al (2015) Barcoding of live human peripheral blood mononuclear cells for multiplexed mass cytometry. J Immunol 194(4):2022–2031. https://doi.org/10.4049/jimmunol.1402661
7. Zunder ER, Finck R, Behbehani GK et al (2015) Palladium-based mass tag cell barcoding with a doublet-filtering scheme and single-cell deconvolution algorithm. Nat Protoc 10(2):316–333. https://doi.org/10.1038/nprot.2015.020
8. McCarthy RL, Mak DH, Burks JK et al (2017) Rapid monoisotopic cisplatin based barcoding for multiplexed mass cytometry. Sci Rep 7(1):3779. https://doi.org/10.1038/s41598-017-03610-2
9. Kleinsteuber K, Corleis B, Rashidi N et al (2016) Standardization and quality control for high-dimensional mass cytometry studies of human samples. Cytometry A 89(10):903–913. https://doi.org/10.1002/cyto.a.22935

Chapter 2

Acquisition, Processing, and Quality Control of Mass Cytometry Data

Brian H. Lee and Adeeb H. Rahman

Abstract

Mass cytometry uniquely combines the principles of mass spectrometry and flow cytometry for high dimensional profiling of immune cells at a single cell level. Using isotopically conjugated antibodies, mass cytometry overcomes the limitations of spectral overlap associated with flow cytometry and allows for deeper single cell characterization of complex biospecimens using more cellular markers. However, the nature of mass spectrometry-based single cell measurements requires specific considerations in acquiring and processing data. This chapter provides an overview of how to optimally acquire mass cytometry data and how to process this data for subsequent analysis and characterization of cell populations.

Key words Mass cytometry, Data acquisition, Data processing, Normalization, Barcoding

1 Introduction

Mass cytometry is able to analyze over 40 different cellular markers in a single sample using time-of-flight inductively coupled plasma mass spectrometry (TOF ICPMS). This technology allows researchers to combine the high-throughput single cell analysis approach of conventional flow cytometry with the high sensitivity and mass resolution of mass spectrometry to perform higher dimensional analyses. While the final processed outputs of mass cytometry data are similar to conventional flow cytometry, the way in which these data are generated results in some specific data processing considerations. This chapter discusses how mass cytometry data are generated and provides methods to optimize the acquisition and processing of data to ensure measurement of cellular markers with high resolution, quality, and reproducibility.

Helen M. McGuire and Thomas M. Ashhurst (eds.), *Mass Cytometry: Methods and Protocols*, Methods in Molecular Biology, vol. 1989, https://doi.org/10.1007/978-1-4939-9454-0_2,

1.1 Ion Cloud Formation and TOF Measurement

A liquid cell suspension stained with metal-conjugated antibodies is first introduced into the "Helios" mass cytometry system (Fluidigm) using a nebulizer, which aerosolizes the liquid cell suspension into single cell-containing droplets. The droplets are pushed into a heated spray chamber, partially vaporizing the sample and directing it into the inductively coupled plasma (ICP) torch. As the aerosolized sample enters the plasma, the cells are completely vaporized, atomized, and ionized into their constituent metal ions, resulting in the formation of an ion cloud. The ion cloud is sampled through a vacuum cone interface into a quadrupole ion deflector where low-mass ion and photons are filtered out, primarily leaving only the high-mass antibody-derived ions of interest. After the ion cloud has been filtered, it is sent to the TOF mass analyzer. Every 13 μs [1], voltage is applied to a push-out plate, pushing the ions into the TOF detector. The detector then analyzes the time of flight based on each ion's mass-to-charge ratio and the signal is converted into a digital value.

1.2 IMD/FCS Generation

After a sample has been acquired, the data are digitized as a raw integrated mass data (IMD) file, which represents a matrix of ion counts for each selected mass channel for every push (Fig. 1a). The Fluidigm software then converts the IMD to a flow cytometry standard (FCS) file, in which the data from individual pushes are integrated into mass-specific ion counts for discrete cellular "events" based on user-adjustable parameters, such as the minimum and maximum number of pushes for event length (typically 10–150 pushes on a Helios mass cytometer) (Fig. 1b). During the conversion, regions of the IMD comprised of undetectable ions (zero data) or pulses that are below the minimum or beyond the maximum event length are removed. Pulses that meet the event threshold criteria are converted to events. Thus, the resulting FCS file contains total integrated ion counts for every selected channel for every event (Fig. 1c) and can be analyzed using FlowJo, Cytobank, or other available third party cytometry analysis software. While events are generally expected to correspond to single cells, specific data processing steps are required to maximize data quality and filter potential debris and cell multiplets prior to data analysis.

2 Materials

2.1 Data

1. IMD and FCS files acquired on a Fluidigm CyTOF system (CyTOF, CyTOF2, or Helios).

2.2 Software

1. Fluidigm CyTOF® Software v6.7.

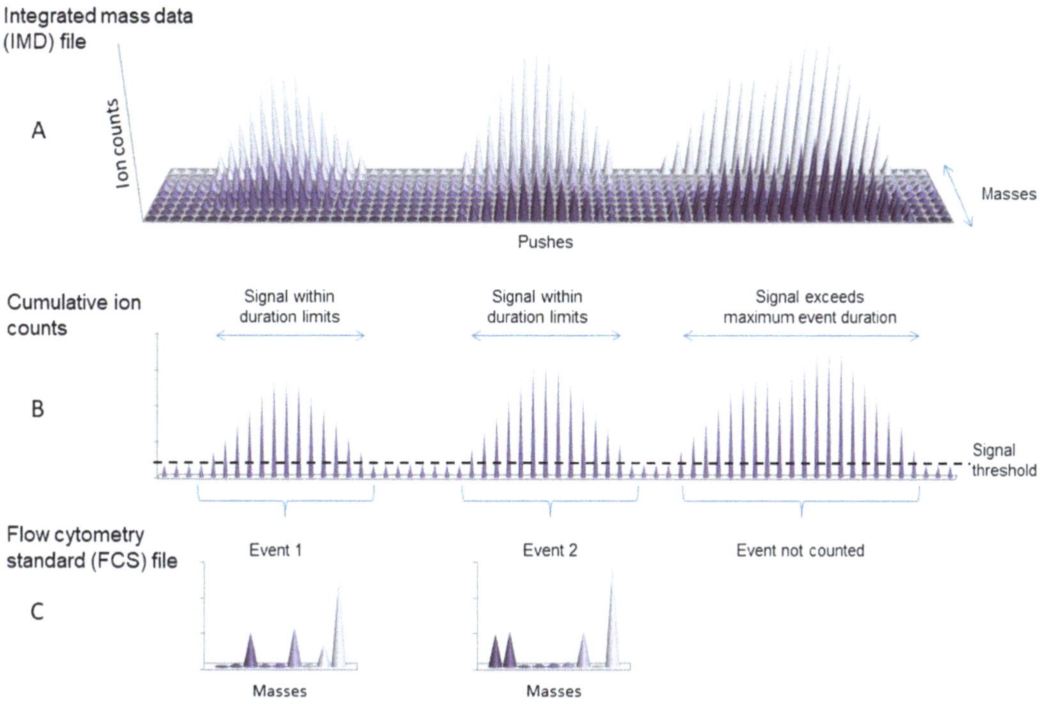

Fig. 1 Signal processing. (**a**) Individual push data from every selected mass channel is integrated into ion counts. (**b**) During conversion to an FCS file, pulses that meet the user-defined event threshold criteria are converted into events in the FCS file, together with the integrated signal for each mass channel. (**c**) Pushes that do not meet the minimum and maximum thresholds are not recorded as events in the FCS file

2. Nolan Lab bead-based normalizer: https://github.com/nolanlab/bead-normalization/releases.
3. Zunder Lab Single-Cell Debarcoder: https://github.com/zunderlab/single-cell-debarcoder.
4. Ismmshimc Cytutils tools for data quality and reproducibility: https://github.com/ismmshimc/cytutils.
5. Cytobank: https://www.cytobank.org/.

2.3 Solutions

1. EQ Four Element Calibration Beads (Fluidigm).
2. CyTOF Tuning Solution (Fluidigm).
3. Ultrapure water.

3 Methods

3.1 Sample Preparation and Acquisition

While this chapter focuses primarily on data processing, it is important to follow certain best practice guidelines when preparing and acquiring samples to ensure high quality data and reproducibility (*see* **Notes 1–3**).

3.2 Initial Sample Quality Control (QC)

Changes in instrument performance can occur between samples and even within a single acquisition of a sample. This can be due to gradual loss of detector sensitivity, build-up of salts or cellular debris in the instrument, or changes in plasma ionization efficiency. Fluidigm's EQ Four Element Calibration Beads are polymer beads that contain known standards of four elements at natural isotopic abundance (cerium, europium, holmium, and lutetium). Several metrics can be evaluated using EQ bead-derived signals to track daily instrument performance and to monitor changes during sample acquisition using conventional cytometry analysis software or automated tools (*see* **Note 4**).

3.2.1 Bead Intensity

1. Identify the EQ bead singlet population by plotting Ce140 vs. DNA and gating on the Ce140+ DNA− events.
2. Determine the median signal intensity of Eu153 on the beads and compare it to a reference bead sample collected immediately following instrument tuning at the start of the day.
3. If the median has dropped by more than a user-defined threshold, retune the detector voltage to an acceptable Eu153 median. We use a 25% drop, or a minimum Eu153 intensity of 1500 as our threshold.

3.2.2 Coefficient of Variance (CV)

1. The CV represents the precision of detection of repeated measurements. Given that beads should have identical ion content, the CV of the beads reflects the amount of error in measurement. High bead CVs would reflect a reduced ability to accurately resolve small changes in protein expression levels between cells (Fig. 2a, b). While minor variations will occur, the instrument should be properly tuned and optimized to minimize this variance.
2. To determine the CV, gate on the EQ Bead population and calculate the CV of Eu153.
3. Determine the CV value and if it is above a user-defined threshold, retune. We use a CV of 13 as our limit.

3.2.3 Oxidation

1. Oxidation of metal isotopes will occur during acquisition, resulting in signal from a given isotope being detected in its +16 dalton (Da) mass channel (e.g., Nd146 oxides will be recorded as Dy162). While oxidation is typically <5% for most isotopes, this can nevertheless be a significant source of isotopic cross-talk and should be considered when designing antibody panels [2] (*see* **Note 5**). Relative amounts of oxide formation are affected by gas flows and plasma characteristics, and instrument performance should be optimized to minimize oxide formation relative to detection sensitivity (*see* **Note 6**). EQ beads can be used to monitor changes in oxide formation during sample acquisition.

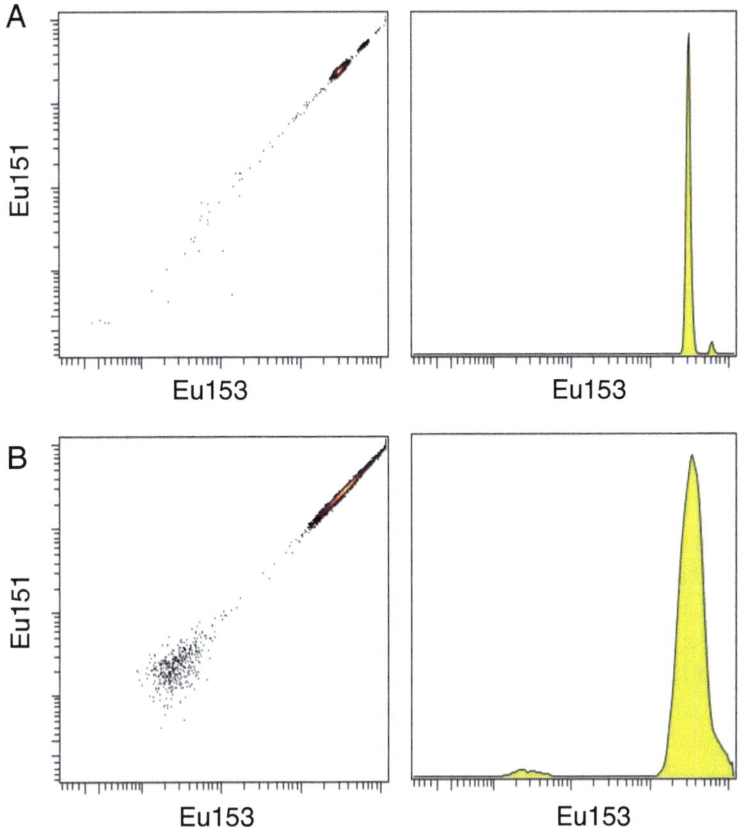

Fig. 2 EQ beads. (**a**) Example of a low CV in the EQ bead population resulting in well-resolved singlet and doublet bead populations. (**b**) Example of a high CV resulting in poor bead resolution, reflecting low precision of repeated bead measurements

2. To determine the oxide ratio calculate the ratio between the dual counts of Ce140 and Gd156 in the EQ beads population.

3.2.4 Signal Stability

1. Cell signal intensity may decrease during acquisition as a result of instrument performance or sample degradation. While instrument performance can be tracked using EQ beads, cell-specific degradation cannot.
2. Identify the Ce140+ DNA− EQ bead population and the Ce140−DNA+ cell population.
3. For each of these populations, plot time vs. a measured parameter (e.g., Eu153 for beads or CD45 for cells) and calculate the slope.
4. A slight negative slope is generally expected, reflecting gradual loss of instrument sensitivity over time. A steeper slope for both cells and beads reflects a more rapid loss of instrument sensitivity over the period of acquisition, while a steeper slope for

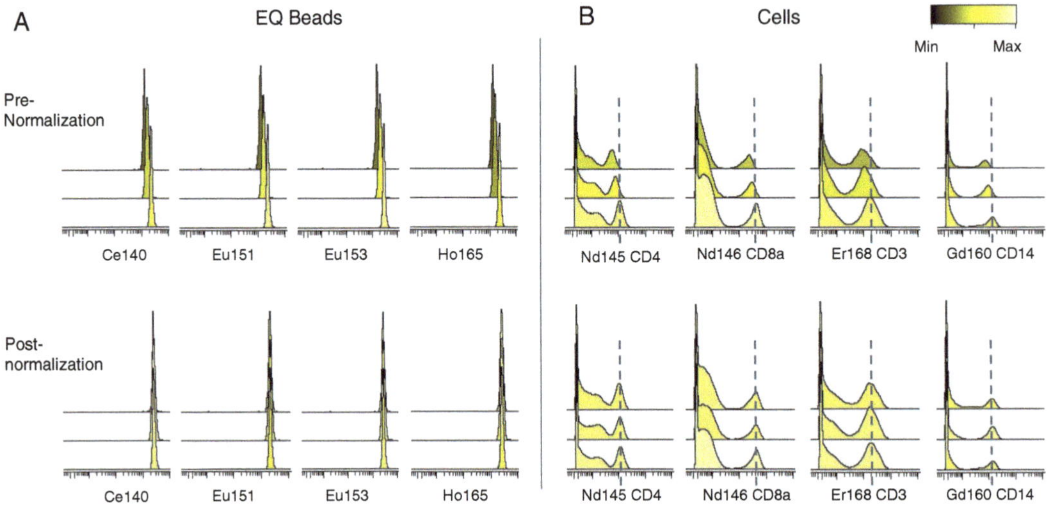

Fig. 3 Data normalization. Three aliquots of the same sample were spiked with EQ beads and acquired on different days on different instruments. (**a**) Initial signal intensities for beads vary depending on instrument performance, but normalization to the Helios global standard passport results in matching bead intensities. (**b**) The cellular signal in the same sample is correspondingly adjusted, resulting in more closely matched median signal intensities. Histograms are colored based on normalized relative median intensity

measured cell parameters relative to the measured bead parameters can indicate sample degradation (e.g., due to poor fixation). Samples with slopes exceeding specific thresholds may be flagged for sample quality issues.

5. Transient spikes in cell recovery or signal intensity over time may be indicative of micro-clogs during acquisition. These time windows should be excluded from analysis since they can be misinterpreted as false positive signals.

3.3 Normalization

Once a sample has been acquired and undergone initial QC, several steps must be taken to process the data for analysis. As mentioned previously, instrument performance can vary within a single sample. To address this issue, EQ bead normalization must be performed to account for this technical variability in order to better represent real biological differences between samples (Fig. 3a, b). Normalization can be accomplished using either a Matlab-based utility developed by the Nolan lab (https://github.com/nolanlab/bead-normalization/releases) or the normalization tool included as part of the Fluidigm CyTOF software. The Fluidigm normalization tool utilizes a "Bead passport" that contains target values for mean dual counts for Ce140, Eu151, Eu153, Ho165, and Lu175 from a specified EQ bead lot number established based on the mass response curve of a reference mass cytometer. Using the normalization tool in the CyTOF software, bead signal is measured throughout an acquisition and a median is calculated within

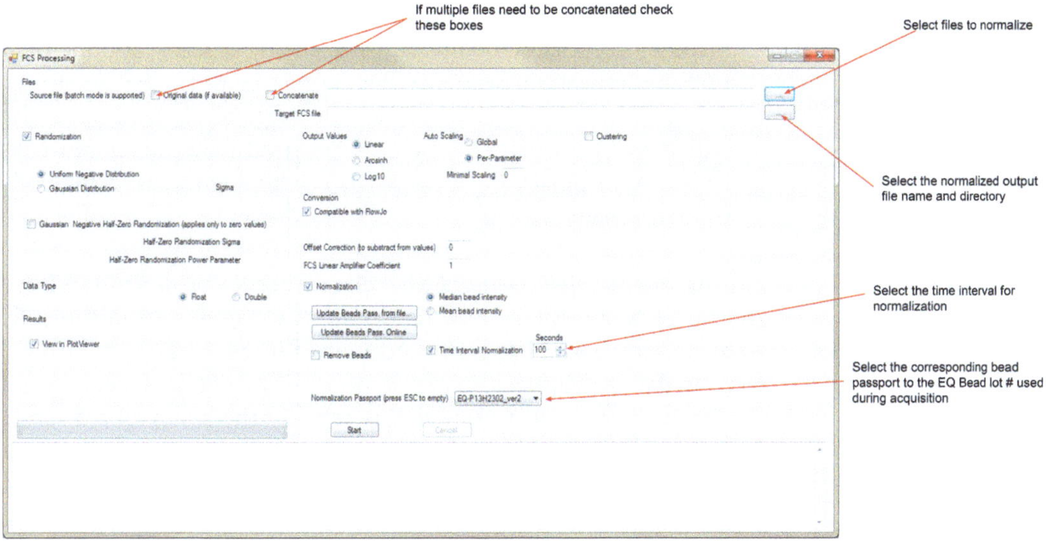

Fig. 4 Fluidigm CyTOF v6.7 normalization and concatenation tool

a user-defined time interval. To adjust bead signal to the global Helios passport standard, a slope is calculated against the global mean in each interval by assuming a linear rate of decay in signal. Every event within the interval is multiplied by the slope, normalizing bead signal to the global passport [3] (*see* **Note 7**).

1. Normalization of the raw FCS files is available in the Fluidigm CyTOF software. Select the "Process" tab in the software and select FCS processing.
2. Select the corresponding bead passport to the EQ bead lot number used during acquisition.
3. Select the time interval that the normalization factor will be applied to. By default, the software will use a 100 s interval (*see* **Note 8**).
4. Create a file name for the output normalized FCS and select the directory to be saved in.
5. Select start to begin normalization (refer to Fig. 4).

3.4 Concatenation

During acquisition a sample may need to be acquired in multiple rounds due to a high volume of sample (*see* **Note 2**), clogs, or unexpected machine shutdown. In order to combine these multiple acquisitions of a single sample, FCS files can be concatenated into a single file during normalization.

1. In the same FCS processing tab used for normalization in the Fluidigm CyTOF software, select the FCS files to concatenate and check the original data (if available) box and set a filename and directory for the concatenated file. Keep in mind that when using the normalization software to concatenate, it will automatically normalize the data (Fig. 4).

2. Cytobank also offers an FCS concatenation tool available on https://support.cytobank.org/hc/en-us/articles/206336147-FCS-file-concatenation-tool. Data must be normalized prior to using this tool. This tool allows the original file number to be added as a parameter in the concatenated FCS file so that cells from the original FCS files can be readily identified in the concatenated file.

3.5 Randomization

During normalization, users can select whether or not to randomize the FCS file. By default, randomization is selected in the software. When visualizing non-randomized integer values in an FCS file, the data will appear "picket-fenced." The blank pickets should not be confused as 0 values as it is actually the result of a discrete integer value being plotted on a continuous scale. To account for this, randomization is applied to every value, redistributing the integer into a randomized continuous number within a 0–1 range. For example, an ion count of 10 will be randomly distributed in the interval between 9 and 10 (excluding 9 and including 10) (Fig. 5a, b).

3.6 Channel Relabeling

When comparing samples across different acquisitions, it is important to keep the target names for each isotope channel consistent. If the channel names differ (e.g., HLA-DR vs. HLA_DR), problems can arise when normalizing, concatenating, or using a cytometry analysis software. There are several tools available to address this issue.

1. The HIMC at Mt. Sinai has developed an R package to rename the channels of multiple FCS files available on https://github.com/ismmshimc/cytutils.
2. The Nolan Lab at Stanford also has an R package available on https://github.com/nolanlab/cytofCore.

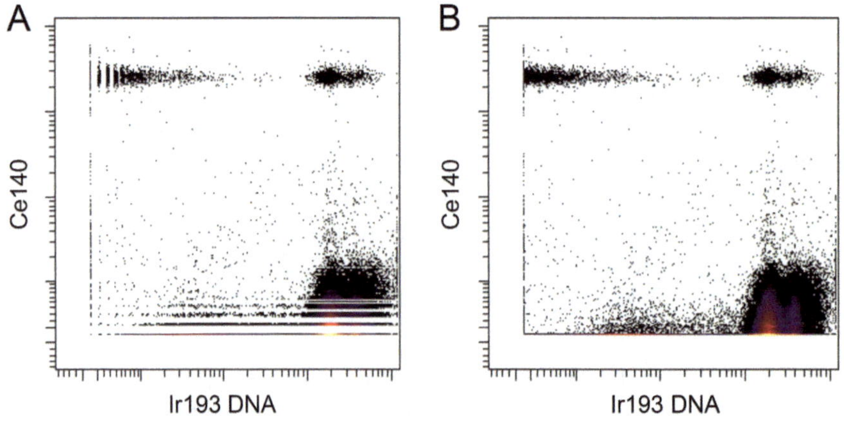

Fig. 5 Randomization. (**a**) Example of a non-randomized FCS file resulting in "picketed" appearance of events in the lowest decade due to their integer values. (**b**) Visualization of the same FCS file after randomization

3. Cytobank's FCS concatenation tool also allows users to edit the output channel names in the final concatenated file. Again, data must be normalized before using this tool.

3.7 Doublet Discrimination

One significant challenge faced in accurately identifying single cell marker co-expression patterns in mass cytometry data is the issue of doublets/multiplets. True cell-cell doublets can occur from cells that adhere to each other during sample preparation or in the instrument fluidics prior to the nebulizer, resulting in a single droplet containing two cells. This will ultimately result in an ion cloud composed of ions from the two cells that may be erroneously misinterpreted as ions from antibodies co-expressed by the same cell. A second form of doublets specific to mass cytometry is an ion cloud fusion doublet, where ion clouds from two separate droplets fuse together prior to entering the TOF chamber, and may again be misinterpreted as signals from a single cell. Ion cloud fusion doublets are related to gas expansion kinetics and are highly dependent on acquisition speed (*see* **Note 3**).

1. Traditional doublet exclusion strategy based on DNA and event length: plot Ir193 (DNA) vs. event length. Events with high DNA intensity and high event length can be interpreted as doublets (*see* **Note 9**) (Fig. 6).
2. Gaussian parameter-based doublet exclusion: FCS files generated using Fluidigm's newest Helios mass cytometer software include several additional parameters for each event that are calculated as part of the IMD to FCS conversion process. These Gaussian parameters can be used to further exclude cell doublets. The four parameters—residual, offset, center and width can be used to describe the shape of the distribution of pushes in

Fig. 6 Doublet discrimination. (**a**) A sample was split and separate aliquots were barcoded using CD45 antibodies conjugated to 154Sm and 159Tb isotopes and combined into a single sample for acquisition, allowing for accurate identification of known doublets based on 154Sm and 159Tb double-positivity. (**b**) Overlay of the known doublet and singlet populations on a biaxial plot of DNA and event length as would be used for traditional doublet exclusion

a pulse. If a pulse follows a normal or bell-shaped distribution it can be considered a high quality pulse and if pulse has an irregular, or bimodal distribution it can be considered a low quality pulse (Fig. 7a, b). The residual effectively represents how well the pushes fit to a normal distribution (Fig. 7c, d). The greater the residual, the larger the difference between the pushes in a pulse and its fitted normal distribution curve [4]. These Gaussian parameters can also be used to identify cell-cell doublets. Plot event length vs. residual: events with high residual and event length can be defined as doublets (*see* **Note 10**) (Fig. 8a, b).

3.8 Debarcoding

Cell barcoding is a technique used to label all the individual cells in a given sample with a unique isotopic label. The barcoded samples can then be pooled together into one tube and stained and acquired as a single sample, thereby eliminating antibody staining variability between samples and variability in instrument performance [5, 6]. Another major advantage of this technique is its ability to robustly identify cell doublets and exclude them from subsequent analysis. Since each sample has a unique combinatorial barcode, any cross-sample doublets will have an invalid barcode and can be excluded. Importantly, this will apply both to cross-sample doublets that result from true cell-cell aggregates or ion cloud fusion events. The greater the number of barcoded samples pooled in a single sample, the greater the likelihood that doublets will result from cells from two different barcoded samples, which can be identified and excluded as cross-sample doublets (Fig. 9).

Barcoded samples can be deconvoluted into their individual constituent samples using Boolean gating, where positive and negative gates are manually drawn for each barcode channel and cells are assigned to their respective samples based on a defined barcode key. Alternatively, a more sophisticated approach utilizes automated single cell deconvolution (SCD) algorithms, which rank the intensity of each barcode channel on a per cell basis and assigns cells to the appropriate sample based on the top channels per cell, and calculates barcode separation (BS) and Mahalanobis distance (MD) features which reflect the resolution of each cell's barcode intensity [5]. While earlier versions of SCD software required users to define a fixed BS and MD value for all samples in a barcode pool, the Zunder Lab's most recent version of the Single Cell Debarcoder

Fig. 7 (continued) (**b**) The distance between each push and the fitted curve is calculated, and (**c**) the residual is determined by effectively taking the sum of the distances between each push to the fitted curve as a chi-square. (**d**) Each event in the FCS file contains four parameters describing the shape of the fitted curve for the corresponding pulse. Relative to singlets, ion cloud fusion doublets are typically characterized by higher residual, slightly reduced width, a left or right shifted center, and a lower offset. Note that a subset of defined cross-sample doublets (identified as in Fig. 6) overlaps with the same region as the singlet population and may represent cell-cell aggregates that existed prior to ionization, which consequently exhibit normal ion cloud characteristics

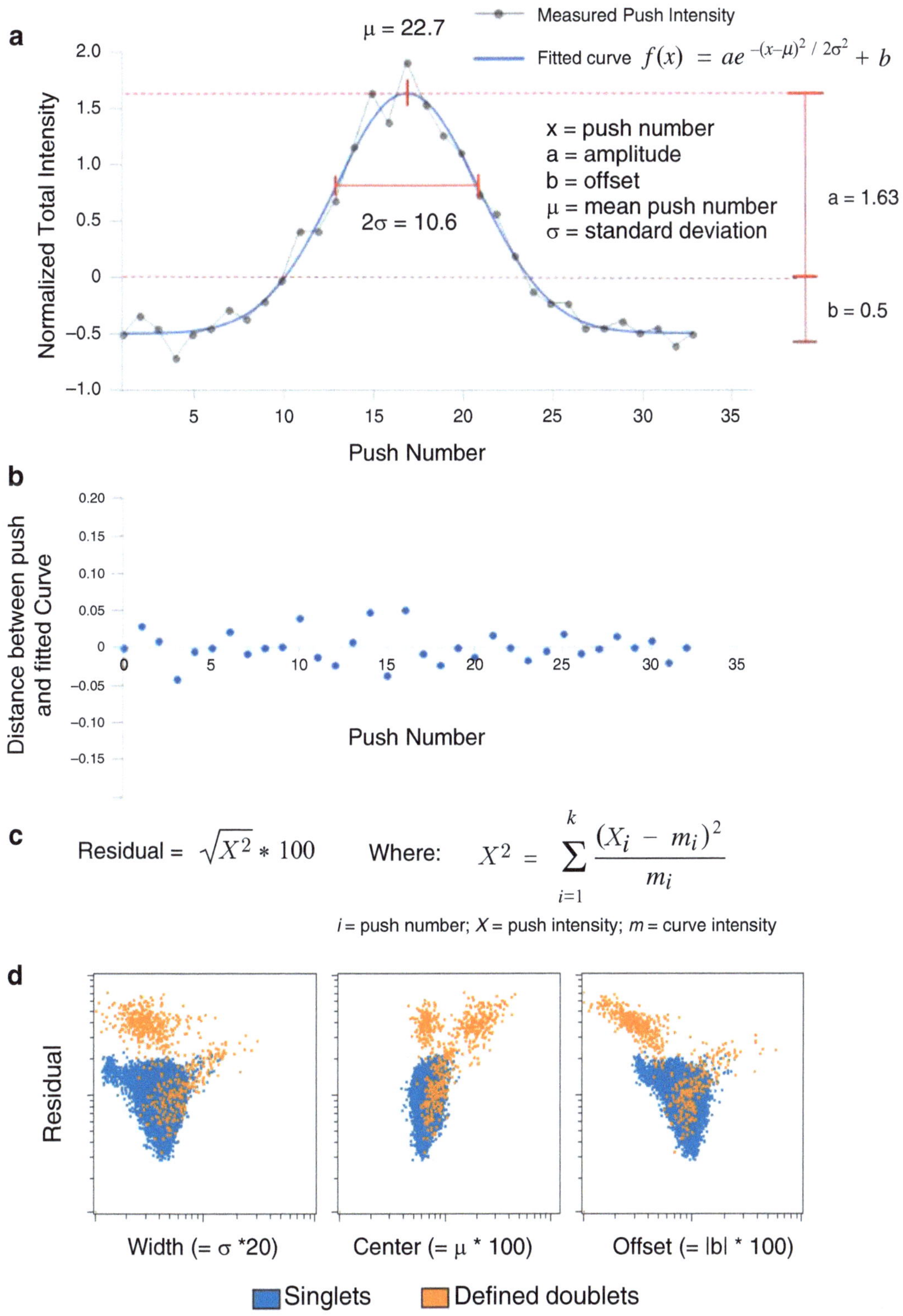

Fig. 7 Gaussian parameter calculations. (**a**) During IMD to FCS conversion, the push data from each pulse are fitted to a Gaussian distribution based on the formula: $f(x) = ae^{-(x-u)^2/2\sigma^2} + b$, where x = push number, a = amplitude above threshold, b = offset relative to baseline, μ = mean push number, σ = standard deviation.

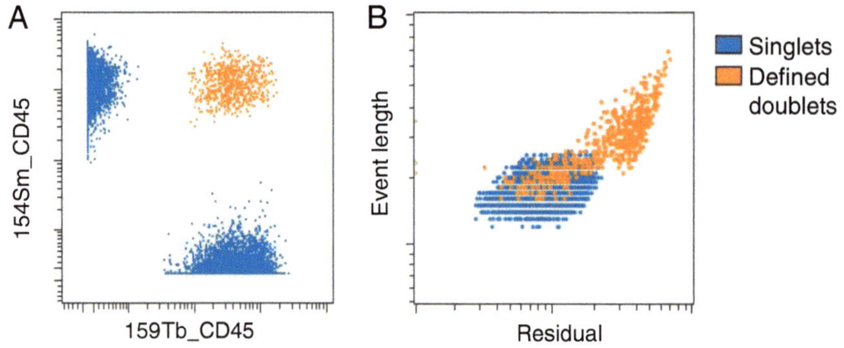

Fig. 8 Utilization of residual parameter for doublet exclusion. (**a**) Samples are barcoded using CD45 antibodies conjugated to 154Sm and 159Tb isotopes as in Fig. 6. (**b**) Overlay of the known doublet and singlet populations on a biaxial plot of residual and event length. Note that while ion cloud fusion doublets can be distinguished, some known doublets overlap with the singlet region and cannot be effectively excluded by manual gating on these parameters

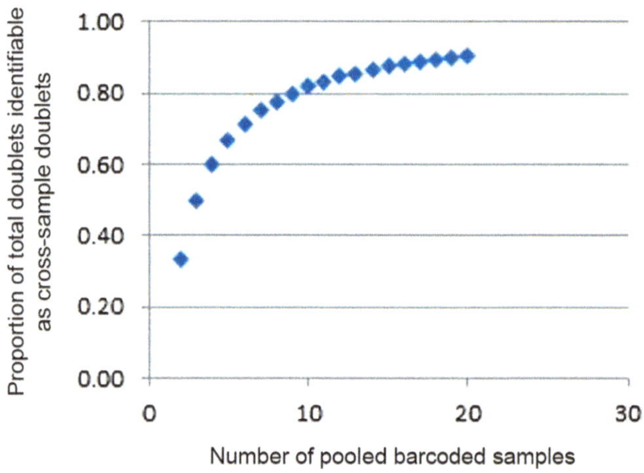

Fig. 9 The proportion of total doublets identifiable as cross-sample doublets increases as the number of pooled barcoded samples increases

exports the barcode separation and Mahalanobis distance as parameters in the debarcoded FCS files, allowing users to more accurately tailor these parameters to each specific barcoded sample [7].

1. Open the Zunder Lab Single Cell Debarcoder available on https://github.com/zunderlab/single-cell-debarcoder.
2. Select the FCS key file and the normalized FCS file.
3. Once uploaded, the software will generate a barcode separation curve. Well-resolved barcodes will typically have a barcode separation curve with a defined plateau and a sharp drop off, while a sloping curve reflects poorly resolved barcodes (Fig. 10a, b).

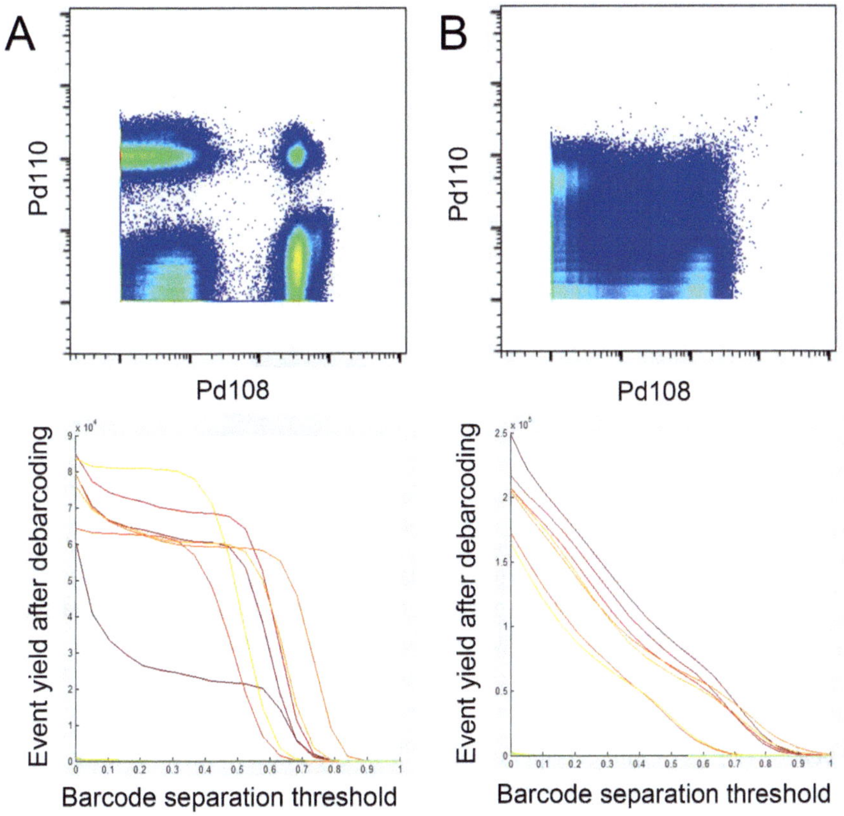

Fig. 10 Barcode resolution. (**a**) Example of well-resolved palladium barcoding resulting in a good barcode separation curve. (**b**) Example of poorly resolved palladium barcodes and an ambiguous barcode separation curve

4. Create a base file name for the debarcoded samples and select the directory to save the debarcoded files. The software will automatically append the sample name in the key file to the base file name.
5. Rather than selecting an optimal BS for the overall file, the Zunder Lab version of the Single Cell Debarcoder will set the default BS separation to zero and export the BS and MD for each single cell as parameters in the debarcoded FCS files. Manual gating can be used to tailor these parameters on a per sample basis to more accurately exclude cross-sample doublets and better resolve the barcoded cells.
6. After gating on the DNA+Ce140− cell population, plot Mahalanobis distance vs. BC separation. Gate on the population with high BC separation and low Mahalanobis distance (Fig. 11).
7. Plotting event length vs. residual can be used to further exclude remaining within-sample doublets (refer to Subheading 3.7).

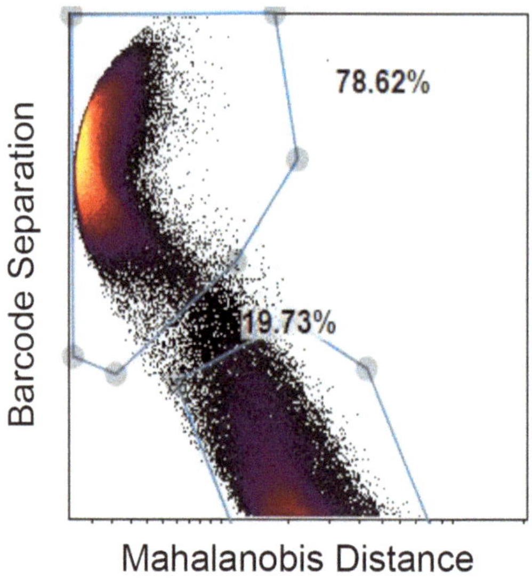

Fig. 11 Example of a manual gating performed on a sample following Zunder Lab SCD software to exclude cross-sample doublets and debris characterized by low barcode separation and Mahalanobis distance

3.9 Live Cell Identification

Unlike conventional flow cytometry, mass cytometry lacks light scatter parameters to distinguish cell events. Instead, nucleic acid intercalators are used to label nucleated cells and distinguish these from non-nucleated debris (*see* **Note 11**). While nucleic intercalators can be used post-fixation/permeabilization to identify all nucleated cells, they can also be applied prior to fixation in which case they will be selectively taken up by dead cells [8]. This allows exclusion of dead cells, which is in important step in the data processing workflow since they can nonspecifically bind antibodies resulting in false positive signals. The gating strategy described below assumes the use of Ir191/193 intercalator for pan-nucleated cell identification and Rh103 nucleic acid intercalators for dead cell identification, though cisplatin may also be used as an alternative reagent for dead cell identification and gated in an analogous fashion [9].

1. Plot Ce140 vs. Ir193 (DNA) and gate on the DNA+ Ce140− population to exclude any residual EQ beads and EQ bead-cell doublets (Fig. 12a).
2. Exclude doublets and gate on the singlet population by plotting event length vs. residual (*see* Subheading 3.7) (Fig. 12b).
3. Within the singlet cell population, plot the viability marker (i.e., Rh103 or cisplatin) used in staining vs. DNA. Any dead cells with a compromised membrane will stain positively for Rh103 or cisplatin. By gating on the Rh103- or Cisplatin- population, live cells can be identified and dead cells excluded (Fig. 12c).

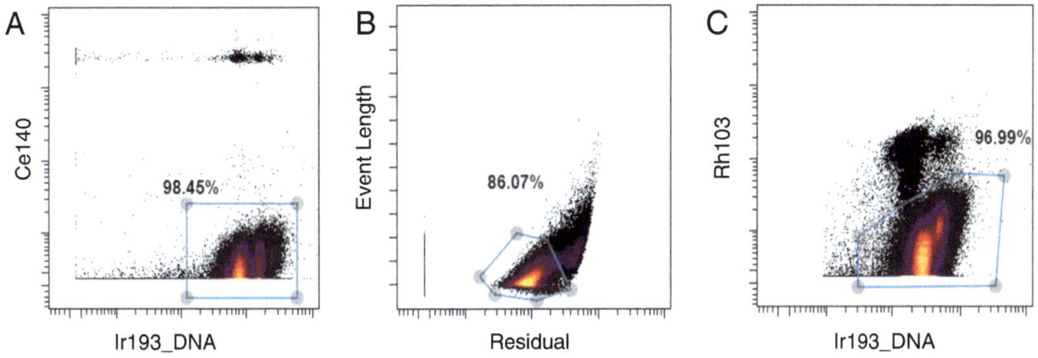

Fig. 12 Example of a gating strategy used to identify live singlet cells. (**a**) EQ beads and EQ bead-cell doublets are excluded. (**b**) Event length vs. residual is plotted and ion cloud fusion doublets are excluded. (**c**) Live cells are identified as negative for the Rh103 viability intercalator

The live singlet population can now be used for downstream analyses such as manual gating, dimensionality reduction, or automated clustering approaches.

4 Notes

1. All mass cytometry samples should be spiked with EQ Four Element Calibration Beads (Fluidigm) prior to data acquisition. These beads allow for monitoring of instrument performance and for normalization of signal intensity to account for fluctuations over time or variations between instruments.
2. Good cell fixation is extremely important to allow cells to withstand the hypotonic stress associated with the final water washes prior to sample acquisition; all stained samples should be post-fixed with freshly diluted 1–4% formaldehyde in phosphate buffered saline (PBS), as described in Chapter 1. Inadequate fixation will result in sample degradation, which may manifest as significant loss of cells during the water wash. Even in appropriately fixed samples, exposure to water for prolonged periods of time will cause cellular degradation [10]. When acquiring large volumes of a sample, prepare multiple aliquots of the sample in isotonic buffer (e.g., PBS containing bovine serum albumin) and do not wash with water until ready for acquisition.
3. Acquiring data at an appropriate event rate is important for data quality. We typically aim for optimal event rates of <300 events/s on a Helios mass cytometer (Fig. 13a). Increasing the event rate will result in a greater number of ion cloud fusion events, resulting in increasing numbers of ion cloud fusion doublets (where ion clouds from two separate cells are fused, resulting in a single event data point containing

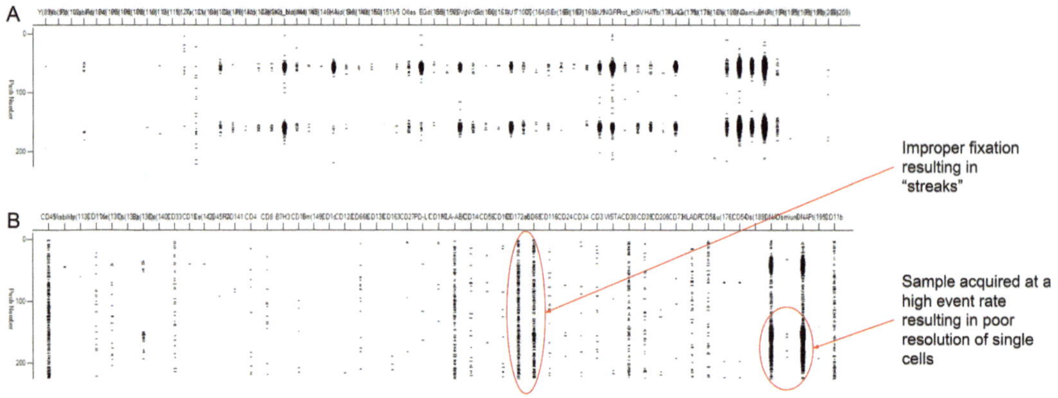

Fig. 13 Data acquisition quality assessment. (**a**) Example of the rain plot in the Fluidigm acquisition software showing an optimal event rate and properly fixed cells resulting in clean separation of ion clouds associated with single cells. (**b**) Example of a high event rate and/or improper fixation resulting in "streaks" and poor resolution of ion clouds associated with single cells

co-mingled signals from both cells), and ultimately loss of single cell resolution. Resolution of single cells may also be impacted by low quality samples that are overstained or poorly fixed, in which case isotope channels may appear as "streaks" in the rain plot rather than well-resolved "droplets" (Fig. 13b).

4. The Human Immune Monitoring Center (HIMC) at the Icahn School of Medicine at Mt. Sinai has developed an automated bead QC application to detect bead performance and to export the results as a CSV file (Fig. 14). This application is freely available at https://github.com/ismmshimc/cytutils.
5. Spillover due to oxidation, isotopic impurity, and abundance sensitivity can cause false positives in mass cytometry data. Takahashi et al. [2] explain practices that can be used to design an antibody panel that reduces the effect of this signal interference for improved data quality. Furthermore, single stained bead controls can also be used to correct for cross-talk between channels similar to conventional fluorescence-based compensation [11].
6. Limiting oxide formation is an important part of daily instrument optimization using CyTOF Tuning Solution (Fluidigm), as described in Chapter 1. Oxide formation during tuning is calculated based on La139 oxides, which are detected in the Gd155 channel, and the balance between oxide formation and detection sensitivity is expressed as the ratio between Tb159 and Gd155. Fluidigm recommends tuning instrument performance using a ratio threshold of 3%. Due to the trade-off nature between oxidation and signal intensity, increasing signal intensity during tuning typically results in a higher oxidation ratio. Given the complications that oxide-induced signal cross-talk

A		B			C	D	E		F	G		H	I		
qc reporter	qc report timestamp	data file name	data file start time	data file end time	total evts	cell evts	Ir 193 median	Ir 193 time corr	bead evts	Eu 153 median	Eu 153 time corr	Eu 153 rCV	Ce 140 median	Gd 156 median	Oxide %
QCToolkit_v170622	11/29/17 20:29:11	171127	11/29 18:53	18:58	190,640	177,252	3,147	-0.034	842	2506	-0.205	7.8	2420	16.0	0.7%
QCToolkit_v170622	11/29/17 20:28:35	171127	11/29 19:01	19:06	232,722	215,532	3,206	-0.022	758	2436	-0.190	7.7	2346	16.6	0.7%
QCToolkit_v170622	11/29/17 20:27:52	171127	11/29 19:09	19:14	267,987	243,149	3,282	-0.027	716	2384	-0.210	7.9	2268	15.4	0.7%
QCToolkit_v170622	11/29/17 20:27:01	171127	11/29 19:19	19:24	293,336	267,832	3,281	-0.030	592	2321	-0.206	7.5	2191	14.6	0.7%
QCToolkit_v170622	11/29/17 20:26:02	171127	11/29 19:28	19:33	317,915	291,739	3,274	-0.042	487	2242	-0.195	7.5	2100	14.4	0.7%
QCToolkit_v170622	11/30/17 00:07:23	171127	11/29 21:44	21:49	50,247	47,459	1,918	-0.039	824	2105	-0.171	7.6	2073	13.2	0.6%
QCToolkit_v170622	11/30/17 00:07:07	171127	11/29 21:52	21:56	77,178	73,440	1,943	-0.064	529	2108	-0.237	8.1	2063	14.5	0.7%
QCToolkit_v170622	11/30/17 00:06:47	171127	11/29 21:58	22:03	135,895	128,477	1,948	-0.060	559	2091	-0.316	8.7	2020	13.9	0.7%
QCToolkit_v170622	11/30/17 00:06:19	171127	11/29 22:05	22:10	186,474	176,567	1,957	-0.058	592	2081	-0.298	8.3	1985	14.1	0.7%
QCToolkit_v170622	11/30/17 00:05:42	171127	11/29 22:13	22:18	193,041	184,736	1,963	-0.036	431	2066	-0.184	8.2	1982	13.3	0.7%
QCToolkit_v170622	11/30/17 00:05:06	171127	11/29 22:21	22:25	205,927	196,050	1,912	0.017	350	2015	0.105	10.3	1881	14.4	0.8%

Fig. 14 Example of the output by the HIMC cytutils automated bead QC application, tracking bead performance across multiple samples. (**a**) Reporter version and timestamp for when the QC was performed; (**b**) start and end time for the file to calculate acquisition duration; (**c**) total events and (**d**) cell event count as identified and recorded based on Ir193+Ce140− events; (**e**) Ir193 median intensity and correlation with time to track cell-associated signal loss; (**f**) bead events count as identified and recorded based on Ce140+1r193− events; (**g**) Eu153 median intensity and correlation with time to track bead-associated signal loss; (**h**) bead CV for Eu153 to track measurement precision; and (**i**) Ce140 and Gd156 intensity and ratio to calculate oxidation

can pose during data analysis, we prefer to sacrifice some signal intensity in favor of minimizing oxidation and consequently use a more stringent oxidation threshold of 2% during tuning.

7. There are some limitations to the bead passport normalization method. Firstly, a relatively limited range of isotopic masses is used to extrapolate the entire mass range, which could theoretically mean that not all channels are normalized effectively. Secondly, normalization cannot recover low intensity signal measurements that drop below the detection threshold due to a drop in instrument sensitivity. In this case, normalization will only amplify the background noise. Thirdly, while normalization can correct for variations in instrument performance, since relies on bead intensity, it cannot correct for fluctuations that result from discrepancies in the behavior between cells and beads, such as sample degradation due to poor fixation. Lastly, normalization cannot correct for pre-analytical variability resulting from variations in sample staining or processing.
8. While smaller time intervals can more accurately account for changes in instrument performance, if the interval is too small, there may not be sufficient bead events present to accurately normalize the data. Time intervals that do not contain a sufficient number of beads are excluded from the normalized file.
9. Traditionally, event length vs. DNA has been used to identify singlets. However, there are shortcomings to this method.

DNA staining intensity is based on the ability of the Ir191/193 intercalator to bind to DNA. However, the density of chromatin can vary between cell types [12] and can affect how effectively the intercalator binds to DNA ultimately affecting the staining intensity (Fig. 15). Furthermore, proliferating cells in S-phase will have higher DNA content, and their DNA expression can fall within the cell doublet region if using the traditional DNA doublet exclusion strategy.

10. In contrast to DNA, Gaussian parameters are purely a reflection of the ion cloud characteristics, which are primarily driven by gas expansion kinetics and are independent of cell type. Using these parameters for doublet discrimination is less sample-dependent when compared to DNA-based doublet discrimination. Importantly however, Gaussian parameters only reflect features of the ion cloud, which are largely independent of cell-specific features, besides total ion content. Therefore, Gaussian parameters can only be used to identify ion cloud fusion doublets and not doublets that occurred due to cell adherence prior to sample ionization.

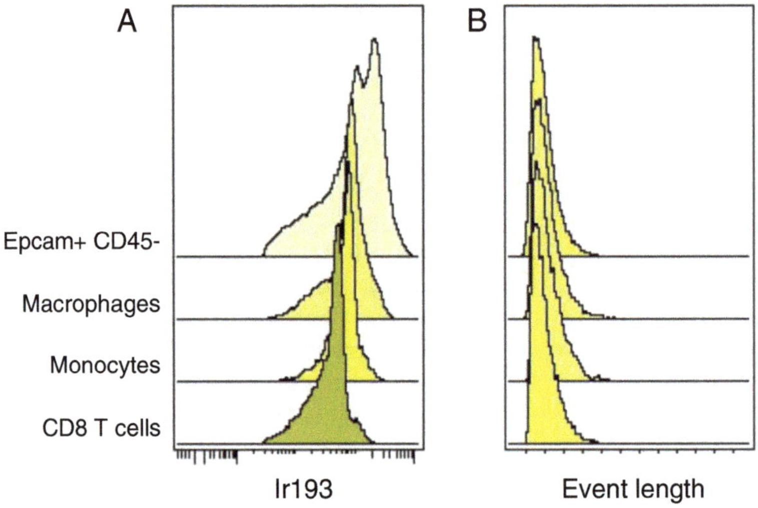

Fig. 15 DNA intercalator profiles across cell types. While cell-cell doublets generally have higher Ir intensity than singlets, it is important to note that DNA staining intensity is influenced by sample preparation and biological differences between cell types, and using an Ir-based doublet exclusion strategy can result in loss of certain cell population. (**a**) Distinct cell types were identified in a suspension from human lung tumors and express different levels of Ir signal. (**b**) By contrast, ion cloud features such as event length are purely a reflection of the ion cloud and are consistent across cell types

11. In flow cytometry, forward scatter and side scatter can be used to measure cell size and granularity. While mass cytometry may not have these tools, cell features independent of antibody staining can still be identified by staining cells with osmium tetroxide [13]. OsO_4 binds to the plasma membrane and is optimally detected in the Os192 channel. Based on staining intensity, OsO_4 can be used to infer cell features such as size and granularity, analogous to traditional light scatter properties.

References

1. Fluidigm | CyTOF 2 User Guide. https://www.fluidigm.com/search?query=user%20guide. Accessed 20 May 2018
2. Takahashi C, Au-Yeung A, Fuh F et al (2017) Mass cytometry panel optimization through the designed distribution of signal interference. Cytometry A 91:39–47. https://doi.org/10.1002/cyto.a.22977
3. Finck R, Simonds EF, Jager A et al (2013) Normalization of mass cytometry data with bead standards. Cytometry A 83:483–494. https://doi.org/10.1002/cyto.a.22271
4. Bagwell B Fluidigm | 6th Annual Mass Cytometry Summit Presentations. http://cn.fluidigm.com/articles/mass-cyto-summit-videos. Accessed 22 May 2018
5. Zunder ER, Finck R, Behbehani GK et al (2015) Palladium-based mass-tag cell barcoding with a doublet-filtering scheme and single cell deconvolution algorithm. Nat Protoc 10:316–333. https://doi.org/10.1038/nprot.2015.020
6. Mei HE, Leipold MD, Schulz AR et al (2015) Barcoding of live human PBMC for multiplexed mass cytometry. J Immunol 194:2022–2031. https://doi.org/10.4049/jimmunol.1402661
7. Fread KI, Strickland WD, Nolan GP et al (2017) An updated debarcoding tool for mass cytometry with cell type-specific and cell sample-specific stringency adjustment. Pac Symp Biocomput 22:588–598. https://doi.org/10.1142/9789813207813_0054
8. Ornatsky OI, Lou X, Nitz M et al (2008) Study of cell antigens and intracellular DNA by identification of element-containing labels and metallointercalators using inductively coupled plasma mass spectrometry. Anal Chem 80:2539–2547. https://doi.org/10.1021/ac702128m
9. Fienberg HG, Simonds EF, Fantl WJ et al (2012) A platinum-based covalent viability reagent for single-cell mass cytometry. Cytometry A 81:467–475. https://doi.org/10.1002/cyto.a.22067
10. Amir E-AD, Guo XV, Mayovska O et al (2018) Average overlap frequency: a simple metric to evaluate staining quality and community identification in high dimensional mass cytometry experiments. J Immunol Methods 453:20–29. https://doi.org/10.1016/j.jim.2017.08.011
11. Chevrier S, Crowell HL, Zanotelli VRT et al (2018) Compensation of signal spillover in suspension and imaging mass cytometry. Cell Syst 6(5):612–620.e5. https://doi.org/10.1016/j.cels.2018.02.010
12. Jensen IM, Ellegaard J, Hokland P (1993) Differences in relative DNA content between human peripheral blood and bone marrow subpopulations—consequences for DNA index calculations. Cytometry 14:936–942. https://doi.org/10.1002/cyto.990140813
13. Stern AD, Rahman AH, Birtwistle MR (2017) Cell size assays for mass cytometry. Cytometry A 91:14–24. https://doi.org/10.1002/cyto.a.23000

Part II

Panel Design and Reagent Preparation

Chapter 3

Visualization of Mass Cytometry Signal Background to Enable Optimal Core Panel Customization and Signal Threshold Gating

Amelia Au-Yeung, Chikara Takahashi, W. Rodney Mathews, and William E. O'Gorman

Abstract

Signal interference or overlap in mass cytometry is minimal compared to flow cytometry but must still be considered for optimal panel design and assay sensitivity. Here we describe a procedure for evaluating signal interference dynamics in the context of a 25-parameter core immunophenotyping panel. Specifically, a mass-minus-many (MMM) approach was used to assess background signals in "empty" or "blank" channels intended for further customization. Through this approach cell type-specific variability in signal background is revealed. Further panel customization can thus be performed with an understanding of cell type and channel-specific background levels to enable rational panel design and the objective delineation of gating thresholds during analysis.

Key words Mass cytometry, CyTOF, Panel design

1 Introduction

Mass cytometry is an established technology for high-dimensional cytometric analysis; however, basic questions surrounding the delineation of signal from background still exist. While autofluorescence does not confound mass cytometry analysis, sources of background caused by signal interference have been described [1, 2], and are primarily caused by the oxidation and isotopic impurity of reporter molecules.

An adaptation of the fluorescence-minus-one control, termed as mass-minus-one (MMO), has previously been used to define gating thresholds that delineate signal from background in mass cytometry data [3]. In practice however, MMO controls are time-consuming to apply in the context of a 40-parameter panel. Here we describe a procedure for implementing a mass-minus-many (MMM) control in the context of customizing a deep immunophenotyping panel

Helen M. McGuire and Thomas M. Ashhurst (eds.), *Mass Cytometry: Methods and Protocols*, Methods in Molecular Biology, vol. 1989, https://doi.org/10.1007/978-1-4939-9454-0_3,

around core markers that define 23 immune cell populations (*see* Table 1 and Fig. 1).

In this approach, channels to be customized are simply left blank thus revealing the background signals that exist due to signal overlap from the core set of markers. Figure 2 demonstrates variability in signal background levels that are both cell type and channel specific. For this particular core panel configuration signal background levels are generally higher for T cells and B cells than for natural killer and myeloid cells. Specifically, the 144Nd channel should not be used for markers that are weakly expressed in T cells and B cells. Similarly, the 158^{Gd} and 164^{Dy} channels should not be used for markers that are weakly expressed on B cells.

An important caveat for the MMM approach is that it does not account for signal interference between the empty channels that are open for customization. Therefore, potential interactions between newly introduced markers should be considered. Table 2 shows signal overlap dynamics between the open channels in this panel configuration. There is less than 1% overlap for all but five channel interactions thus minimizing the potential impact of channel crosstalk between newly introduced markers. Lastly, once all open channels have been filled, the effects of full panel occupancy on cell population resolution should be assessed to ensure signal overlap from newly introduced markers does not compromise the core gating strategy.

MMM controls can also be used to objectively define gating thresholds, which is particularly important for signals that are continuous with background. Figure 3 shows how an MMM control can be used to define signal thresholds that should be customized on a per cell type basis in order to reduce the frequency of false-positive events. Target antigens producing signals that are continuous with background were intentionally chosen in order to illustrate the difficulty in setting objective gates in the absence of a control sample.

In summary, following the establishment of an adequate core panel, further panel customization can be performed rationally by visualizing signal background levels in blank channels through an MMM control. Subsequently, this control is also useful for objectively defining gating thresholds.

2 Materials

Prepare all solutions fresh with sterile technique. Follow all waste disposal regulations when disposing waste materials.

1. 50 mL Leucosep™ tubes (Greiner Bio-One International, *see* **Note 1**).

Table 1
Core panel (in bold) configuration and antibody clone and vendor information

Mass	Metal	Target	Source	Vendor	Clone	Titration
89	Y	**CD45**	Fluidigm		HI30	2.0 μL
113	In	Open				
115	In	**CD57**	Custom	Biolegend	HCD57	200 μg/mL
140	Ce	EQ beads	Fluidigm			
141	Pr	Open				
142	Nd	CCR4	Custom	R&D	205410	200 μg/mL
143	Nd	**CD127**	Fluidigm		A019D5	1.0 μL
144	Nd	Open				
145	Nd	**CD4**	Fluidigm		RPA-T4	1.0 μL
146	Nd	**IgD**	Fluidigm		IA6-2	1.0 μL
147	Sm	Open				
148	Nd	Open				
149	Sm	CD45RO	Fluidigm		UCHL1	1.0 μL
150	Nd	Open				
151	Eu	**CD123**	Fluidigm		6H6	1.0 μL
152	Sm	**γδTCR**	Custom	Life Technologies	SA6.E9	200 μg/mL
153	Eu	TIM3	Custom	R&D	344823	400 μg/mL
154	Sm	**CD3**	Fluidigm		UCHT1	1.0 μL
155	Gd	**CD27**	Fluidigm		L128	1.0 μL
156	Gd	CXCR3	Fluidigm		G025H7	1.0 μL
157	Gd	**CD19**	Custom	Biolegend	HIB19	200 μg/mL
158	Gd	Open				
159	Tb	**CD11c**	Fluidigm		Bu15	1.0 μL
160	Gd	**CD56**	Custom	Miltenyi	REA196	200 μg/mL
161	Dy	**CD66**	Custom	BD	B1.1	200 μg/mL
162	Dy	**Foxp3**	Fluidigm		PCH101	1.0 μL
163	Dy	**CD20**	Custom	Biolegend	2H7	200 μg/mL
164	Dy	Open				
165	Ho	Open				
166	Er	**FcεRI**	Custom	Biolegend	AER-37 (CRA-1)	200 μg/mL
167	Er	**CCR7**	Fluidigm		G043H7	4.0 μL

(continued)

Table 1
(continued)

Mass	Metal	Target	Source	Vendor	Clone	Titration
168	Er	**CD8**	Fluidigm		SK1	1.0 μL
169	Tm	**CD25**	Fluidigm		2A3	1.0 μL
170	Er	**CD45RA**	Fluidigm		HI100	0.1 μL
171	Yb	**CD1c**	Custom	Biolegend	L161	200 μg/mL
172	Yb	**CD14**	Custom	Biolegend	M5E2	400 μg/mL
173	Yb	TIGIT	Fluidigm		MBSA43	3.0 μL
174	Yb	**HLA-DR**	Fluidigm		L243	1.0 μL
175	Lu	PD-1	Fluidigm		EH12.2H7	3.0 μL
176	Yb	**CD38**	Custom	Biolegend	HIT2	200 μg/mL
191	Ir	Nucleic acid	Fluidigm			
192	Pt	Cisplatin	Fluidigm			
193	Ir	Nucleic acid	Fluidigm			
195	Pt	Cisplatin	Fluidigm			
209	Bi	**CD16**	Fluidigm		3G8	1.0 μL

2. Ficoll-Paque™ PLUS: Ficoll™ PM400, 1.077 g/mL density, Sodium diatrizoate (GE Healthcare).
3. PBS: 144 mg/L Monopotassium phosphate, 9000 mg/L Sodium chloride, 795 mg/L Disodium phosphate, Calcium free, Magnesium free.
4. 50 mL conical centrifuge tubes.
5. Lyse™ buffer (BD Pharm).
6. Cell-ID™ Cisplatin (Fluidigm, *see* **Note 2**).
7. 5 mL 12 × 75 mm tubes.
8. MaxPar® Cell Staining Buffer (Fluidigm).
9. Human TruStain FcX (BioLegend).
10. Paraformaldehyde fixative: 16% stock solution (10×), dilute to 1.6% (1×) working dilution with PBS.
11. Foxp3/Transcription Factor Staining Buffer Set (eBioscience).
12. Isotope conjugated antibodies (*see* **Note 3**, including Table 3).
13. Cell-ID™ Intercalator-Ir, working solution 125 nM (Fluidigm).
14. MilliQ Water.

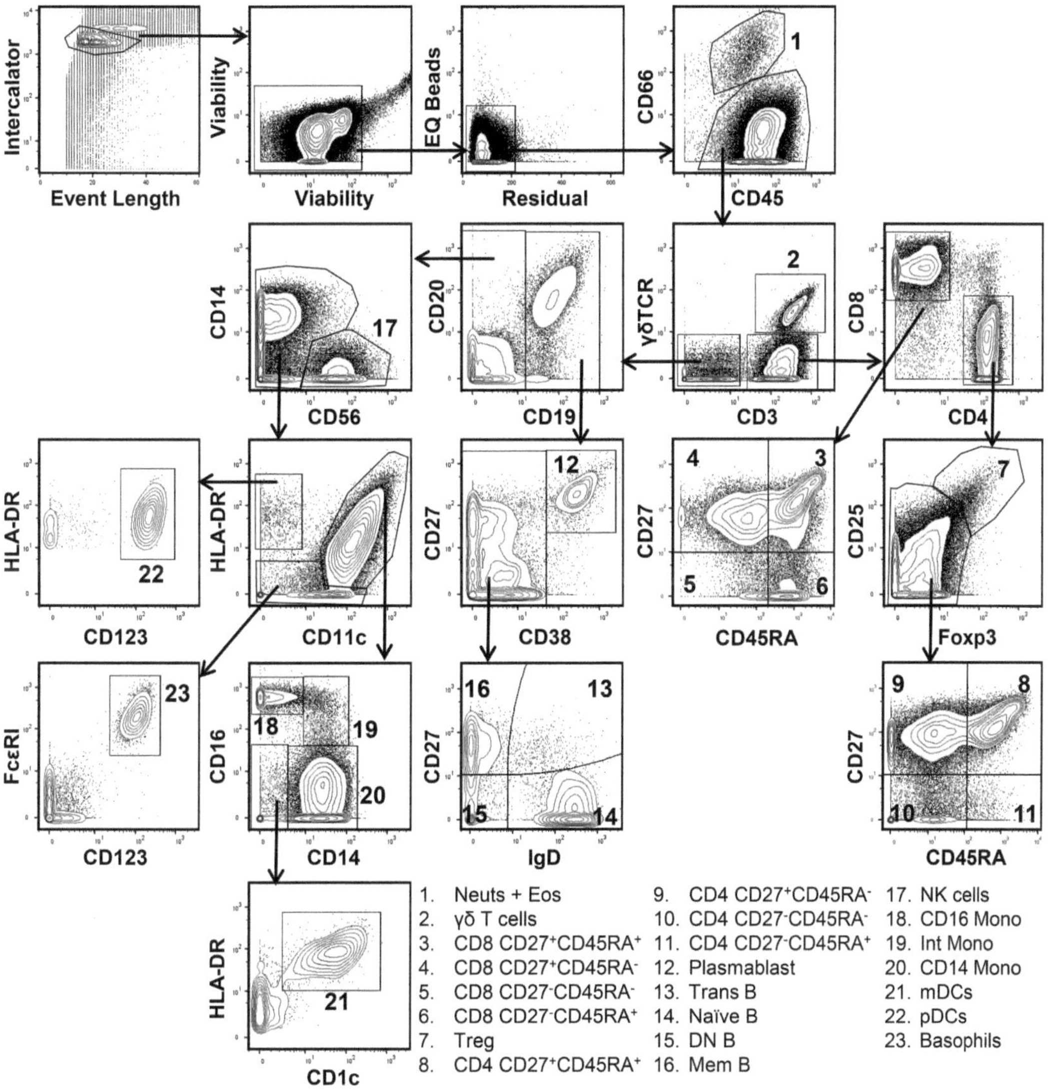

Fig. 1 Gating strategy to identify human immune cell subpopulations. Healthy PBMCs were stained with cisplatin (to assess viability) and 24 cell surface markers (*see* Table 1 for extended details on panel contents). Cells were then fixed and permeabilized prior to FoxP3 intracellular staining. Following overnight intercalator staining cells were acquired on a Helios mass cytometer. Twenty-one surface and intracellular markers were used to define one NK cell, two granulocyte, five B cell, ten T cell, and five myeloid cell populations

15. EQ Four Element Calibration Beads (Fluidigm).
16. 5 mL 12 × 75 mm tubes with 35 μm cell strainer snap cap.
17. MATLAB® (MathWorks®, Natick, Massachusetts) normalizer [4].
18. FlowJo® software (FlowJo, Ashland, Oregon).

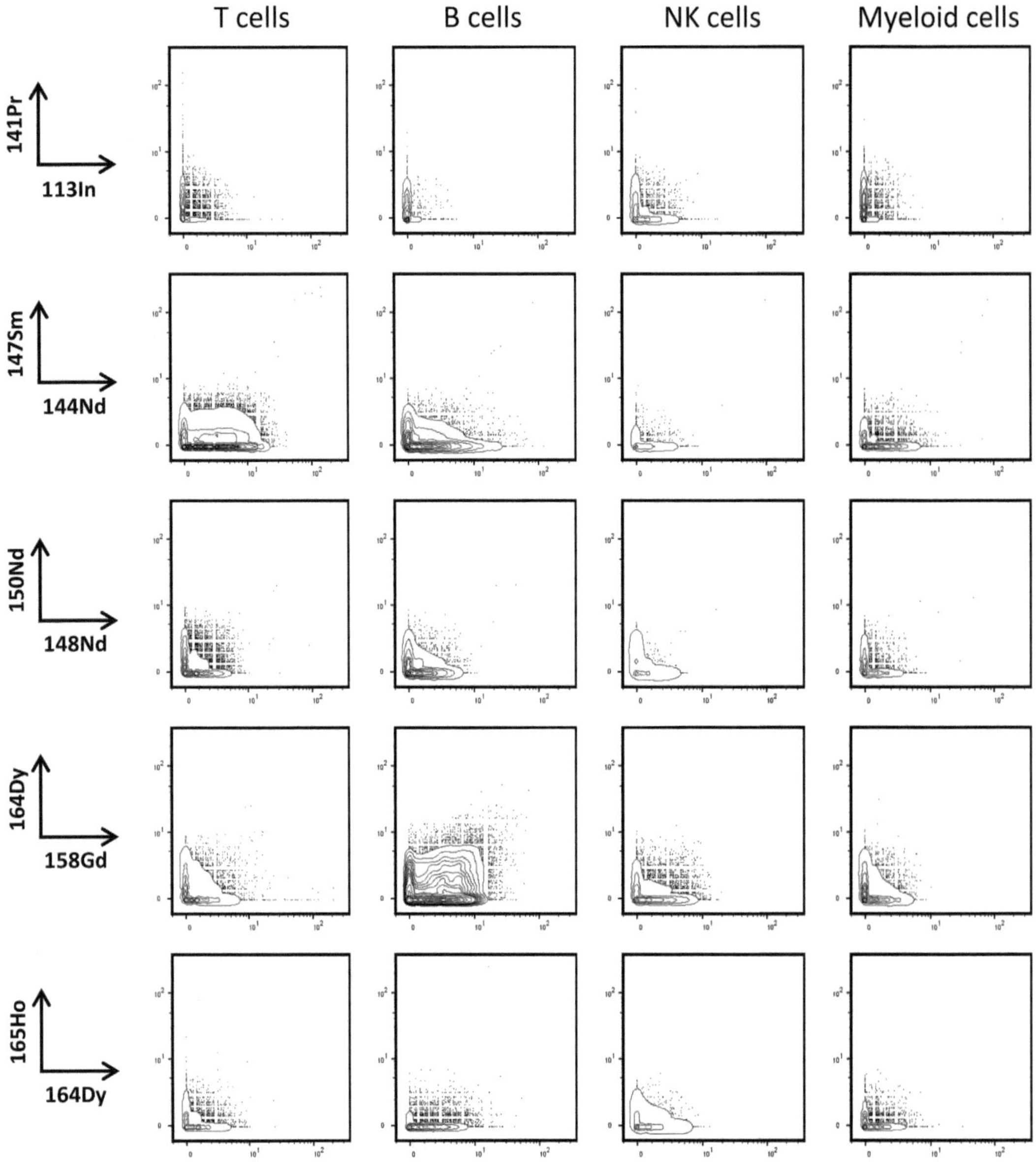

Fig. 2 Signal background visualization using a mass-minus-many (MMM) control. Cells were processed as in Fig. 1

3 Methods

Follow all procedures on ice unless otherwise specified. All samples were acquired using the Helios™, a CyTOF® System.

3.1 PBMC Isolation

1. Healthy whole blood was drawn into sodium heparin blood collection tubes.
2. Dilute whole blood 1:1 with PBS.

Table 2
Signal interference matrix adapted from "Guidelines for Mass Cytometry Panel Design" white article (DVS Sciences, WP13-01_1012)

	Mass Channel														
Mass Tag	113	141	142	144	147	148	149	150	153	156	158	164	165	173	175
113 In	**93**														
141 Pr		**100**													
142 Nd			**100**	0.3							*3.0*				
144 Nd			0.2	**100**											
147 Sm					**100**	*2.2*	*0.5*	0.2							
148 Nd			*0.5*	*0.5*		**100**		*0.5*				*3.0*			
149 Sm					0.2	*0.8*	**100**	*1.4*					0.3		
150 Nd			*0.5*	*0.7*		0.3		**100**							
153 Eu									**100**						
156 Gd										**100**	*1.0*				
158 Gd										0.2	**100**				
164 Dy												**100**			
165 Ho													**100**		
173 Yb														**100**	
175 Lu															**100**

Truncated version of signal overlap matrix displaying signal interference dynamics between channels open for customization

3. Load diluted blood into 50 mL Leucosep™ conical tubes containing Ficoll-Paque™ PLUS.
4. Centrifuge for 15 min at 800 × *g* at room temperature (RT), with brakes turned off.
5. Harvest PBMCs by pipetting buffy coat layer into a new 50 mL conical centrifuge tube.
6. Wash the PBMCs with PBS and subsequently centrifuge for 10 min at 250 × *g* at RT.
7. Lyse red blood cells using Lyse buffer, if needed (*see* **Note 4**).

3.2 Surface Staining of Cells

1. Resuspend PBMC in an appropriate volume of PBS to obtain a cell concentration of 10^7 cells/mL.
2. Incubate cells with Fluidigm's viability reagent, Cell-ID™ Cisplatin at a final concentration of 5 μM at RT for 5 min.
3. Quench Cisplatin by washing with 5× volume of MaxPar® Cell Staining Buffer and centrifuge at 300 × *g* for 5 min.

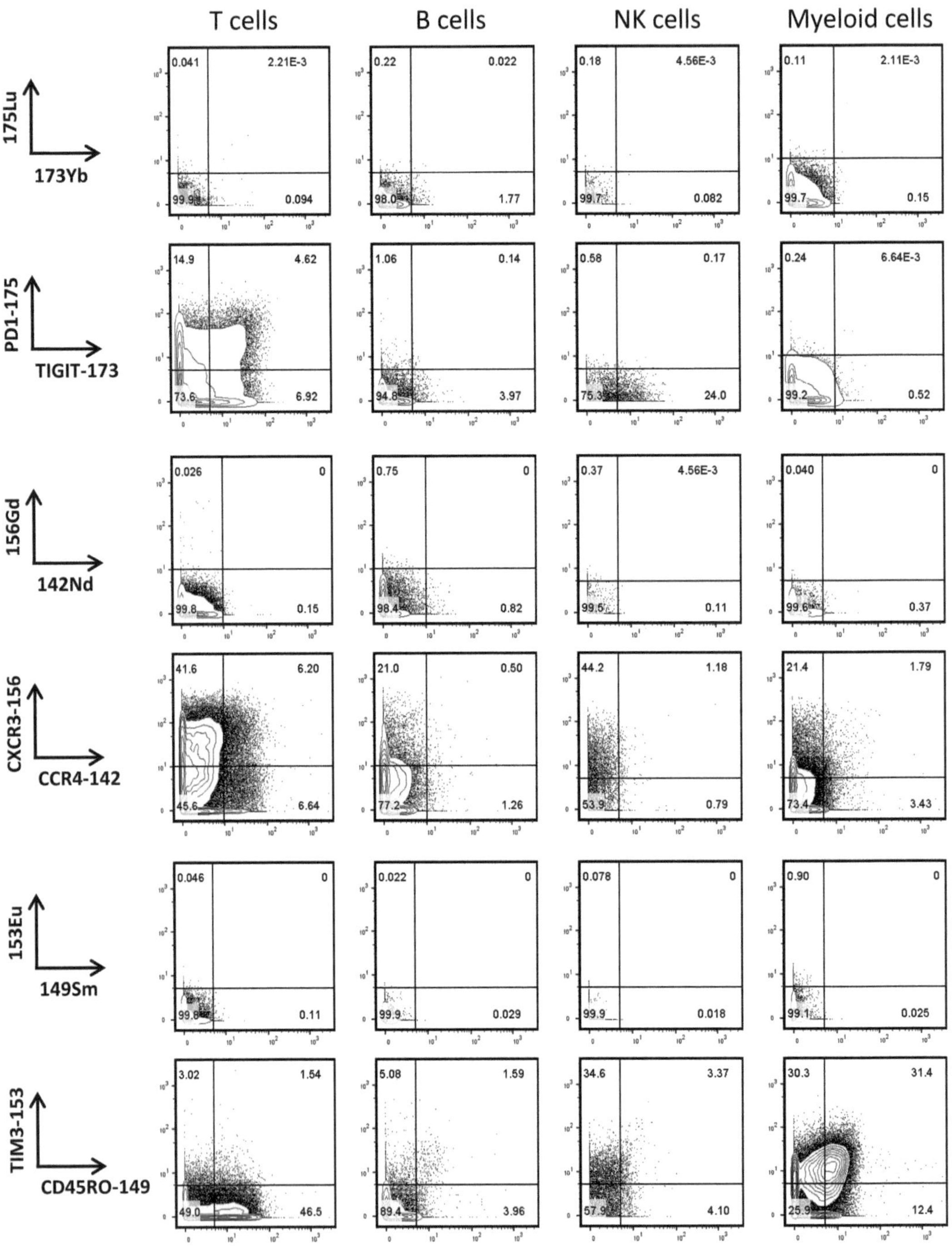

Fig. 3 Signal threshold gating using an MMM control on a per cell type basis to set gates with minimal false-positive events

Table 3
Reference isotope impurities for metal isotopes sourced directly from Trace Sciences International

	Atomic mass (Da)	
Isotope-113	**113**	115
Enrichment (%)	93.1	6.9
Isotope-115	113	**115**
Enrichment (%)	0.17	99.83

Isotope-157	156	**157**	158
Enrichment (%)	1.9	92.3	5.4

4. Resuspend in a final concentration of 30 million cells/mL in staining buffer.
5. Transfer three million cells to 5 mL 12 × 75 mm tubes and incubate with 5 μL of Human TruStain FcX™ for 10 min to block Fc receptor binding.
6. Add 50 μL master antibody cocktail containing all metal-conjugated core panel antibodies to samples for cell surface staining and incubate for 30 min (*see* **Note 5**).
7. Wash cells once with 4 mL MaxPar® Cell Staining Buffer and centrifuge at 300 × *g* for 5 min.
8. Proceed to stain for intracellular markers.

3.3 Permeabilization and Intracellular Labeling of Cells

1. Resuspend cells in 1 mL of FoxP3 staining fixation/permeabilization solution for 45 min (*see* **Note 6**).
2. Wash with 3 mL of permeabilization buffer at 800 × *g* for 5 min, then resuspend cells in 50 μL of permeabilization buffer.
3. Stain cells for intracellular targets by addition of 50 μL intracellular antibody cocktail.
4. After 30 min incubation, wash cells with 4 mL MaxPar® Cell Staining Buffer and centrifuge at 800 × *g* for 5 min.
5. Stain cells overnight at 4 °C in 1 mL of Cell-ID™ Intercalator-Ir reagent with prepared paraformaldehyde fixative (*see* **Note 7**).

3.4 Acquisition on the CyTOF® Instrument

1. Wash cells with 3 mL of MaxPar® Cell Staining Buffer and centrifuge at 800 × *g* for 5 min.
2. Wash with 4 mL of MilliQ Water.
3. Resuspend cells in 1 mL MilliQ Water and count cells.

4. Add an additional 3 mL of MilliQ Water and wash cells at 800 × *g* for 5 min.
5. Resuspend pelleted cells with MilliQ Water containing EQ™ Four Element Calibration Beads according to cell count to achieve 8×10^5 cells/mL.
6. Filter cells using a 12 × 75 mm tube with a 35 μm nylon mesh cell strainer cap prior to introduction into the Helios™, a CyTOF® System (*see* **Note 8**).

3.5 Data Processing and Normalization

1. All FCS files were normalized using the MATLAB® normalizer [4] and analyzed using FlowJo® software.

4 Notes

1. Use of Leucosep tubes with porous barrier allows for easy and efficient separation of lymphocytes and monocytes. Add 1:1 dilution of PBS: whole blood for optimal separation and buffy coat layer.
2. Handle with care as cisplatin is a mutagenic and carcinogenic agent.
3. All isotope sources are from Fluidigm with the exception of 113In, 115In, and 157Gd from Trace Sciences International. *See* Table 3 for the purities of Trace Sciences isotopes.
4. Prepare 1:10 lyse buffer with water to lyse red blood cells in isolated PBMCs, if bloody. Warming up the solution will help with the lysing of the red blood cells, as per manufacturer's instructions.
5. Create a master mix cocktail of all antibodies to ensure all samples will receive the same cocktail with less variability between samples. Do not allow master mix cocktails to sit overnight, prepare fresh.
6. Prepare fresh each time for best results.
7. Prepare 1.6% of paraformaldehyde with PBS and store in the dark, can last for 1 week diluted. Use Cell-ID™ Intercalator-Ir at a final concentration of 125 nM.
8. Always remember to filter prior to acquiring on the CyTOF instrument to avoid clogging issues. Filtering twice using a new filter can help to avoid clogging further. Back flushing of the sample line can help get rid of debris in the PSI system and lines.

References

1. Ornatsky OI, Kinach R, Bandura DR et al (2008) Development of analytical methods for multiplex bio-assay with inductively coupled plasma mass spectrometry. J Anal At Spectrom 23:463–468
2. Takahashi C, Au-Yeung A, Fuh F et al (2017) Mass cytometry panel optimization through the designed distribution of signal interference. Cytometry A 91:39–47
3. Nicholas KJ, Greenplate AR, Flaherty DK et al (2015) Multiparameter analysis of stimulated human peripheral blood mono-nuclear cells: a comparison of mass and fluorescence cytometry. Cytometry A 89:271–280
4. Finck R, Simonds EF, Jager A et al (2013) Normalization of mass cytometry data with bead standards. Cytometry A 83(5):483–494. https://doi.org/10.1002/cyto.a.22271

Chapter 4

Method for Tagging Antibodies with Metals for Mass Cytometry Experiments

Stephen Gregory Chang and Cynthia J. Guidos

Abstract

Mass cytometers are time-of-flight (TOF) mass spectrometer-coupled flow cytometers (known as CyTOFs) that quantify the abundance of metal-tagged antibodies (Abs) or other cellular probes within single cell suspensions or laser-ablated tissue sections. While many strategies exist for covalently crosslinking to proteins, the Fluidigm MaxPar® process is currently the most widely used and involves first loading a metal-chelating polymer with an elementally and isotopically enriched metal. Once the chelation sites have been filled, a maleimide moiety on the polymer is reacted with the free thiol groups on the partially reduced monoclonal immunoglobulin G (IgG) Ab to form an irreversible covalent bond. Here we describe modifications to the Fluidigm MaxPar® protocol that increase the quantity of Ab per reaction up to 150 μg, introduce an initial Ab quality control step, utilize metal labels not included in the Fluidigm catalog, and provide an option to perform two reactions in one centrifugal filter.

Key words Metal-chelating polymers, MaxPar, Mass cytometry, CyTOF, Helios, Fluidigm

1 Introduction

Mass cytometry permits a much higher degree of "parameterization" than fluorescence-based flow cytometry and allows researchers to quantify expression of >40 markers/cell in order to resolve complex developmental trajectories and the diverse cell subsets underpinning immune-mediated and other diseases [1]. However, the current third generation "Helios" CyTOF has 135 detection channels, so as additional metals and chemistries become available for conjugating to Abs, many more markers per cell can be measured. To perform mass cytometry, cells or tissue sections are first labeled with a cocktail of metal-tagged antibodies specific for cell surface, cytoplasmic or nuclear markers of cell lineage, differentiation/activation status, proliferation, or viability. In some cases, researchers choose to pre-label individual cell samples with metal tag barcodes [1], allowing multiplexed staining and analysis of multiple samples in a single tube, which limits technical variation

Helen M. McGuire and Thomas M. Ashhurst (eds.), *Mass Cytometry: Methods and Protocols*, Methods in Molecular Biology, vol. 1989, https://doi.org/10.1007/978-1-4939-9454-0_4,

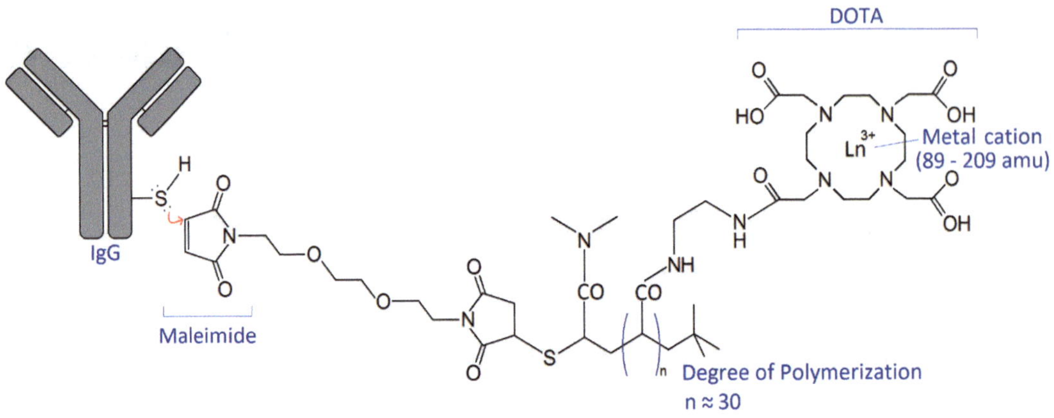

Fig. 1 Schematic of the MaxPar® X8 metal-chelating polymer. Adapted from US Patent 20080003616 A1 and Lou et al. [3, 7], depicting a thiol group of a partially reduced IgG reacting with the maleimide moiety of the polymer

and increases throughput of sample acquisition on the CyTOF. In this article we provide methods that can be used to label metal-tagged antibodies (Abs) with lanthanide or other metals (atomic mass 75–209) for use in mass cytometry.

The Fluidigm MaxPar® metal conjugation kit includes a proprietary polymer that contains approximately 30 repeats of the metal chelator 1,4,7,10-Tetraazacyclododecane-1,4,7,10-tetraacetic acid (DOTA) to bind Ln and other "+3" metal ions [2]. MaxPar® kits that contain either the linear X8 or branching DN3 polymer formats are available [3, 4], but most Abs label well using the X8 polymer. The DOTA polymer also contains a maleimide moiety that forms a stable thioether linkage with thiols of partially reduced IgG at neutral pH (Fig. 1). On average, each IgG molecule will be tagged with two to four polymers [2]. The advantage of this strategy is that intra-chain disulfide bonds between cysteine residues of the IgG fragment crystallizable (Fc) region can be selectively reduced by Tris(2-carboxyethyl)phosphine (TCEP) to form free thiols [5]. Targeting the Fc region decreases the likelihood that the metal-loaded polymers will sterically hinder the antigen binding site. A disadvantage of this approach is that there are fewer available thiol binding sites for the metal-chelating polymer compared to isothiocyanato-benzene (SCN-Bn) chemistries which target primary amines on lysine residues and the N-terminus of each subunit [6]. However, the partial reduction of IgG may diminish shelf life of some metal-tagged Abs compared to those tagged with fluorophores through primary amines [unpublished data]. Here we demonstrate modifications to the Fluidigm protocol PRD002 to increase the quantity of starting material by 50% to conjugate up to 150 μg of IgG per reaction and incorporate an Ab quality control step. We have also used our method to tag Abs with Bismuth, Yttrium, and Indium, metal tags not currently available from Fluidigm as conjugation reagents.

2 Materials

1. MaxPar® Antibody Labeling Kit (Fluidigm) (*see* **Notes 1** and **2**).
2. Stock Ab: Carrier-free (no BSA or gelatin) IgG, 100–150 μg per reaction.
3. 3 kDa Amicon® Ultra-0.5 mL Centrifugal Filters.
4. 50 kDa Amicon® Ultra-0.5 mL Centrifugal Filters.
5. Two microcentrifuges.
6. 37 °C water bath with floating tube rack (55 °C if using indium).
7. 4 mM TCEP solution: 8 μL of 0.5 M Bond-Breaker® TCEP stock solution (Thermo Scientific) with 992 μL of R-Buffer from MaxPar® Antibody Labeling Kit.
8. Stabilization buffer: Antibody Stabilizer PBS (CANDOR Bioscience) with 0.09% Sodium azide.
9. NanoDrop™ or other microvolume spectrophotometer.
10. 50 mM Lanthanide solution: depending which tag you want to conjugate, use isotopically enriched metal salt solutions, provided at 50 mM ranging from 141 praseodymium to 176 ytterbium (Fluidigm) or natural abundance Bismuth(III) nitrate (Millipore Sigma 467839), natural abundance yttrium(III) nitrate (Millipore Sigma 217239), or natural abundance indium(III) chloride (Millipore Sigma 203440).
11. Screw-Cap Microtubes.

3 Methods

All steps are carried out at room temperature using filter pipette tips unless otherwise stated. Furthermore, all steps described below should be performed in sequence without pausing between sections.

3.1 Quantifying the Antibody

1. Centrifuge the stock Ab vial for 10 s at 10,000 RCF in a microcentrifuge.
2. Determine Ab concentration by NanoDrop™ set to IgG using 2 μL of stock Ab, blanking against PBS (*see* **Note 3**).

3. Calculate the volume of Ab solution required to add up to 150 μg for one reaction or up to 300 μg for two reactions (referred to as double reactions).

3.2 Loading the Maxpar® Polymer with Metal Ions

1. Briefly centrifuge the lyophilized polymer tube, resuspend in 95 μL of L-buffer, and mix by pipetting. For double reactions transfer the 95 μL polymer solution to a second lyophilized polymer tube.
2. Add 5 μL of the 50 mM Lanthanide solution for single reactions, or 10 μL for double reactions, to the tube containing the polymer in 95 μL of L-buffer from the previous step (*see* **Note 4**).
3. Vortex and centrifuge polymer metal solution and incubate in a 37 °C water bath for 30–40 min. During this incubation, proceed to the next stage.

3.3 Assessing Antibody Integrity

1. Transfer up to 150 μg of Ab (or up to 300 μg for double reactions) into a 50 kDa centrifugal filter. If the volume is greater than 500 μL, add the remainder after the following centrifugation step and repeat if necessary.
2. Centrifuge the 50 kDa filter containing the Ab for 4 min at 12,000 RCF.
3. Transfer the 50 kDa filter to a new microcentrifuge tube, keeping the previous filtrate for protein quantification (*see* **Note 5**).

3.4 Reducing the Antibody

1. Add 400 μL of R-buffer to the 50 kDa filter containing the concentrated Ab and centrifuge for 10 min at 12,000 RCF.
2. During the previous step's centrifugation, prepare the 4 mM TCEP solution.
3. Discard the filtrate and resuspend the Ab in 100 μL of the previously prepared 4 mM TCEP solution per reaction (200 μL for double reactions), mix gently by pipetting avoiding contact with the filter membrane, then incubate in a 37 °C water bath for 30 min (*see* **Note 6**).

3.5 Purifying the Polymer

1. Add 200 μL of L-buffer to a 3 kDa filter.
2. Recover the metal-loaded polymer solution from the water bath and transfer to the 3 kDa filter.
3. Centrifuge the 3 kDa filter for 25 min at 12,000 RCF (*see* **Note 7**).
4. Add 400 μL of C-buffer and centrifuge again for 30 min at 12,000 RCF.

3.6 Purifying the Antibody

1. At precisely 30 min after adding the TCEP, remove the Ab from the water bath and add 300 μL of C-buffer (*see* **Note 8**).

2. Centrifuge the 50 kDa filter containing the Ab for 4 min at 12,000 RCF.
3. Discard the filtrate and add 400 μL of C-buffer to the Ab retentate in the 50 kDa filter, repeat steps 2 and 3.
4. Centrifuge the 50 kDa filter for 10 min at 12,000 RCF.

3.7 Conjugation

1. Resuspend the polymer retentate in the 3 kDa filter with 60 μL C-buffer.
2. Transfer the resuspended polymer to the partially reduced Ab retentate in the 50 kDa filter.
3. Mix gently by pipetting.
4. Incubate in a 37 °C water bath for 2 h.

3.8 Buffer Exchange and Storage

1. Recover the polymer-conjugated Ab from the water bath.
2. Add 300 μL of W-buffer and centrifuge the 50 kDa filter for 4 min at 12,000 RCF.
3. Discard the filtrate, add 400 μL of W-buffer, and centrifuge for 4 min at 12,000 RCF (repeat twice more, increasing the centrifugation time to 10 min the second time).
4. Resuspend the ~20 μL of concentrated Ab with 80 μL of W-buffer (or 180 μL for double reactions), mix thoroughly and gently by pipetting. The total volume will be ~100 μL per reaction.
5. Quantify the protein concentration in the resuspended retentate by NanoDrop™ (*see* **Note 9**).
6. Centrifuge for 10 min at 12,000 RCF to concentrate the Ab a final time; the concentrated volume will be ~20 μL.
7. Calculate the volume required to dilute the concentrated Ab to 0.5 mg/mL and resuspend the concentrated Ab in this volume of stabilization buffer (*see* **Note 10**).
8. Elute the conjugated Ab in stabilization buffer from the 50 kDa filter by inverting the filter into a new microcentrifuge tube and centrifuging for 2 min at 1000 RCF, or pipette directly out of the filter.
9. Transfer to a screw-cap microtube and store at 4 °C.

4 Notes

1. While the contents of the included buffers are proprietary, a publication by Fluidigm scientists and collaborators suggests they are likely as described [7] and listed here:
   - R-buffer—100 mM phosphate buffer (pH 7.2) with 2.5 mM EDTA

- C-buffer—20 mM Tris-buffered saline (TBS, pH 7.0)
- L-buffer—20 mM ammonium acetate (pH 6.0)
- W-buffer—20 mM Tris-buffered saline (TBS, pH 7.0)
- Maxpar® Metal-Chelating Polymer
- Lanthanide solution—50 mM metal salt in L-buffer (*see* **Note 2**)

2. Fluidigm provides a choice of 35 different isotopically enriched 50 mM metal salt solutions as part of their MaxPar® conjugation kit that range from 141 praseodymium to 176 ytterbium, and are working to increase their catalog. Additional isotopically enriched metal salts can be purchased from Trace Sciences International (Canada and USA) or BuyIsotope (Sweden) and natural abundance metal salts can be purchased from Millipore Sigma.
3. The Fluidigm protocol recommends users quantify the Ab concentration by "blanking against the buffer they are suspended in." While this would be ideal, this is not practical as in a given day we are likely conjugating antibodies from several different vendors and a detailed description of buffer components or reference solution to blank is not always available. For calculating the volume required to load up to 150 μg per reaction we will use the lower concentration value of either the NanoDrop blanked against PBS with an extinction coefficient of 210,000 $M^{-1}cm^{-1}$ or what is printed on the label.
4. Prepare a 50 mM solution of the metal salts by dissolving in L-buffer. If using bismuth(III) nitrate, solubility in L-buffer will be poor and undissolved solids will be present; therefore after vortexing thoroughly, centrifuge at maximum RCF for 2 min and add 25 μL of this supernatant per reaction instead of 5 μL. Abs conjugated to yttrium(III) nitrate are expected to have lower signal intensity since mass channel 89 has the lowest sensitivity on the mass response curve of the instrument. Fluidigm sells higher intensity 89Y conjugated antibodies but they are using a different proprietary polymer in order to increase metal quantity and improve signal. If using Indium(III) chloride, add the 50 mM solution to the polymer, mix by pipetting and then incubate at 55 °C instead of 37 °C. Maleimide ring opening may occur at higher temperatures; if that is a concern, longer incubations at 37 °C may be preferable but we have not explored this option. Some precipitation of the polymer may occur but our experience has been that the conjugation results are still satisfactory. Palladium(III) nitrate completely precipitates the X8 polymer rendering it ineffective by this protocol. It is important that the metal cations are loaded onto the polymer and the excess washed away prior to coming in contact with the IgG since metal cations can

independently bind to the IgG which can cause cleavage of the protein due to hydrolysis or oxidation [8]. Finally, inspect the polymer tube for cracks that may occur during shipping and attempt to vissualize the lyophilized polymer by holding the tube up to a light source while adding the L-buffer.

5. If the NanoDrop quantification of the first spin filtrate shows that >3% of the A280 nm absorbing material has passed through the filter, this is indicative that there may be an issue with the quality or purity of the starting material. More Ab can be added to bring the quantity of the retained Ab up to approximately 150 μg (or 300 μg for double reactions). However, if the starting material showed an excessively higher than expected initial protein concentration, i.e., 1.0 mg/mL when the label stated 0.5 mg/mL and approximately 50% passed through the filter, do not be alarmed. In this case, it is likely the result of UV absorbing buffer components, for instance in buffers used for some lots sold by BD Biosciences, and the appropriate quantity of starting material will be approximated in the retentate if the lower of the two concentration values was used. Ab formulations containing glycerol will initially filter at a slower rate and require additional rounds of centrifugation with R-buffer prior to advancing to the reduction steps.
6. When mixing solutions containing Ab by pipetting, avoid generating bubbles. The TCEP reduction step is precisely timed; do not attempt more than eight tubes per batch. TCEP is used since it does not contain sulfhydryls that would otherwise interfere with the maleimide reaction chemistry.
7. Unchelated metal ions will pass through the 3 kDa filter while retaining the larger ~10kDa polymer.
8. Move quickly to minimize the time from the end of the 30 min incubation to starting the centrifuge. If conjugating multiple antibodies at once, open only one tube at a time to prevent cross contamination.
9. The Fluidigm protocol suggests expected recoveries will be in the order of 60%. Losses can occur when A280 absorbing starting material such as peptide fragments pass through the 50 kDa filter, either initially or during later steps resulting from an over-reduction of the IgG into constituent subunits. Proteins will also bind to the centrifugal filter membrane and losses can occur as a result of residual volume on pipettes. Our experience has been that "BioLegend MaxPar Ready" formulations, in which the vendor has performed the initial buffer exchange step, typically yield an average recovery of 80% with less variability than lots that have not previously been buffer exchanged on 50 kDa Amicon® filters prior to the initial

quantification. If loss is occurring during the initial centrifugations, the "Assessing Antibody Integrity" steps described in Subheading 3.3 provide the opportunity to detect and correct this deficiency before costly reagents are consumed. Losses can also occure from mechanical insult to the filter membrane with a pipette tip which will comprimize the filters ability to retain the Ab, or from excessive delay after centrifugation steps.

10. Stabilization buffer in the final volume is essential to prolonging shelf life. We recommend re-titrating conjugated antibodies 1 year after conjugation or sooner if the marker intensity shows signs of diminishing relative to initial titration values. The stabilization buffer contains a significant concentration of protein that prevents accurate quantification of the Ab concentration. Perform NanoDrop™ quantification and/or SDS-PAGE analysis if desired prior to adding the stabilization buffer.

Acknowledgments

This protocol was prepared with the support of the Ontario Institute for Cancer Research through funding provided by the Government of Ontario, The SickKids Centre for Advanced Single Cell Analysis (CASCA), and discussions with Fluidigm Corporation.

References

1. Spitzer MH, Nolan GP (2016) Mass cytometry: single cells, many features. Cell 165(4):780–791. https://doi.org/10.1016/j.cell.2016.04.019
2. Tanner SD, Bandura D, Ornatsky O et al (2008) Flow cytometer with mass spectrometer detection for massively multiplexed single-cell biomarker assay. Pure Appl Chem 80(12):2627–2641. https://doi.org/10.1351/pac200880122627
3. Loboda AV, Bandura DR, Baranov VI (2016) Mass cytometry apparatus and methods. US Patent 20160195466 A1
4. Winnik M, Nitz M, Baranov V et al. (2008) Polymer backbone element tags US Patent 20080003616 A1
5. Burns JA, Butler JC, Moran J et al (1991) Selective reduction of disulfides by tris(2-carboxyethyl)phosphine. J Organomet Chem 56(8):2648–2650. https://doi.org/10.1021/jo00008a014
6. Mei HE, Leipold MD, Schulz AR et al (2015) Barcoding of live human PBMC for multiplexed mass cytometry. J Immunol 194(4):2022–2031. https://doi.org/10.4049/jimmunol.1402661
7. Lou X, Zhang G, Herrera I et al (2007) Polymer-based elemental tags for sensitive bioassays. Angew Chem Int Ed Eng 46(32):6111–6114. https://doi.org/10.1002/anie.200700796
8. Glover ZK, Basa L, Moore B et al (2015) Metal ion interactions with mAbs: part 1 pH and conformation modulate copper-mediated site-specific fragmentation of the IgG1 hinge region. MAbs 7(5):901–911. https://doi.org/10.1080/19420862.2015.1062193

Chapter 5

Scalable Conjugation and Characterization of Immunoglobulins with Stable Mass Isotope Reporters for Single-Cell Mass Cytometry Analysis

Felix J. Hartmann, Erin F. Simonds, Nora Vivanco, Trevor Bruce, Luciene Borges, Garry P. Nolan, Matthew H. Spitzer, and Sean C. Bendall

Abstract

The advent of mass cytometry (CyTOF®) has permitted simultaneous detection of more than 40 antibody parameters at the single-cell level, although a limited number of metal-labeled antibodies are commercially available. Here we present optimized and scalable protocols for conjugation of lanthanide as well as bismuth ions to immunoglobulin (Ig) using a maleimide-functionalized chelating polymer and for characterization of the conjugate. The maleimide functional group is reactive with cysteine sulfhydryl groups generated through partial reduction of the Ig Fc region. Incubation of Ig with polymer pre-loaded with lanthanide ions produces metal-labeled Ig without disrupting antigen specificity. Antibody recovery rates can be determined by UV spectrophotometry and frequently exceeds 60%. Each custom-conjugated antibody is validated using positive and negative cellular control populations and is titrated for optimal staining at concentrations ranging from 0.1 to 10 μg/mL. The preparation of metal-labeled antibodies can be completed in 4.5 h, and titration requires an additional 3–5 h.

Key words CyTOF, Mass cytometry, Conjugation, Antibody, Immunoglobulin, IgG, Lanthanide, Bismuth, Isotope, Titration

1 Introduction

1.1 Background

Multicellular biological systems require the interplay of a diversity of cellular phenotypes. The advent of monoclonal antibody technology allowed the precise definition of many cellular phenotypes, first in the immune system, and more recently in solid tissues and tumors. Cellular phenotypes are most often defined by combinations of extracellular surface antigens. Many of these markers, such as those best understood in the immune system, are gained and lost during developmental maturation and in response to environmental stimuli. As the knowledge of cellular roles, particularly among immune cell subtypes, has become increasingly detailed, phenotypic definitions have begun to involve intracellular

Helen M. McGuire and Thomas M. Ashhurst (eds.), *Mass Cytometry: Methods and Protocols*, Methods in Molecular Biology, vol. 1989,
https://doi.org/10.1007/978-1-4939-9454-0_5,

regulatory proteins or protein modifications that can act as proxies for cellular function: these include transcription factors, cytokines, and post-translational modifications of signaling proteins (e.g., phosphorylation, acetylation, cleavage).

Multiparametric single-cell fluorescence cytometry platforms are unique in their ability to measure multiple features per cell on thousands or millions of cells per experiment—allowing for quantitative capturing of subtle or wholesale shifts in cell subset frequencies and marker expression across diverse cellular phenotypes. The absolute number of cellular components that can be measured simultaneously on each cell is limited in fluorescence-based cytometry by constraints inherent in light-based measurements and emission spectral overlap of available fluorophores. This restriction in the number of simultaneous measurements has limited the scope of inquiry regarding the biological system under study such as human immune system states and cancers.

Recognizing the need for increased simultaneous measurement and quantification on a per cell basis, the Tanner group at the University of Toronto created a new detection modality, in which antibodies were tagged with stable heavy metal isotopes and quantified using a technology called inductively coupled plasma mass spectrometry (ICP-MS) [1]. By combining sensitive and highly multiparametric ICP-MS immunoassay technology with a single-cell acquisition source, Tanner and colleagues produced the first mass cytometry platform, which was later released commercially as the CyTOF® [2].

To date, this technology has been employed to analyze multiple facets of biology, biochemistry, and molecular regulation at the single-cell level. Using the protocols described herein our groups have created numerous customized antibody panels, including those for studying human hematopoiesis and regulatory cell signals [3]; cell and context specific kinase inhibitor activity in a high-throughput assay [4]; broad facets of cell cycle across the human hematopoietic compartment [5]; cellular apoptosis and necrosis [6]; activation profiles of virus-specific cytotoxic T cells; and comparison of regulatory phosphorylation kinetics governing T cell receptor activation across different populations [7].

Using CyTOF®, more than 40 antibody-based parameters can be analyzed simultaneously at the single-cell level [3, 4, 8], though reagents for its implementation are currently not as widespread as fluorophore-conjugated materials. Additionally, the pursuit of new biology combined with the enormity of possible combinations of measurement reagents will likely always necessitate the creation of novel, custom-conjugated antibodies. Designing an optimal panel to investigate relevant biological questions of interest requires custom conjugation of purified antibodies with heavy metal ions such that bound antibodies can be detected by inductively coupled plasma time-of-flight mass spectrometry (ICP-TOF-MS).

To standardize and optimize this process, we have developed a protocol for labeling of purified monoclonal and polyclonal antibodies designed to maximize conjugation efficiency while maintaining desired binding affinity.

This antibody labeling protocol utilizes MaxPar® chelating polymers commercially available from Fluidigm. These water-soluble polymers contain a sulfhydryl-reactive bismaleimide group and a trivalent metal cation-chelating 1,4,7,10-tetraazacyclododecane-1,4,7,10-tetraacetic acid (DOTA) or diethylene triamine pentaacetic acid (DTPA) group. The binding affinities of DOTA and DTPA for lanthanide metal reporters approach 10^{-16} M [9]. Importantly, an advantage of this platform is that there are 44 mass channels within the optimal mass cytometry measurement range that can be occupied by stable isotopes of transition metals, mainly the lanthanide series. These transition metals readily form stable trivalent salts in solution with an oxidation state of (III) (Table 1). Consequently, antibody reagents for all 44 mass channels can be created using the protocol described herein.

The process of antibody conjugation has three fundamental steps:

1. The MaxPar® polymer is incubated in a trivalent metal salt solution to facilitate chelation with high efficiency.
2. To generate sites for maleimide labeling, the purified, carrier-free immunoglobulin, type G (IgG) is incubated in a low concentration of tris-carboxyethyl phosphine (TCEP) to preferentially reduce disulfide bonds within the F_c region, ideally without compromising the antigen specificity of the F_{ab} region [10, 11].
3. Lastly, the chelated polymer and partially reduced antibody are mixed and incubated to achieve full conjugation.

Notably, antibodies, as well as many buffer components such as phosphates, have a propensity to precipitate out of solution containing even low concentrations of free lanthanide ion; therefore, it is critical to chelate metal ions and wash the polymer thoroughly prior to addition of the reduced antibody. Furthermore, the tertiary structure of Fab regions of antibodies incubated under harsh reducing conditions may become altered due to disruption of structural disulfide bonds. As such, the concentration of Tris-carboxyethyl phosphine (TCEP) in reducing buffer and the length of incubation are critical, and the protocol below should be followed precisely as described. To maximize conjugation efficiency, we utilize a series of buffers optimized for metal loading, antibody reduction, polymer coupling to the antibody, and subsequent washing. It is also essential to ensure that carrier protein (i.e., BSA) is removed from the antibody solution as the carrier may compete for free maleimide groups of the polymer. If the monoclonal or

Table 1
A summary of compatible trivalent stable elemental isotopic reporters, commercial sources, and relative sensitivities and performance notes and common interferences in mass cytometry assays

Isotopic mass	Element (symbol)[a]	Commercial availability[b]	Relative sensitivity[c]	Most common interference[d] *Ba oxide or hydroxide (contaminant)
113	Indium (In)	P	0.1	^{113}In impurity from another In
115	Indium (In)	P	0.1	^{115}In impurity from another In
139	Lanthanum (La)	N	0.3	^{138}Ba (contamination) +1 signal bleed
140	Cerium (Ce)	P	0.3	
141	Praseodymium (Pr)	M, N	0.3	
142	Neodymium (Nd)	M, P	0.4	^{142}Nd impurity from another Nd
143	Neodymium (Nd)	M, P	0.4	^{143}Nd impurity from another Nd
144	Neodymium (Nd)	M, P	0.5	^{144}Nd impurity from another Nd
145	Neodymium (Nd)	M, P	0.5	^{145}Nd impurity from another Nd
146	Neodymium (Nd)	M, P	0.5	^{146}Nd impurity from another Nd
147	Samarium (Sm)	M, P	0.6	^{147}Sm impurity from another Sm
148	Neodymium (Nd)	M, P	0.6	^{148}Sm impurity from an Sm reporter
149	Samarium (Sm)	M, P	0.6	^{149}Sm impurity from another Sm
150	Neodymium (Nd)	M, P	0.7	^{150}Sm impurity from an Sm reporter
151	Europium (Eu)	M, P	0.7	*
152	Samarium (Sm)	M, P	0.7	*; ^{152}Gd impurity from a Gd reporter
153	Europium (Eu)	M, P	0.8	*
154	Samarium (Sm)	M, P	0.8	*
155	Gadolinium (Gd)	M, P	0.8	*; ^{139}La Oxide; ^{155}Gd impurity from another Gd
156	Gadolinium (Gd)	M, P	0.9	^{140}Ce Oxide; ^{156}Gd impurity from another Gd
157	Gadolinium (Gd)	P	0.9	^{141}Pr Oxide; ^{157}Gd impurity from another Gd
158	Gadolinium (Gd)	M, P	0.9	^{142}Nd Oxide; ^{158}Gd impurity from another Gd
159	Terbium (Tb)	M, N	1	^{143}Nd Oxide
160	Gadolinium (Gd)	M, P	1	^{144}Nd Oxide; ^{160}Dy impurity from a Dy reporter
161	Dysprosium (Dy)	M, P	1	^{145}Nd Oxide; ^{161}Dy impurity from another Dy

(continued)

Table 1 (continued)

Isotopic mass	Element (symbol)[a]	Commercial availability[b]	Relative sensitivity[c]	Most common interference[d] *Ba oxide or hydroxide (contaminant)
162	Dysprosium (Dy)	M, P	1	^{146}Nd Oxide; ^{162}Dy impurity from another Dy
163	Dysprosium (Dy)	M, P	1	^{147}Sm Oxide; ^{163}Dy impurity from another Dy
164	Dysprosium (Dy)	M, P	1	^{148}Nd Oxide; ^{164}Dy impurity from another Dy
165	Holmium (Ho)	M, N	1	^{149}Sm Oxide
166	Erbium (Er)	M, P	1	^{150}Nd Oxide; ^{166}Er impurity from another Er
167	Erbium (Er)	M, P	1	^{167}Er impurity from another Er
168	Erbium (Er)	M, P	1	^{152}Sm Oxide; ^{168}Er impurity from another Er
169	Thulium (Tm)	M, N	1	
170	Erbium (Er)	M, P	0.9	^{154}Sm Oxide; ^{170}Yb impurity from a Yb reporter
171	Ytterbium (Yb)	M, P	0.9	^{155}Gd Oxide; ^{171}Yb impurity from another Yb
172	Ytterbium (Yb)	M, P	0.9	^{156}Gd Oxide; ^{172}Yb impurity from another Yb
173	Ytterbium (Yb)	M, P	0.8	^{157}Gd Oxide; ^{173}Yb impurity from another Yb
174	Ytterbium (Yb)	M, P	0.8	^{158}Gd Oxide; ^{174}Yb impurity from another Yb
175	Lutetium (Lu)	M, P	0.8	^{159}Tb Oxide
176	Ytterbium (Yb)	M, P	0.8	^{160}Gd-oxide; ^{176}Lu impurity from a Lu reporter
197	Gold (Au)	N	0.3	High nonspecific binding
203	Thallium (Tl)	P	0.5	^{203}Tl impurity from another Tl
205	Thallium (Tl)	P	0.5	^{205}Tl impurity from another Tl
209	Bismuth (Bi)	N	0.5	^{193}Ir Oxide (DNA intercalator)

[a]Most abundant elemental isotope with that mass. Other stable elemental isotopes with the same mass may exist

[b]*M* MaxPAR labeling kit, *P* source (III) purified chloride or nitrate isotope from supplier (e.g., Trace Sciences International), *N* natural (III) chloride or nitrate element is >99.9% monoisotopic

[c]Based on the original mass cytometry specifications [2]. Transmission efficiency varies between instruments. This should serve only as a guide

[d]Most significant source(s) of signal interference, if any, in mass cytometry assays. Other isotopic contaminants and ion adducts (i.e., oxidation) may not be listed. For a more complete list *see* Ornatsky et al. [21] and Table 2

polyclonal antibody preparation to be used includes carrier protein, these proteins may be removed by Melon gel (Pierce) or Protein G column purification. Using purified antibodies, the coupling reaction of polymer to partially reduced antibody approaches completion, resulting in covalent binding of around six polymers to each IgG.

1.2 Experimental Considerations

The overriding advantage of mass cytometry over other biological analysis platforms is its ability to acquire single-cell, 40–50 parameter data with negligible cross-channel overlap. Although this cytometric analysis is best known as a method for analyzing immunologic subpopulations [3, 8], the utility of high-parameter single-cell analysis goes well beyond blood and bone marrow—it is useful in virtually any biological system that can be prepared as a single-cell suspension. For example, the multiparametric and quantitative data afforded by mass cytometry is valuable in studies of solid-tissue stem cell differentiation, cellular heterogeneity in tumors [12, 13], coordinated antitumor immunity across an entire organism [14], defining cellular phenotypes in autoimmunity [15], tumor-stroma interactions, stochasticity and kinetics in cell culture models of apoptosis and cell cycle [5, 6], network analysis of the interplay between phosphorylation events, and as a highly parallelized drug screening platform capable of monitoring many kinase targets simultaneously [4].

The common theme of most of these applications is the need for simultaneous detection of many cellular components at a single-cell level to enable the fine-grained discrimination of many heterogeneous cell types, to efficiently multiplex many functional assays in a single analysis, or to gain insights into correlated phenomena and emergent properties that can only be mined from quantitative, high-dimensional data [16, 17]. These high-dimensional single-cell approaches create a challenge and an opportunity to leverage the mutual information encoded in these experiments to make new models of human systems and take unsupervised approaches to cellular identification and disease classification [12, 18–20].

One minor limitation is the nature of the immunoglobulins used with success in this protocol. IgGs from mouse, rat, rabbit, goat, and sheep have been tested extensively and are compatible with the protocol described here. Attempts to use type E and type M immunoglobulins (Igs) have been less successful presumably due to disulfide bonds in these antibody types which are required to maintain structure and binding activity. We have also occasionally observed that labeling Igs from an Armenian hamster background decreases binding activity.

1.3 Experimental Design

When designing a mass cytometry experiment, the researcher must first take into consideration the technological limitations of the current generation of mass cytometry instruments: (i) The expected data capture rate is only ~30–50% of the input number of cells for each sample depending on the staining protocol and cell introduction system. (ii) The cell acquisition rate should be limited to 500–1000 cells/s to maximize data quality and avoid double cell events. (iii) This is a destructive method. Cells are vaporized as they enter the plasma ionization source; thus no cell recovery is possible. (iv) Lastly, investigator-specific reagents must be labeled and tested as detailed in the protocol herein. Although many experiments have been designed to circumvent these limitations, there are some experiments that are simply not well suited for mass cytometry analysis. Mainly, experiments that involve an extremely rare cell population such as hematopoietic progenitors [3] or antigen-specific T cells [8] require either very large numbers of cells, some sort of pre-enrichment with the use of carrier cells to be discriminated in silico, or both.

Mass cytometry and the use of elemental isotopic reporters overcome many of the limitations inherent with fluorescent or colorimetric reporter technologies, namely spectral overlap and background signal inherent in biological samples. Still, there are sources of interference that can confound analysis, namely oxide formation, isotopic contamination, and, to a lesser extent, spectral overlap ("signal bleed") when strong signals are adjacent to weaker ones on the mass scale (Fig. 1). Isotopic reporter signals that differ by >10^2 may cause visible spectral bleed from the higher-abundance ion into the lower-abundant mass channel next to it on the mass scale. For example, a small percentage of a strong ^{169}Tm signal may be present in the ^{170}Er measurement window (Fig. 1a) despite no expected signal. Because the magnitude of this "signal bleed" is only a very small proportion of the overall interfering signal (^{169}Tm), if the expected signal in the channel targeted by the interference (^{170}Er) is not 100 times (two orders of magnitude) lower than the interfering signal, it is not expected to confound the analysis. Additionally, most of this spectral overlap in time-of-flight mass spectrometry, as in the CyTOF mass cytometer, tails in the M + 1 direction (i.e., the ^{169}Tm has a higher propensity to be measured as ^{170}Er rather than the other way around).

The two primary sources of convoluting reporter signal interference are isotopic impurities (Fig. 1b) and oxide formation during ionization (Fig. 1c). Isotopic impurities are a result of incomplete purification from the naturally occurring element, which can be a mixture of stable isotopes. As a general rule, the most common isotopic contaminants are in those elements that have +1 or −1 mass adjacent isotopes. For example, the contaminants in 97% pure ^{145}Nd are likely ^{144}Nd and ^{146}Nd. In contrast, ^{151}Eu can be obtained in 99% purity with little or no contaminating ^{153}Eu (the only other stable Eu isotope) because the additional mass resolution allows for more efficient purification. A profile of

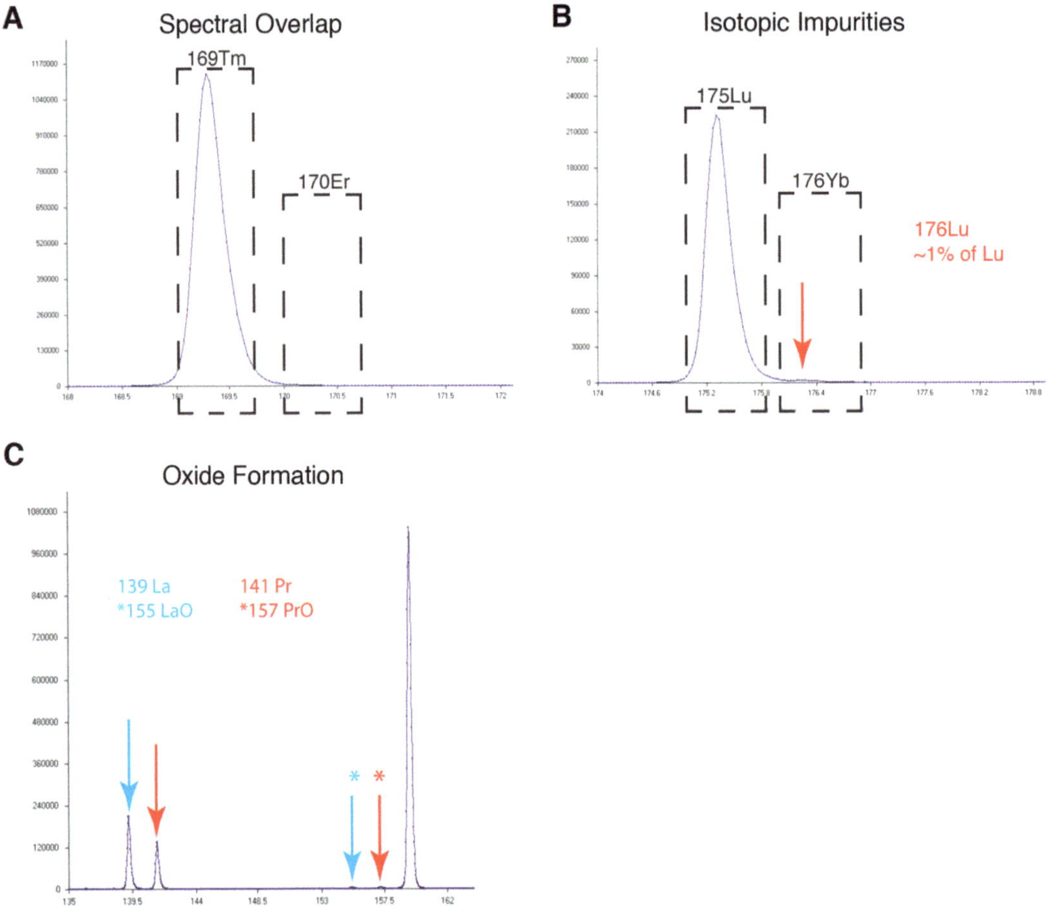

Fig. 1 Sources of signal interference in mass cytometry assays as demonstrated by analyzing a mixture of metal standard (^{139}La, ^{141}Pr, ^{159}Tb, ^{169}Tm, and ^{175}Lu) on a CyTOF mass cytometer. (**a**) Spectral overlap from the ^{169}Tm standard into the ^{170}Er measurement channel. ^{169}Tm is naturally mono-isotopic with no expectation of a constituent with mass 170. (**b**) Natural Lu, used to make the standard in this experiment, contains ~99% ^{175}Lu and 1% ^{176}Lu. A distinct peak corresponding to the ^{176}Lu isotopic "contaminant" can be seen in the measurement window that would be used for ^{176}Yb labeled reagents (red arrow). The dashed black boxes in (**a**) and (**b**) represent the measurement window where all signals within that range would be attributed to the indicated elemental isotope. (**c**) La and Pr oxide formation is indicated by the blue and red arrows, respectively. Here the resulting "+16 Da" oxide interference peak is indicated by a "*" of a matching color

the contaminating isotopes is generally available from the supplier of the isotopically enriched material. Other heavy elements (La, Pr, Tb, Ho, and Tm) are natural or near natural (>99.9% pure) mono-isotopes; therefore isotopic contamination is also not a concern in these cases. These natural mono-isotopes and series of mass adjacent isotopic reporters are also noted in Table 1 and were reviewed in Ornatsky et al. [21].

Oxide, and to a much lesser extent hydroxide, formation occurs during analysis when incomplete vaporization/ionization results in the formation of adducts between the elemental isotopic reporter with oxygen or a hydroxide, thus adding 16 or 17 atomic mass units (amu) to the reporter mass, respectively [22]. For example, ^{139}La, one of the most likely elements to oxidize, will also have a small signal in the ^{155}Gd channel resulting from the oxidation product (^{139}La+^{16}O) (Fig. 1c). While oxidation creates this M + 16 artifact for certain metal isotope reporters, the mass cytometer is tuned such that this occurrence is typically 3% or less of the total signal. The level of oxidation is dependent on the oxide bond strength of the atomic ion of the reporter element and the local temperature of the plasma. The relative likelihoods of oxide or hydroxide formation for common mass cytometry elements are summarized in Table 2. The most significant sources of interference due to oxide formation combined with common isotopic interferences are summarized in Table 1 and should be considered in the context of expected signal levels when selecting panels of elemental isotopic reporters for mass cytometry analyses. For example, antibodies against low-abundance antigens should not be conjugated to an isotope that will receive considerable oxidation from a reporter metal used to label an antibody targeting a high-abundance, or "bright," marker.

Finally, the CyTOF utilizes TOF mass measurement in conjunction with a series of mass filters in order to remove overly abundant ions inherent in biological samples as well as the ions from argon plasma that can be detrimental to the detector. Current mass cytometry instrumentation can be tuned with a mass window of approximately 130 amu. While this allows the detection of stable isotopic reporters with masses between ~80 and 238 amu, the peak sensitivity typically lies between 160 and 170 amu and drops off sharply towards the low mass end and more gradually towards the higher end. The relative sensitivity of various isotopic mass reporters is summarized in Table 1 and should be considered in the context of the expected antigen expression level on a cell when assigning a mass reporter to an antibody. Specifically, low-abundance or "weak" antigens are often best measured using reporter isotopes in the peak sensitivity range. However, because these channels can receive bleed from the oxidation of lighter isotopes, effective panel design aims to optimize these factors to minimize confounding signal as is routinely performed for fluorescence-based methods.

2 Materials

2.1 General Considerations

Barium is a commonly occurring element in many detergent products and is frequently found at high concentrations in bottles cleaned in laboratory dishwashers. For optimum sensitivity and

Table 2
The expected relative occurrence of oxide (M + O) and hydroxide (M + OH) ion adducts during ICP-MS analysis of metals commonly present in mass cytometry assays

	Relative to LaO occurrence[a]	
Element (symbol)	Oxide level (M + ^{16}O)	Hydroxide level (M + ^{16}O^{1}H)
Barium (Ba)	0.04	0.05
Lanthanum (La)	1.0	0.09
Cerium (Ce)	1.0	0.07
Praseodymium (Pr)	0.7	0.05
Neodymium (Nd)	0.58	0.03
Samarium (Sm)	0.11	0.01
Europium (Eu)	0.02	–
Gadolinium (Gd)	0.30	0.08
Terbium (Tb)	0.25	0.02
Dysprosium (Dy)	0.11	0.01
Holmium (Ho)	0.10	0.00
Erbium (Er)	0.09	0.00
Thulium (Tm)	0.03	0.00
Ytterbium (Yb)	0.01	0.00
Lutetium (Lu)	0.07	0.01

The frequency of occurrence has been normalized to the ratio of LaO/La acquired under the same conditions

[a]Based on the published ratios of oxide (M + O/M) or hydroxide (M + OH/M) measured by ICP-MS analysis [22], normalized to the expected ratio of LaO/La in the same experiment and accounting for the relative mass sensitivities as reported in Table 1. This frequency of LaO occurrence is a common metric in tuning oxidation levels in mass cytometry experiments. For example, if the level of LaO was found to be 2% of the total La signal then the expected GdO level would only be 0.6% of the total Gd signal. Note that the abundance of oxide ions also depends on local plasma temperature and thus may deviate slightly from these values.

lifetime of the CyTOF® instrument, wash solutions used in the immunostaining protocol must be nearly free of barium (<1 ppb). Many commercially available biological reagents (e.g., 500 mL liquid bottles of liquid GIBCO DPBS) contain high amounts of barium and should be avoided in later steps of the staining protocol. The recipes below are specifically designed to avoid barium contamination, but it is recommended that each laboratory test every wash solution in the workflow for barium contamination before running the first set of samples on the CyTOF.

2.2 Reagents

1. 100–500 μg immunoglobulin.
2. Tris-carboxyethyl phosphine (TCEP), neutral pH (0.5 M in 10 μL aliquots) (Thermo Scientific).
3. MaxPar X8 or DN3 Antibody Labeling Kit (Fluidigm): MaxPar X8 or DN3 Antibody Labeling Reagent (one test per 0.1 mg of Ig), L-Buffer (suitable substitution: 20 mM ammonium acetate, pH 6), R-Buffer (suitable substitution: 0.1 M phosphate buffer with 2.5 mM EDTA, pH 7.2), C-Buffer (suitable substitution: *tris*-buffered salt with 1 mM EDTA, pH 7.5), W-Buffer (suitable substitution: *tris*-buffered salt, pH 7.5), trivalent metal lanthanide solution (0.05 M stocks of XCl3 or $X(NO_3)_3$ in L-buffer, where X is the elemental metal isotope).
4. Modified C-Buffer: 150 mM *tris*, 150 mM NaCl, 1 mM EDTA, pH 7.5.
5. 10× PBS: 320 g of NaCl, 8 g of KCl, 46 g of $Na_2HPO_4{\cdot}7H_2O$, and 8 g of KH_2PO_4 in 3 L of ddH_2O. Bring solution to pH 7.4 using concentrated aqueous NaOH. Bring volume to 4 L with ddH_2O. To create a 1× stock mix one part of 10× stock with nine parts ddH_2O.
6. FACS buffer: 500 mL PBS, 2.5 g BSA (final concentration: 0.5% wt/vol), and 100 mg sodium azide NaN_3 (final concentration: 0.02% wt/vol). Store at 4 °C for up to 4 months.
7. Antibody stabilization buffer: 0.1% (wt/vol) (NaN_3) in Antibody Stabilizer solution. Store at 4 °C for up to several years.
8. 16% Paraformaldehyde (PFA) ampules (wt/vol in water).
9. DNA intercalator: 1 mL PBS, 100 μL of 16% PFA, and 0.25 μL of Ir-Intercalator (500 μM stock concentration, Fluidigm).
10. Methanol, store at 4 °C.
11. Positive and negative control cells (or cell populations) of interest.
12. Monensin Solution (1000×, Biolegend).
13. Brefeldin A Solution (1000×, Biolegend).

2.3 Equipment

1. Amicon Ultra-0.5 mL Centrifugal Filter Unit with Ultracel-3 membrane (Millipore).
2. Amicon Ultra-0.5 mL Centrifugal Filter Unit with Ultracel-50 membrane (Millipore).
3. Amicon Ultrafree Durapore 0.1 μm PVDF 0.5-mL centrifugal filters (Millipore).
4. Screw-top Eppendorf tubes.
5. Pipettes.
6. Filter pipette tips.

7. Polystyrene round bottom test tubes (FACS tubes).
8. Pasteur pipettes.
9. Vacuum flask connected to vacuum line.
10. Water bath heated to 37 °C.
11. Room temperature centrifuge with rotor for microtubes.
12. Refrigerated centrifuge with rotor for FACS tubes.
13. A low volume (μL) UV/Vis spectrophotometer.
14. CyTOF mass cytometer.
15. Flow cytometry analysis software (we use Cytobank [23], www.Cytobank.org).

3 Methods

3.1 Antibody Conjugation

The following conjugation procedure has some critical timing steps that are important to the success of the protocol. These steps are noted in text and summarized in a workflow diagram in Fig. 2. In the main text, we describe the conjugation of 100 μg of immunoglobulin. Due to the inevitable loss of a certain amount of protein during the conjugation procedure, conjugation of <100 μg in one reaction is not recommended. However, conjugation of multiples of 100 μg is possible and might be desirable in cases in which frequent usage is anticipated and where potential batch effects between different conjugations are to be avoided. In order to conjugate >100 μg (we have tested conjugation of up to 500 μg), the presented protocol has to be slightly modified which will be pointed out and referred to in the Notes section. Conjugation of antibodies to Bismuth (^{209}Bi) requires slightly adjusted reagents and again, this will be referred to in the Notes section.

3.1.1 Preloading the Chelating Polymer (40 Min)

1. Per 100 μg of immunoglobulin (*see* **Note 1**) to be conjugated, spin one tube containing the MaxPar chelating polymer for 10 s in a microcentrifuge. Polymer is difficult to see by eye and can escape from the tube easily upon opening. Spinning beforehand ensures that the reagent is at the bottom of the tube.

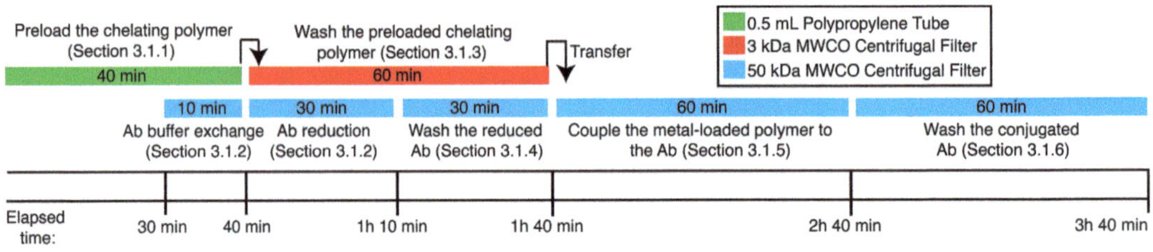

Fig. 2 A workflow summarizing the timing and coordination of steps for the conjugation of a purified immunoglobulin with a sulfhydryl-reactive polymer pre-loaded with metal isotope reporters

2. Reconstitute the MaxPar reagent in 95 μL of L-Buffer per labeling reaction using a filter pipette tip (for conjugation of >100 μg of antibody *see* **Note 2** and for conjugation of antibodies to ^{209}Bi *see* **Note 3**).
3. Add 5 μL of 0.05 M stock of metal solution to the polymer solution (final concentration: 2.5 mM). Vortex briefly to mix (for conjugation of >100 μg of antibody *see* **Note 4**).
4. Incubate at room temperature (RT) for 40 min, vortexing every 10 min. In order to ensure that the polymer and antibody are ready for conjugation simultaneously, proceed to Subheading 3.1.2 after approximately 30 min of the incubation has elapsed (*see* Fig. 2).

3.1.2 Buffer Exchanging and Reducing the Antibody (40 Min)

1. Add 300 μL of R-buffer to a 50-kDa MWCO micro-filter device (*see* **Note 5**).
2. Add 100 μg of antibody (*see* **Note 6**) to the R-buffer in the 50-kDa MWCO micro-filter device (*see* **Note 7**).
3. Reduce volume by spinning at 12,000 × *g* for 10 min at RT. Discard the flow-through. Final volume should be 20 μL or less before proceeding.
4. Mix 8 μL of TCEP stock with 992 μL of R-buffer (final concentration: 4 mM TCEP).
5. Add 100 μL of the diluted TCEP solution to the concentrated antibody in the 50-kDa MWCO micro-filter device. Tap the tube by hand to mix. Mixing too vigorously by vortexing at high speeds can compromise the structural integrity of the antibody when mixed with the mild reducing agent TCEP.
6. Incubate covered for 30 min at 37 °C. The antibody should not be left in TCEP for more than 30 min; longer incubation may result in full reduction of disulfide bonds necessary for the structural integrity of the protein.

3.1.3 Washing Pre-loaded MaxPar Labeling Reagent (60 Min)

1. Following the 40 min incubation, add 200 μL of C-buffer to the metal-loaded polymer. Pipette the mixture or briefly vortex the column to mix (for conjugation to ^{209}Bi *see* **Note 8**).
2. Transfer the mixture to the 3-kDa MWCO micro-filter device.
3. Reduce the volume by centrifugation at 12,000 × *g* for 25 min at RT. Discard the flow-through.
4. Add 300 μL of C-buffer to the 3-kDa MWCO micro-filter. Pipette the mixture or briefly vortex the column to mix.
5. Reduce the volume by centrifugation at 12,000 × *g* for 30 min at RT. Discard the flow-through. Final volume should not exceed 20 μL. A higher volume could result in excess free metal concentration and induce antibody precipitation.

3.1.4 Washing the Partially Reduced Antibody (30 Min)

1. Following the 30 min incubation (Subheading 3.1.2), collect the partially reduced antibody from the 37 °C incubator. Add 300 μL of C-buffer to the partially reduced antibody in the 50-kDa MWCO micro-filter device. Pipette the mixture or briefly vortex the column to mix.
2. Reduce the volume by centrifugation at 12,000 × *g* for 10 min at RT. Discard the flow-through.
3. Add an additional 400 μL of C-buffer to the 3-kDa MWCO micro-filter device. Pipette the mixture or briefly vortex the column to mix.
4. Reduce the volume by centrifugation at 12,000 × *g* for 10 min at RT. Discard the flow-through.

3.1.5 Coupling Metal-Loaded Polymer to Partially Reduced Antibody (1 H)

1. Remove all micro-filter devices from the centrifuge. Resuspend the metal-loaded polymer in 60 μL of C-buffer in the 3-kDa MWCO micro-filter device using a pipette equipped with a filter tip.
2. Transfer the contents of the 3-kDa MWCO micro-filter device into the corresponding 50-kDa MWCO micro-filter device containing the partially reduced antibody of choice. Pipette the mixture or briefly vortex the column to mix.
3. Incubate at 37 °C for at least 60 min. Incubation time can be extended up to 2 h, though the reaction should approach completion after 60 min.

3.1.6 Washing and Recovering the Conjugated Antibody (1 H)

1. Add 250 μL of W-buffer to the antibody conjugation mixture in the 50-kDa MWCO micro-filter device. Pipette the mixture or briefly vortex the column to mix.
2. Centrifuge at 12,000 × *g* for 10 min at RT. Discard the flow-through. Volume should not exceed 20 μL after spin.
3. Add 400 μL of W-buffer to the antibody conjugation mixture in the 50-kDa MWCO micro-filter device. Pipette or briefly vortex to mix.
4. Centrifuge at 12,000 × *g* for 10 min at RT. Discard flow-through. Volume should not exceed 20 μL after spin.
5. Repeat **steps 3** and **4** twice.
6. Add 50 μL of W-buffer to the 50-kDa MWCO micro-filter device. Pipette to mix and rinse the walls of the column.
7. Invert the micro-filter device into a new collection tube (supplied with the AMICON filters).
8. Centrifuge at 1000 × *g* for 2 min at RT.
9. Gently remove micro-filter device from collection tube. Add an additional 50 μL of W-buffer to the 50-kDa MWCO micro-filter device. Pipette to mix and rinse the walls of the column.

10. Invert the micro-filter device into the same collection tube.
11. Centrifuge at 1000 × *g* for 2 min at RT.
12. Using a pipette equipped with a filter tip, transfer the conjugated antibody to a screw-top polypropylene tube (to prevent evaporation) for long-term storage. The antibody can be stored in W-buffer at 4 °C for at least a week. For long-term storage, the antibody should be diluted to the appropriate concentration as determined by titration in antibody stabilization buffer supplemented with NaN_3. Ideally, the antibody should be diluted such that 1–2 μL is sufficient for a 100 μL cell staining reaction. This facilitates the creation of low volume staining cocktails containing more than 20 antibodies.

3.2 Quantification of Conjugated Antibody (15 Min)

1. Set up a low volume spectrophotometer according to the manufacturer's instructions. Measure the absorbance of W-buffer at 280 nm and use as a "blank" for subsequent measurements.
2. Against the blank, measure the absorbance of the conjugated antibody at 280 nm. Calculate the concentration of antibody present in solution. For mammalian IgG, an A_{280} of 1.38 absorbance units corresponds to a concentration of 1 mg/mL. The expected recovery is >60% of conjugated antibody. For quantification of Bismuth-conjugated antibodies *see* **Note 9**.
3. Within a week of labeling proceed to antibody validation and titration in order to select the appropriate storage concentration. If in doubt, dilute the conjugated antibody to a concentration of 0.2 mg/mL in antibody stabilization buffer supplemented with NaN_3 for long-term storage. The antibody can be stored in antibody stabilization solution at 4 °C for 6 months or more. However, before using the conjugated antibody for an experiment on the mass cytometer, the antibody must be validated and titrated as follows.

3.3 Staining Samples for Validation and Titration of Metal-Conjugated Antibody (3–5 H)

1. Here we focus on the validation and titration of antibodies against stable cell surface molecules. Cases in which conjugated antibodies recognize inducible intracellular modifications or secreted molecules (i.e., cytokines) will be referred to in **Note 10**.
2. Obtain a suspension of single cells that are expected to contain the marker of choice (positive control) as well as cells that are not (negative control). Choosing appropriate cell populations is critical to validate that the antigen specificity of the newly conjugated antibody has not been altered. For example, when validating an antibody against anti-human CD3, Jurkat T cells could serve as an appropriate positive control, and Nalm-6 pre-B cells could serve as an appropriate negative control. In

this same example, human peripheral blood mononuclear cells (PBMCs) could be used and CD3-positive T cells could be identified with a CD2 or CD5 stain and CD3 negative B cells with a CD19 or CD20 stain. For more examples, *see* the expected results in Fig. 3.

3. Add 1 mL of cell culture media (user's choice based on cell line or cell type) or FACS buffer at 37 °C to two FACS tubes labeled positive and negative. Add 10×10^6 positive control cells or 10×10^6 negative control cells to the FACS tubes, respectively. If using a single control (i.e., when using cells containing known positive and negative cell types; e.g., PBMCs) use a single tube. If the recognized epitope is stimulation dependent, *see* **Note 10**.
4. Add 111 μL of 16% paraformaldehyde to the 1 mL solution (to achieve a final concentration of 1.6%), and pipette thoroughly to mix.
5. Incubate for 10 min at RT.
6. Centrifuge at 500 × *g* for 5 min at 4 °C. Aspirate supernatant using a vacuum equipped Pasteur pipette. Vortex to resuspend cells. Leaving the cells in minimal residual volume greatly enhances the efficiency of these wash steps. We prefer to aspirate the supernatant rather than decanting FACS tubes. This also maximizes cell recovery. Immediately continue to the following washing steps to prevent over-fixation of cells.
7. Add 3.3 mL of FACS buffer to each tube.
8. Add 500 μL of the above cell mixture into each of six FACS tubes, labeled "stain 1" to "stain 6" (resulting in 6× tubes of the "positive" cells and 6× tubes containing "negative" cells).
9. Centrifuge at 500 × *g* for 5 min at 4 °C. Aspirate supernatant using a Pasteur pipette equipped with a fresh pipette tip, leaving cells in 60 μL residual volume.

Fig. 3 (continued) antibodies were simultaneously titrated from 0.25 to 8 μg/mL on human peripheral blood mononuclear cells that were counterstained with antibodies against CD45-^{154}Sm, CD3-^{170}Er, CD20-^{147}Sm, CD33-^{158}Gd, and CD16-^{165}Ho in order to identify positive and negative cell populations. (**a**) From data analysis at cytobank.org, the following populations were identified: (1) CD33-positive myeloid cells, (2) CD20-positive B cells, (3) CD3-positive T cells, and (4) CD16-positive NK cells. (**b**) Histogram overlays of CD95, CD21, and CD14 expression levels in the gated populations from (**a**) were created at cytobank.org. Histogram color scale indicates minimum (black) and maximum (yellow) median counts for each given antibody. (**c**) Dot plots summarizing the median counts of representative positive (blue) and negative (red) cell controls for each of the titrated antibodies across the range of 0.25–8 μg/mL. In (**b**) and (**c**) the red asterisk indicates the ideal antibody concentration with maximum signal-to-noise (the ratio of positive control to negative control signal) and lowest antibody concentration where the positive signal begins to saturate

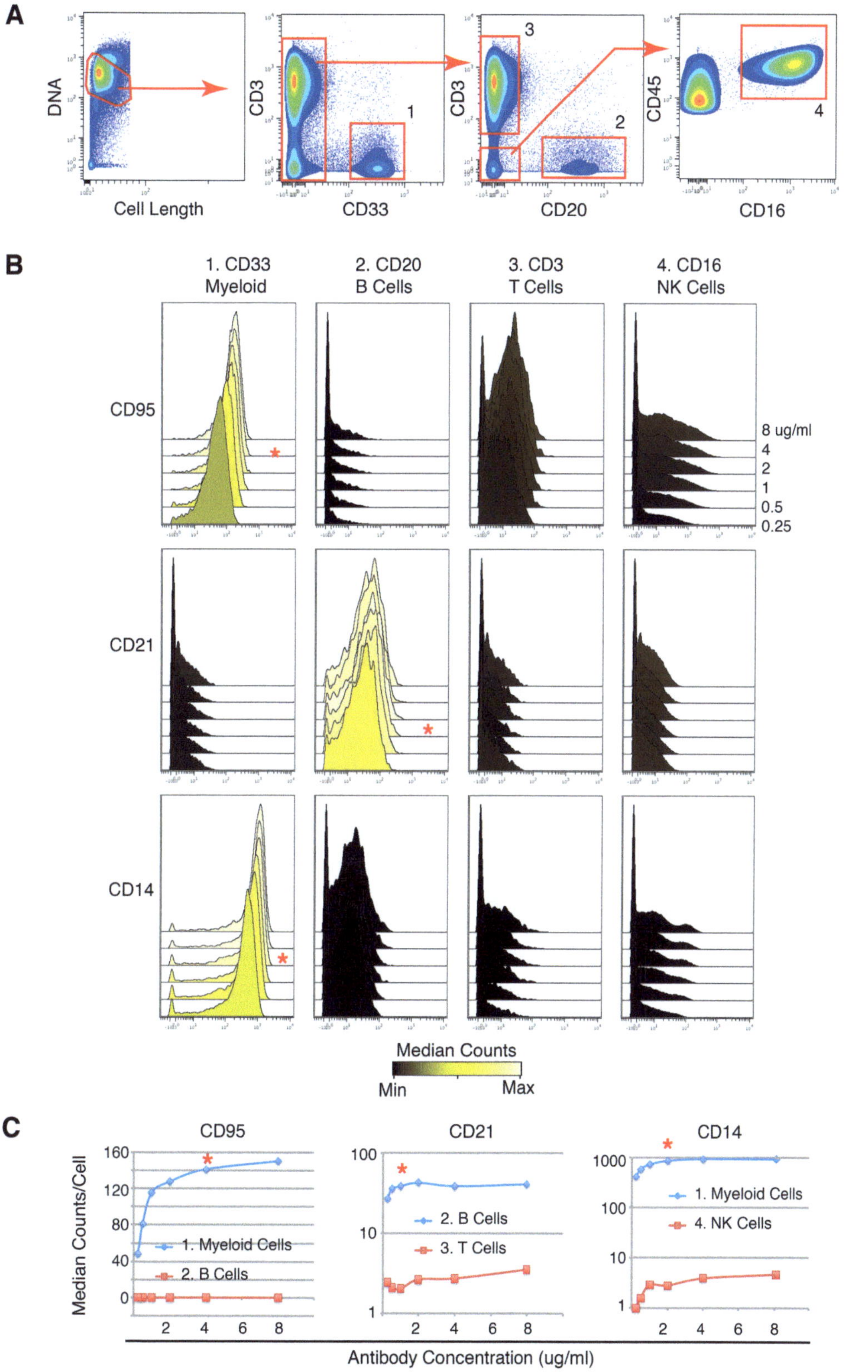

Fig. 3 Anticipated results for the titration of metal reporter conjugated antibodies by CyTOF mass cytometry. Using the protocol described herein, antibodies against human CD95, CD21, and CD14 were labeled with ^{164}Dy, ^{152}Sm, and ^{148}Nd, respectively. Using the surface staining protocol and a methanol permeabilization, these

10. If surface antibody staining is being performed, but not being titrated, *see* **Note 11** and proceed to **step 16**. Otherwise, proceed to **step 11**.
11. Create an antibody staining cocktail by diluting 4 μg of each antibody for titration to a total volume of 200 μL in FACS buffer (final concentration: 20 μg/mL per antibody) and label this tube "dilution 1" (*see* **Note 12**). This dilution assumes a staining volume of 100 μL per sample. If adjustments are required for smaller or larger staining volumes, *see* **Note 13**.
12. Add 100 μL FACS buffer to 5× empty tubes, labeled "dilution 2" to "dilution 6."
13. Perform a six-step, two-fold serial dilution: Add 100 μL of the "dilution 1" solution (20 μg/mL solution from **step 11**) to 100 μL of FACS buffer in the "dilution 2" tube and mix (final concentration: 10 μg/mL). Subsequently add 100 μL of the solution in the "dilution 2" tube to 100 μL of FACS buffer in the "dilution 3" tube (final concentration: 5 μg/mL). Repeat for all dilution tubes.
14. Add 40 μL of "dilution 1" to the 60 μL cell suspension in the "stain 1" tube, for both positive and negative cell types. Repeat for "dilution 2" and "stain 2," up to "dilution 6" and "stain 6." The final staining concentrations for each tube will be: 8 μg/mL in "stain 1," 4 μg/mL in "stain 2," 2 μg/mL in "stain 3," 1 μg/mL in "stain 4," 0.5 μg/mL in "stain 5," and 0.25 μg/mL in "stain 6."
15. If titrating some antibodies while holding the concentration of others "constant" (in order to identify control cell populations), prepare a separate, low volume (ideally <10 μL per sample) mixture of the "constant" antibodies, and add to the samples.
16. Incubate for 30 min at RT.
17. Add 3 mL of FACS buffer to each tube.
18. Centrifuge at 500 × *g* for 5 min at 4 °C. Aspirate supernatant using a Pasteur pipette equipped with a fresh pipette tip.
19. Vortex to resuspend cells. If the cell pellet is not resuspended the cells will clump upon addition of methanol for cell permeabilization. Add 1 mL of 4 °C methanol to each disrupted cell pellet. Vortex to mix and incubate on ice for 10 min.
20. Add 2 mL of FACS buffer to each tube.
21. Centrifuge at 500 × *g* for 5 min at 4 °C. Aspirate supernatant using a Pasteur pipette equipped with a fresh pipette tip. Vortex to resuspend cells.
22. Add 3 mL of FACS buffer to each tube. Centrifuge at 500 × *g* for 5 min at 4 °C.

23. Aspirate supernatant, leaving cells in 60 μL residual volume.
24. If staining and/or titrating intracellular antibodies, prepare an intracellular antibody staining cocktail and stain the cell samples according to **steps 10–18**.
25. Add 1 mL of Ir DNA intercalator solution to each FACS tube and mix thoroughly (*see* **Note 14**).
26. Incubate for at least 20 min at RT. Cells can be stored in DNA intercalator solution for up to 3 days at 4 °C before acquisition on mass cytometer.
27. Add 2 mL of FACS buffer to each tube.
28. Centrifuge at 500 × *g* for 5 min at 4 °C. Aspirate supernatant using a Pasteur pipette equipped with a fresh pipette tip. Vortex to resuspend the cells.
29. Add 3 mL of ddH_2O to each tube.
30. Centrifuge at 500 × *g* for 5 min at 4 °C. Aspirate supernatant using a Pasteur pipette equipped with a fresh pipette tip. Vortex to resuspend the cells.
31. Repeat **steps 29** and **30**.
32. Place cell pellets on ice. Just prior to analysis, resuspend cells in ddH_2O at a concentration of $1–2 \times 10^6$ cells per mL.
33. Filter with a cell strainer and analyze each tube on the mass cytometer. At this stage, 1× EQ beads can be added to the sample to enable later data normalization. Collect cell events at a rate no faster than 500–1000 cells/s. If the cells are too concentrated, the instrument can clog and the data quality may suffer due to overlapping cell boundaries. Adjust cell dilution with ddH_2O accordingly.

3.4 Gating Strategy and Data Analysis

Figure 3 provides an example of anticipated results and analysis of titration data.

1. In .FCS file browsing software, gate on single cells (parameters: cell_length, Ir-191 or Ir-193) for all samples (Fig. 3a). This gate may have to be tailored for each sample as the intensity of the Ir DNA intercalator staining can vary depending on a number of conditions such as cell type and number of cell in the staining reaction.
2. View different cell populations or control samples as stacked histograms or dot plots to visually validate epitope specificity of the antibody (Fig. 3b).
3. Select the ideal antibody concentration based on the greatest overall signal in the positive control and signal-to-noise when compared to the negative control. This can be accomplished by comparing the channel medians (50th percentile) (Fig. 3c) or, if focusing on an outlier population, the 95th percentile of

the positive and negative samples at the different concentrations. Alternative approaches such as maximum separation index (SI) could also be employed [24].

3.5 Troubleshooting

1. Troubleshooting advice regarding antibody conjugation and titration can be found in Table 3. All potential problems will be observed during antibody recovery or during data analysis.

3.6 Anticipated Results

1. To demonstrate the utility of the protocol described herein and provide representative data for a mass cytometry antibody titration, antibodies against human CD95, CD21, and CD14 were conjugated to ^{164}Dy, ^{152}Sm, and ^{148}Nd, respectively (Fig. 3).
2. When choosing negative and positive control cell populations for an antibody titration, we prefer to utilize cells from a similar source as those to be interrogated experimentally, which, in this case, were human peripheral blood mononuclear cells (PBMC). The expectation was that CD21, complement component receptor 2, would be uniformly expressed on peripheral B cells, and CD14, the LPS co-receptor, would be expressed on the majority of monocytes. Similarly, we expected that CD95, the FAS receptor, would be expressed by most monocytes, at lower levels on some T cells and NK cells, but not by resting B cells, which constitute the majority of those present in unstimulated healthy human PBMC [25, 26].
3. The staining protocol with sample fixation and methanol permeabilization was used to stain human PBMC, which was expected to contain both positive and negative cell types for these antibody targets. A mixture of anti-CD95, CD21, and CD14 were titrated according to the above described method. A separate mixture with anti-CD45, CD3, CD20, CD33, and CD16 was prepared and added to each sample in **step 15** of Subheading 3.3 in order to identify control cell populations while simultaneously titrating anti-CD95, CD21, and CD14. Six different antibody concentrations were evaluated using six samples, as the lack of crosstalk between reporter signals allowed counterstaining all significant cell populations while titrating all three antibodies at the same time. Approximately, 10^5 cell events were acquired per sample on the CyTOF mass cytometer, and all data was uploaded to and analyzed at cytobank.org.
4. Single cells were first identified by visualizing a biaxial plot of the iridium DNA intercalator signal versus cell length, a metric of the number of individual mass scans integrated to form the cell event as previously described [3]. Based on the expression of the counterstaining antibodies, the following cell populations were subsequently identified and gated: (1) CD33-positive myeloid

Table 3
A troubleshooting guide for common issues observed following custom antibody conjugation

Problem	Possible reason	Solution
Antibody recovery too low post-conjugation (<50%)	Precipitation	Likely induced by excess lanthanide metal exposure or antibody denaturation due to reduction. Ensure complete washes of the polymer and metal with post-centrifuge filter volumes of <20 μL to achieve desired dilution factor
	Antibody starting concentration	Measure concentration of starting antibody stock by A280 to ensure concentration is as stated by manufacturer
	Defective 50 kDa MWCO filter	Repeat labeling protocol with a new filter and fresh antibody preparation
Antibody recovery too high post-conjugation (>95%)	Carrier protein	Check manufacturer specifications for carrier protein (BSA, gelatin, "protein stabilizer"). Obtain carrier free stock or purify antibody away from carrier
	Antibody starting concentration	Measure concentration of starting antibody stock by A280 to ensure concentration is as stated by manufacturer
No antibody staining detected	Poor staining	Repeat staining steps with fresh cells
	Antibody integrity destroyed	After staining cells with conjugated antibody, perform secondary stain with fluorescently labeled anti-IgG antibody and run on flow cytometer (e.g., LSR II from BD)
	Mass cytometer malfunction	Check mass cytometer performance using CyTOF calibration beads
Weak antibody staining detected	Poor staining	Repeat staining steps with fresh cells
	Improper titration range	Repeat staining steps using increased antibody concentration
No difference between positive and negative controls	Improper choice of controls	Check antigen expression using fluorescently labeled antibodies on flow cytometer (e.g., LSR II from BD)
	Loss of antigen specificity	After staining cells with conjugated antibody, perform secondary stain with fluorescently labeled anti-IgG antibody and run on flow cytometer (e.g., LSR II from BD)
No difference between titration steps	Improper titration range	Repeat staining steps using decreased antibody concentration
High background	Improper titration range	Repeat staining steps using decreased antibody concentration
	Loss of antigen specificity	After staining cells with conjugated antibody, perform secondary stain with fluorescently labeled anti-IgG antibody and run on flow cytometer (e.g., LSR II from BD)

cells, (2) CD20-positive B cells, (3) CD3-positive T cells, and (4) CD16-positive NK cells (Fig. 3a).

5. To assess the titration of CD95, CD21, and CD14 for each control cell populations, histogram overlays were created for each control population (Fig. 3b). To more quantitatively visualize the differences between positive and negative control cell populations, selected based on known biology, dot plots summarizing the median counts for each antibody concentration from 0.25 to 8 μg/mL were made (Fig. 3c). Depending on the antibody, linear (CD95) or log (CD21 and CD14) scaled plots best revealed differences across different concentrations. In Fig. 3b, c, the red asterisks indicate the optimal antibody concentrations selected for these particular preparations. These concentrations were selected based on signal-to-noise ratio (the ratio of positive control to negative control signal) and on the lowest antibody concentration where the positive signal was beginning to saturate (plateau at higher concentrations). Alternative mechanisms of titration analysis could include using the 5th or 95th percentile, as opposed to the median, to compare situations where outlier cell events in each population are a primary concern when assessing resolution of positive and negative controls. Separation index is a good example of this and is described by Bigos et al. [24].
6. Although not completely necessary, it is helpful to have the antibody at close to saturating concentration in order to buffer the effects of staining volume differences as well as differences in cell staining numbers. For example, for CD95 (Fig. 3c) at the lower concentrations of antibody there was a linear relationship between the measured counts per cell and the concentration of the antibody. If the antibody were applied at one of these lower concentrations, changes in cell staining volumes, a common source of experimental variation, would have inversely proportional effects on the resulting stain intensity. Variations in cell numbers have a similar effect. Both of these situations are particularly problematic when attempting to perform comparative single-cell measurements in a relatively quantitative fashion. Also of note in Fig. 3c is that all three antibodies had selected titration concentrations that were below the maximum possible positive signal. In the cases shown here, the signal-to-noise ratio was not significantly higher at the selected concentration than at higher concentrations and the positive signal had begun to saturate, evidenced by the titration curve plateau. For these reasons, and to conserve custom-labeled antibody, the lower titration point in each case was selected.

4 Notes

1. Immunoglobulin should be free of a cysteine-containing carrier protein in solution (e.g., BSA). Notably, the cysteine content of gelatin is variable, from 10 to 100-fold lower than that of BSA by mass to completely absent. We have conjugated antibodies in the presence of Porcine gelatin though it is not recommended as it makes the Ig recovery difficult to quantify. If a carrier is present, a purification procedure should be performed before attempting the conjugation. Small molecule preservatives (e.g., trehalose, glycerol, and sodium azide) or standard buffer salts are compatible additives.
2. For the conjugation of >100 μg of the same immunoglobulin to a given heavy metal isotope, use one tube of MaxPar polymer per 100 μg of antibody. You will need to add 5 μL of metal stock per tube of polymer. Resuspend the first polymer in a volume "X" of L-buffer such that "X" = 100 μL − ((Amount of antibody in μg/100 μg) × 5 μL). E.g. for the conjugation of 500 μg, the first polymer tube is to be resuspended in 75 μL of L buffer. Subsequently, resuspend the next four tubes of polymer with the L-buffer from the first. Complete the preloading step by adding 25 μL of metal stock.
3. If antibodies are to be conjugated with ^{209}Bi, replace L buffer with 1% HNO_3 in this step. Note that the Bi stock will also be in 1–10% HNO_3 to maintain solubility.
4. For the conjugation of >100 μg of immunoglobulin, add 5 μL of stock metal solution per 100 μg of protein.
5. To confirm the integrity of the filter column, add 400 μL of ddH_2O and centrifuge at 12,000 × *g* for 30 s. The volume of the flow-through should be around 150 μL. Spin for 5 min to get rid of excess ddH_2O and discard flow-through. During the conjugation procedure, make sure to never touch the column membrane with the pipet tip to avoid scratches which can lead to lower antibody recovery.
6. Optional: To ensure that the volume of antibody solution corresponds to the intended quantity (within a tolerance of ±10%), confirm the protein concentration of the antibody stock solution by measuring the absorbance at 280 nm as described in Subheading 3.2.
7. MWCO micro-filter devices are designed to hold up to 500 μL. If the desired amount of antibody requires addition of more than 200 μL of solution, pre-concentrate the antibody first in the same 50-kDa MWCO micro-filter device by spinning at 12,000 × *g* for 8 min at RT until all of the antibody solution has been added to the column and the final volume is

200 µL. If these additional steps are required to pre-concentrate the antibody, this step should begin earlier (*see* Fig. 2).

8. If antibodies are to be conjugated with ^{209}Bi, replace C-buffer with modified C-buffer in this step.
9. Bismuth-ion chelator complexes interfere with the measured absorbance at 280 nm and thus cannot be easily quantified using this method. Alternatively, bicinchoninic acid assays (BCA) can be used to determine the protein content.
10. In case the recognized epitope is stimulation dependent, proceed as follows: add 1 mL of cell culture media at 37 °C to two FACS tubes, labeled stimulated and unstimulated. Add stimulation cocktail of choice to one tube. Selecting the appropriate stimulation condition is essential for validating and titrating the conjugated antibody. The stimulation should result in selective induction of the activated form of the signaling molecule of interest. Please refer to these for additional information on antibody targets, appropriate stimulation, and timing—reviewed in [27]. For secreted molecules such as cytokines, add 1× brefeldin A and monensin to the cell culture media prior to addition of cells to medium. More details on cytokine production and measurement by mass cytometry can be found in Newell et al. [8]. Add 10 × 10^6 cells to each FACS tube. Incubate cells at 37 °C for the appropriate time for monitoring of the selected cellular target. Because distinct signaling pathways are activated at different times following stimulation, the incubation time should be adjusted accordingly. Some cellular signaling events are very time sensitive. Be ready to proceed to fixation immediately.
11. Dilute the appropriate amount of antibody or antibodies for each staining reaction in FACS buffer at 40 µL per sample (assuming a final staining volume of 100 µL per sample), then add to each 60 µL sample suspension and vortex to mix. Then proceed to the incubation (**step 16**).
12. For an antibody stock solution of 0.2 mg/mL, 20 µL will contain 4 µg.
13. This dilution is designed to produce a final concentration of 8 µg/mL for each antibody in a total staining volume of 100 µL for the first titration step. Therefore, if staining will be performed in a larger volume or the antibody was diluted to a different initial concentration, adjust these guidelines accordingly. Also, this titration is designed for two series (i.e., positive and negative controls) of six concentrations. If only using one series or more than two series of control cell lines, the amount of antibody can be scaled and subsequently diluted accordingly.

14. The reagent should be made fresh for each use. The amount of Ir intercalating reagent can also be reduced (to 0.15–0.2 μL per mL) for more sensitive instruments and larger cells, or increased (to 0.5 μL per mL) for staining cells that were not permeabilized (not described herein).

Acknowledgments

We are grateful to Scott Tanner for providing useful feedback on this manuscript and to Fluidigm for supplying the chelating polymer that was used to optimize the protocol described here. F.J.H received support from the Swiss National Science Foundation (SNF Early Postdoc. Mobility), the Novartis Foundation for medical-biological research (Research Fellowship), and EMBO (Long-Term Fellowship). Also, DP5OD023056, R21AG057224 from the NIH and the Parker Institute for Cancer Immunotherapy to M.H.S. and S10OD018040 Shared Instrumentation Grant to UCSF. E.F.S. is supported by a Damon Runyon Cancer Research Foundation Fellowship (DRG-2190-14). G.P.N. is supported by the NIH grants R01CA184968, 1R01GM10983601, 1R01NS08953301, 1R01CA19665701, R01HL120724, 1R21CA183660, R33CA0183692, 1R33CA183654-01, U19AI057229, 1U19AI100627, U54-UCA149145A, N01-HV-00242, HHSN26820100034C, and 5UH2AR067676; the NIH Northrop-Grumman Corporation Subcontract 7500108142; FDA grant HHSF223201210194C; DOD grants OC110674 and W81XWH-14-1-0180; NWCRA Entertainment Industry Foundation; and Bill and Melinda Gates Foundation grant OPP1113682. S.C.B. is supported by the Damon Runyon Cancer Research Foundation Fellowship (DRG-2017-09), the NIH grants 1DP2OD022550-01, 1R01AG056287–01, 1R01AG057915-01, 1-R00-GM104148-01, 1U24CA224309-01, 5U19AI116484-02, 1U24CA224309-01, The Bill and Melinda Gates Foundation, and a Translational Research Award from the Stanford Cancer Institute.

Competing Financial Interests: S.C.B. and E.F.S. have been paid consultants for the company Fluidigm Sciences, the manufacturers that produced some of the reagents and instrumentation described in this manuscript.

References

1. Baranov VI, Quinn Z, Bandura DR et al (2002) A sensitive and quantitative element-tagged immunoassay with ICPMS detection. Anal Chem 74:1629–1636. https://doi.org/10.1021/ac0110350

2. Bandura DR, Baranov VI, Ornatsky OI et al (2009) Mass cytometry: technique for real time single cell multitarget immunoassay based on inductively coupled plasma time-of-flight mass spectrometry. Anal Chem 81:6813–6822. https://doi.org/10.1021/ac901049w

3. Bendall SC, Simonds EF, Qiu P et al (2011) Single-cell mass cytometry of differential immune and drug responses across a human hematopoietic continuum. Science 332:687–696. https://doi.org/10.1126/science.1198704
4. Bodenmiller B, Zunder ER, Finck R et al (2012) Multiplexed mass cytometry profiling of cellular states perturbed by small-molecule regulators. Nat Biotechnol 30:858–867. https://doi.org/10.1038/nbt.2317
5. Behbehani GK, Bendall SC, Clutter MR et al (2012) Single-cell mass cytometry adapted to measurements of the cell cycle. Cytometry A 81:552–566. https://doi.org/10.1002/cyto.a.22075
6. Fienberg HG, Simonds EF, Fantl WJ et al (2012) A platinum-based covalent viability reagent for single-cell mass cytometry. Cytometry A 81:467–475. https://doi.org/10.1002/cyto.a.22067
7. Mingueneau M, Kreslavsky T, Gray D et al (2013) The transcriptional landscape of alphabeta T cell differentiation. Nat Immunol 14:619–632. https://doi.org/10.1038/ni.2590
8. Newell EW, Sigal N, Bendall SC et al (2012) Cytometry by time-of-flight shows combinatorial cytokine expression and virus-specific cell niches within a continuum of CD8+ T cell phenotypes. Immunity 36:142–152. https://doi.org/10.1016/j.immuni.2012.01.002
9. Lou X, Zhang G, Herrera I et al (2007) Polymer-based elemental tags for sensitive bioassays. Angew Chem Int Ed Eng 46:6111–6114. https://doi.org/10.1002/anie.200700796
10. Seegan GW, Smith CA, Schumaker VN (1979) Changes in quaternary structure of IgG upon reduction of the interheavy-chain disulfide bond. Immunology 76:907–911. https://doi.org/10.1073/pnas.76.2.907
11. Liu H, Chumsae C, Gaza-Bulseco G et al (2010) Ranking the susceptibility of disulfide bonds in human IgG1 antibodies by reduction, differential alkylation, and LC-MS analysis. Anal Chem 82:5219–5226. https://doi.org/10.1021/ac100575n
12. Amir ED, Davis KL, Tadmor MD et al (2013) viSNE enables visualization of high dimensional single-cell data and reveals phenotypic heterogeneity of leukemia. Nat Biotechnol 31:545–552. https://doi.org/10.1038/nbt.2594
13. Gibbs KD, Jager A, Crespo O et al (2012) Decoupling of tumor-initiating activity from stable immunophenotype in HoxA9-Meis1-driven AML. Cell Stem Cell 10:210–217. https://doi.org/10.1016/j.stem.2012.01.004
14. Spitzer MH, Carmi Y, Reticker-Flynn NE et al (2017) Systemic immunity is required for effective cancer immunotherapy. Cell 168:487–502.e15. https://doi.org/10.1016/j.cell.2016.12.022
15. Hartmann FJ, Bernard-Valnet R, Quériault C et al (2016) High-dimensional single-cell analysis reveals the immune signature of narcolepsy. J Exp Med 213:2621–2633. https://doi.org/10.1084/jem.20160897
16. Arrell DK, Terzic A (2010) Network systems biology for drug discovery. Clin Pharmacol Ther 88:120–125. https://doi.org/10.1038/clpt.2010.91
17. Sachs K (2005) Causal protein-signaling networks derived from multiparameter single-cell data. Science 308:523–529. https://doi.org/10.1126/science.1105809
18. Aghaeepour N, Jalali A, O'Neill K et al (2012) RchyOptimyx: cellular hierarchy optimization for flow cytometry. Cytometry A 81:1022–1030. https://doi.org/10.1002/cyto.a.22209
19. Qiu P, Simonds EF, Bendall SC et al (2011) Extracting a cellular hierarchy from high-dimensional cytometry data with SPADE. Nat Biotechnol 29:886–891. https://doi.org/10.1038/nbt.1991
20. Mair F, Hartmann FJ, Mrdjen D et al (2016) The end of gating? An introduction to automated analysis of high dimensional cytometry data. Eur J Immunol 46:34–43. https://doi.org/10.1002/eji.201545774
21. Ornatsky OI, Kinach R, Bandura DR et al (2008) Development of analytical methods for multiplex bio-assay with inductively coupled plasma mass spectrometry. J Anal At Spectrom 23:463. https://doi.org/10.1039/b710510j
22. Dulski P (1994) Interferences of oxide, hydroxide and chloride analyte species in the determination of rare earth elements in geological samples by inductively coupled plasma-mass spectrometry. Fresenius J Anal Chem 350:194–203. https://doi.org/10.1007/BF00322470
23. Kotecha N, Krutzik PO, Irish JM (2010) Web-based analysis and publication of flow cytometry experiments. Curr Protoc Cytom Chapter 10:Unit10.17. https://doi.org/10.1002/0471142956.cy1017s53
24. Bigos M (2007) Separation index: an easy-to-use metric for evaluation of different configurations on the same flow cytometer. Curr Protoc Cytom Chapter 1:Unit1.21. https://doi.org/10.1002/0471142956.cy0121s40

25. Medvedev AE, Johnsen AC, Haux J et al (1997) Regulation of Fas and Fas-ligand expression in NK cells by cytokines and the involvement of Fas-ligand in NK/LAK cell-mediated cytotoxicity. Cytokine 9:394–404. https://doi.org/10.1006/cyto.1996.0181
26. Garrone P, Neidhardt EM, Garcia E et al (1995) Fas ligation induces apoptosis of CD40-activated human B lymphocytes. J Exp Med 182:1265–1273
27. Krutzik PO, Trejo A, Schulz KR, Nolan GP (2011) Phospho flow cytometry methods for the analysis of kinase signaling in cell lines and primary human blood samples. Methods Mol Biol 699:179–202. https://doi.org/10.1007/978-1-61737-950-5_9

Chapter 6

Titration of Mass Cytometry Reagents

Caryn van Vreden, Paula Niewold, Helen M. McGuire, and Barbara Fazekas de St. Groth

Abstract

Mass cytometry (MC) is a powerful research tool enabling high-dimensional analysis of single cells in suspension and within tissue sections following laser ablation. Here we describe the procedure of titrating metal-conjugated antibodies, to ensure that optimal levels of staining are achieved while minimizing nonspecific signals that may occur at high concentrations.

Key words Mass cytometry, CyTOF, Helios, Antibody titration, Conjugation, Best practices

1 Introduction

Mass cytometry (MC) is a powerful research tool enabling high-dimensional analysis of single cells in suspension and within tissue sections following laser ablation. Using lanthanide metals in place of fluorophores as reporters, MC allows for the detection of over 40 parameters with minimal overlap. This approach generates information dense datasets, making it particularly suitable for investigation of rare samples. Despite these advantages, adopting this technology requires significant start-up funds and technical expertise, and these barriers prove prohibitive for many users. Although commercially validated metal-conjugated antibodies are available, researchers routinely conjugate antibodies in-house to obtain reagents not commercially available and to reduce cost. Titrating commercially available and in-house metal-conjugated antibodies not only aids in cost reduction but also ensures that optimal levels of staining are achieved while minimizing nonspecific signals that may occur at high concentrations.

Helen M. McGuire and Thomas M. Ashhurst (eds.), *Mass Cytometry: Methods and Protocols*, Methods in Molecular Biology, vol. 1989, https://doi.org/10.1007/978-1-4939-9454-0_6,

2 Materials

1. Staining buffer: 0.5% Bovine Serum Albumin (BSA), 0.02% sodium azide, 2 mM Ethylenediaminetetraacetic acid (EDTA), in phosphate-buffered saline (PBS).
2. Human or Mouse Fc-block (such as Human TruStain FcX™, Biolegend).
3. Heparin.
4. Eppendorf tubes 1.5 mL.
5. Cisplatin (Fluidigm).
6. Fixative: Methanol and/or FoxP3/Transcription Factor Staining Buffer set (eBioscience).
7. Metal-conjugated antibodies (*see* **Note 1**).
8. Ultrafree-MC VV Centrifugal filters 0.1 μm.
9. Cell ID™ Intercalator (Fluidigm).
10. Paraformaldehyde (PFA): diluted to 4% with PBS.
11. EQ four element calibration beads (Fluidigm).

3 Method (*See* Note 2)

Titrations require ~1–2 × 10^6 cells per titration condition; therefore prepare ~1.4 × 10^7 cells for seven different conditions: gating antibodies only, control vial, and titration concentrations of 8, 4, 2, 1, and 0.5 μg/mL. The protocol used will depend on the type of cells being titrated (*see* **Notes 3** and **4**).

3.1 Sample Preparation

The protocol for sample preparation will differ depending on the cell preservation format. The most common source of samples at a MC facility are cryopreserved peripheral blood mononuclear cells (PBMC) (*see* **Note 5**).

1. Cryopreserved PBMC: These can be prepared by rapidly thawing cryotubes in a 37 °C waterbath and resuspending in 10 mL pre-warmed (37 °C) RPMI + 10% FCS. Centrifuge for 5 min at 500 × *g* at room temperature (RT).
2. Fixed cells: Whole blood that has been preserved in Smart Tube Buffer (STB). This involves defrosting samples at 10 °C followed by a series of lysing steps [1].

3.2 Pre-staining Steps

1. Transfer all the cells to a 15 mL conical tube and add 10 mL of complete media. Centrifuge for 5 min at 500 × *g* at RT. Discard the supernatant. If cells do not require stimulation to express the appropriate antigens, go to Subheading 3.3.

2. Cell stimulation (*see* **Note 6**): Some reagents may require that cells undergo stimulation or activation in order to express the antigen of interest. A common method involves incubating cells at 37 °C for 3–4 h with 50 ng/mL phorbol 12-myristate 13-acetate (PMA) and 500 ng/mL ionomycin in the presence of 1× brefeldin A. However, the specific stimulation conditions depend on the reagent being titrated [1].
3. Following stimulation cells need to be washed 1× by adding complete media and centrifuging for 5 min at 500 × *g* at RT. Discard supernatant.

3.3 Pre-titration Staining Steps for Mass Cytometry

1. Cell viability (*see* **Note 7**): Cisplatin, which enters cells with disrupted membranes, is used to discriminate live and dead cells in MC. This step can be skipped for prefixed cells such as STB-prepared samples. Unfixed cells (7–14 × 10^6) can be stained by adding ~1 mL of Cisplatin diluted in RPMI without FCS or in PBS at a final concentration of 1 μM (1:4000 dilution of 5 mM stock) for 3 min at RT. Stop the reaction by adding 9 mL of staining buffer. Centrifuge cells for 5 min at 500 × *g* at RT and discard supernatant.
2. Blocking step if required (*see* **Note 8**). Perform surface blocking by adding Fc block and/or heparin block according to manufacturer's instructions [2] and incubating for 20–30 min at RT.
3. Wash 1× by adding 10 mL cold staining buffer and centrifuging for 5 min at 500 × *g* at 4 °C.
4. Prepare enough of the gating antibody cocktail for 7–14 × 10^6 cells or 7 tests (cells should still be combined in one tube, *see* **Note 7**). Note that the antibody cocktail needs to be filtered to remove possible aggregates (*see* **Note 9**). Add the gating antibody cocktail to an Ultrafree-MC VV Centrifugal filter tube, and centrifuge for 4 min at 12,000 × *g* at 4 °C to collect flow through. Multiple rounds of spinning may be required, if the total volume of the cocktail is greater than the capacity of the tube.
5. Resuspend the combined cells in the filtered flow through gating antibody cocktail and incubate for ~30 min at 4 °C.
6. Meanwhile, prepare the serial dilutions of antibodies to be titrated.

3.4 Preparing Serial Dilution of Antibodies for Testing (See Note 4*)*

For chemokine receptor staining: due to recycling of chemokine receptors, staining is often optimal at RT or 37 °C (*see* **Note 10**) [3]. Antibodies that require staining at 37 °C or RT should be stained first, using the serial dilution procedure described below (*see* **Note 11**).

1. Prepare a total of 150 μL of the titrating antibody cocktail at a concentration of 8 μg/mL of each antibody in a 0.1 μm filter as described previously by first adding staining buffer followed by antibodies.
2. Spin the solution for 4 min at 12,000 × *g* and collect the flow through.
3. Perform a serial dilution by adding 75 μL of the 8 μg/mL antibody cocktail to 75 μL of staining buffer.
4. Mix well and repeat four times to obtain 4, 2, 1, and 0.5 μg/mL solutions.
5. In addition to the five antibody concentrations, prepare the positive control antibody cocktail at this step.

3.5 Staining Cells with Serial Dilution Series

1. Wash the combined cells 1× by adding 10 mL cold staining buffer and centrifuging for 5 min at 500 × *g* at 4 °C.
2. Resuspend cells in 3.5 mL of staining buffer, mix well, and transfer 500 μL to each of 7 Eppendorf tubes so that there are equal numbers of cells in each tube (~1–2 × 10^6).
3. Top each tube up to 1 mL with staining buffer and centrifuge for 5 min at 500 × *g* at 4 °C.
4. Discard supernatant.
5. First perform staining of reagents with specific temperature requirements such as chemokine receptors.
6. Add 50 μL of each of the five dilutions or the control antibody to the separate tubes containing 1–2 × 10^6 cells. The negative control tube will receive 50 μL of staining buffer only.
7. Incubate at the desired temperature (RT or 37 °C) for 20–30 min.
8. Wash cells by adding 1 mL of staining buffer, centrifuging at 4 °C for 5 min at 500 × *g*.
9. Discard the supernatant.
10. Perform the staining of all other surface antibodies by adding 50 μL of each antibody dilution (prepared as described in Subheading 3.4) to the relevant tube, incubate for ~30 min at 4 °C.
11. Wash cells by adding 1 mL of staining buffer and centrifuge at 4 °C for 5 min at 500 × *g*.
12. Discard supernatant.

3.6 Intracellular Staining Steps

The method of intracellular staining will depend on the type of antibody being titrated (*see* **Note 12**). Cytokine staining requires permeabilization with a saponin-containing reagent [4], whereas transcription factors such as FoxP3 require specialized fixation and permeabilization available in kits like the eBioscience™ FoxP3/Transcription Factor Staining Buffer Set. Optimized staining of phosphorylated targets requires fixation with methanol [1].

1. Following the appropriate fixation procedure (follow manufacturer's instructions) and blocking step (*see* **Note 13**), prepare a serial dilution of intracellular antibodies in the appropriate buffer, as described in Subheading 3.4, and incubate at 4 °C for 40–60 min. Note that transcription factor antibody cocktails should be prepared in Perm/Wash buffer instead of staining buffer, as per manufacturer's instructions.
2. Wash cells by adding 1 mL of staining buffer and centrifuge at 4 °C for 10 min at 1000 × *g*.
3. Discard supernatant.

3.7 Post Staining Steps

1. For fixed samples: Store samples overnight in 100–200 μL 4% PFA and Cell ID™ Intercalator at a dilution of 1:4000.
2. For unfixed samples (i.e., samples that did not require intracellular staining): Store overnight in 4% PFA and perform the Cell ID™ Intercalator staining on the day of acquisition by adding Perm/Wash buffer with Cell ID™ Intercalator (1:4000 dilution) at a 1:1 ratio with the PFA. Incubate at 4 °C for 30 min.
3. Wash samples: Add 1 mL of staining buffer to samples and centrifuge for 4 °C for 10 min at 1000 × *g*. Discard supernatant. Repeat this step 3× with MilliQ water and count cells during the last wash step.
4. Resuspend samples in MilliQ water and 1X EQ four element calibration beads (1:10 dilution) at a concentration of 8×10^5 cells/mL and pass through a cell strainer snap cap into a polystyrene test tube (*see* **Note 14**).
5. Acquire at least 100,000 events on the mass cytometer (*see* **Note 15**).

3.8 Assessment of Titration Results

There are a few factors to consider when deciding on the optimal titration concentration to use for a new reagent.

1. Ensure that the signal of your new antibody is higher than background and has equal or better separation between background and positive staining than your control antibody (Fig. 1).
2. When the concentration of antibody is too high, a reduction of positive staining and increase in background signal can result. We recommend a concentration of 4 μg/mL for anti-CD14-160Gd and 1 μg/mL for anti-CD11c-167Er in the example Fig. 1, as staining has reached saturation point without increasing background signal significantly.

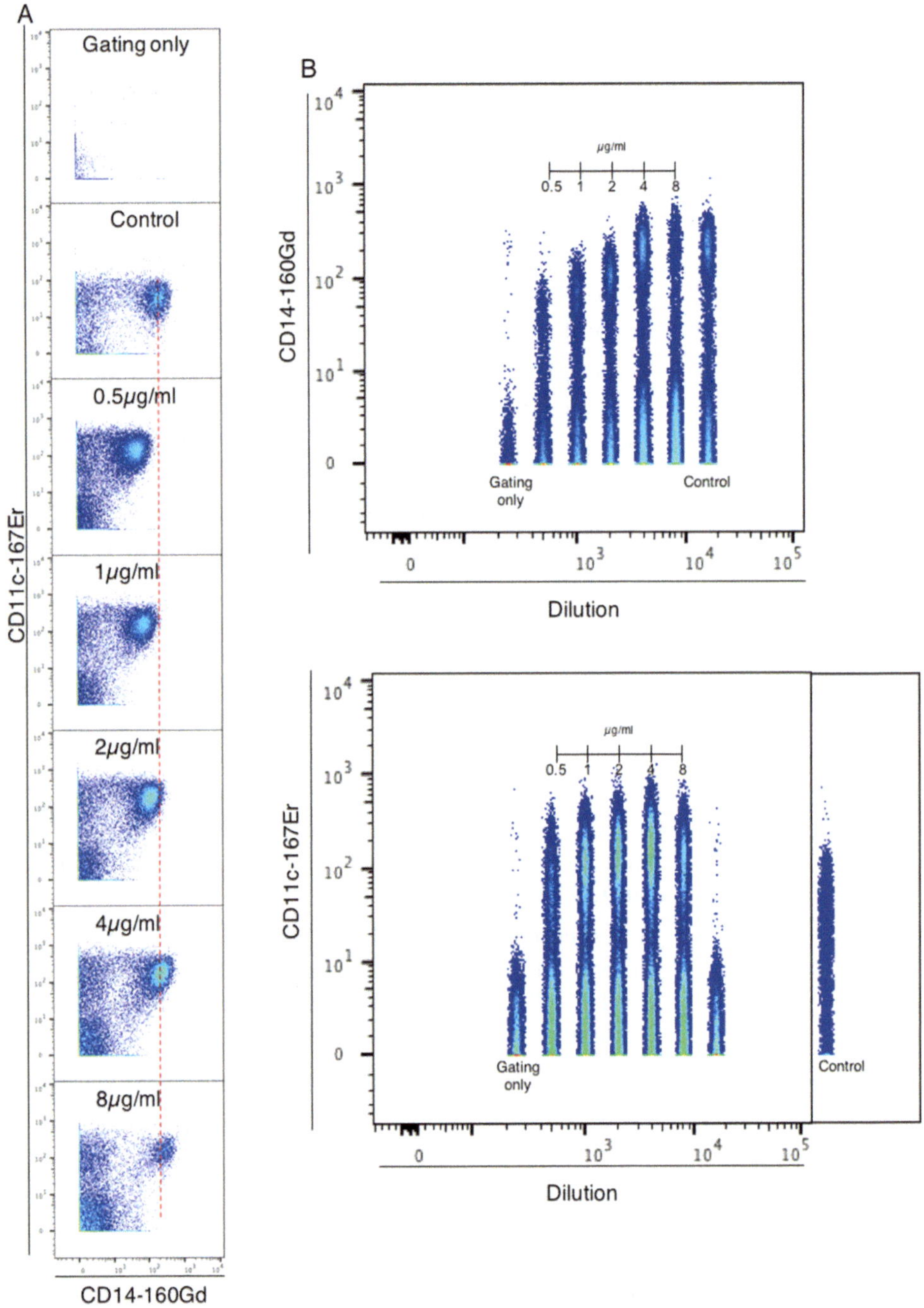

Fig. 1 Example of antibody titration of anti-human CD14-160Gd and anti-human CD11c-167Er. (**a**) Shows 2D plots of CD14 vs. CD11c staining of PBMC. (**b**) Shows the concatenated images of all the anti-human CD14-160Gd and CD11c-167Er concentrations. This is useful when determining at which concentration antibody binding reaches saturation and background increases. Note that for the CD11c-167Er titration the control was on a different metal

3. Check background staining on cells that should not express the antigen, i.e., CD14 signal on $CD3^+$ cells.
4. Reagents should be re-titrated every 2 years or prior to large experiments if they are over a year old.

4 Notes

1. Titrations: In-house metal-conjugated antibodies are generally produced at a concentration between 400 and 500 μg/mL. For a stock concentration of 400–500 μg/mL we recommend performing a serial dilution of 8, 4, 2, 1, and 0.5 μg/mL. If this range is not sufficient, antibodies can always be titrated within a higher or lower range.
2. Consistency: To ensure reproducibility in staining, antibodies should be titrated under conditions that are as close as possible to experiment-day staining conditions. Minimizing variations in protocols ensures reproducibility, predictable staining results, and reduces costs.
3. Gating antibodies: Being able to correctly phenotype cells is a crucial aspect of titration as this allows identification of the cell subset of interest and determination of the level of nonspecific staining or signal spillover. Therefore, the antibody cocktail used for gating should include markers that will allow precise identification of cells of interest. It should also include markers that identify cell subsets that should not stain with the antibodies that are being titrated and which will therefore serve as internal negative controls.

   Control antibodies: Including antibodies that act as positive reference (e.g., a previously validated antibody) is particularly important in a shared resource laboratory setting where there is a shared reagent bank continuously being restocked. Having an existing control with which to compare new reagents allows you to track batch-to-batch variability, improves reproducibility, and ensures that reagent quality remains at a high standard. Thus, wherever possible, aim to include a separate tube of cells that are stained with the gating antibody cocktail plus a control antibody for each new reagent being titrated.
4. Combining antibodies: Co-titrating multiple antibodies that will be used under the same conditions saves time and money, as long as the gating antibody cocktail is carefully designed to allow optimal identification of all the relevant cell populations.
5. Fixation: The state of fixation can impact the level of staining. For example, when staining fresh-frozen PBMC vs. Smart Tube Buffer (STB)-fixed whole blood, comparable staining of some antibodies might require a slightly altered concentration to achieve equivalent separation (Fig. 2).

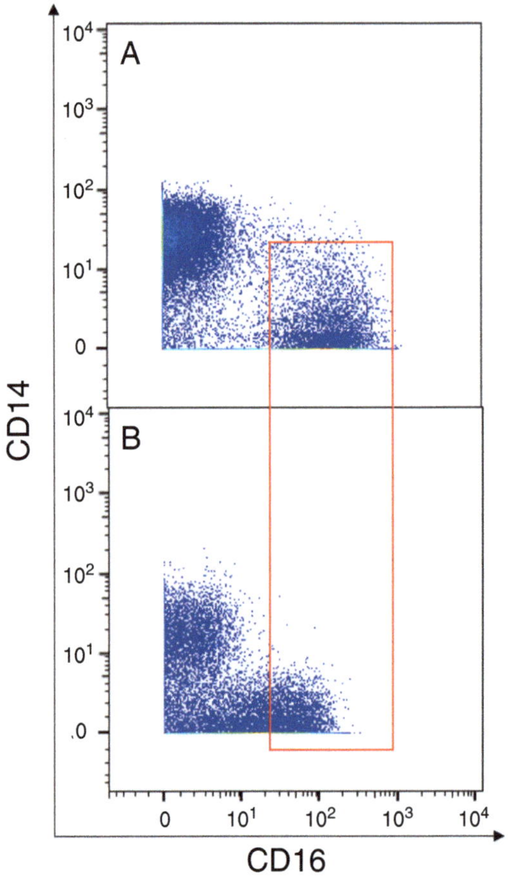

Fig. 2 Comparison of CD16 staining on fresh-frozen PBMC (*a*) and STB-fixed whole blood (*b*). *a* and *b* were stained with the same antibody cocktail and acquired on the same day. The reduced CD16 signal of STB-fixed whole blood highlights the need for titration of reagents under the particular conditions to be used in an experiment

6. Stimulation: When titrating reagents for antigens that are expressed after stimulation, ensure that you include a negative control (unstimulated). Combining barcoded stimulated and unstimulated samples prior to performing a titration is the most accurate method of comparison. Note that the stimulation cocktail should be appropriate for the reagent being titrated [1].
7. Cisplatin Live/Dead and gating antibody staining can be done with all the cells (i.e., prior to splitting cells into vials for different staining conditions) combined in one tube to ensure uniform staining of all titration samples.
8. Blocking: Ensure that all appropriate blocking steps are followed prior to titrating new antibodies. Blocking of Fc receptors with anti-CD16/32 and other nonspecific binding can

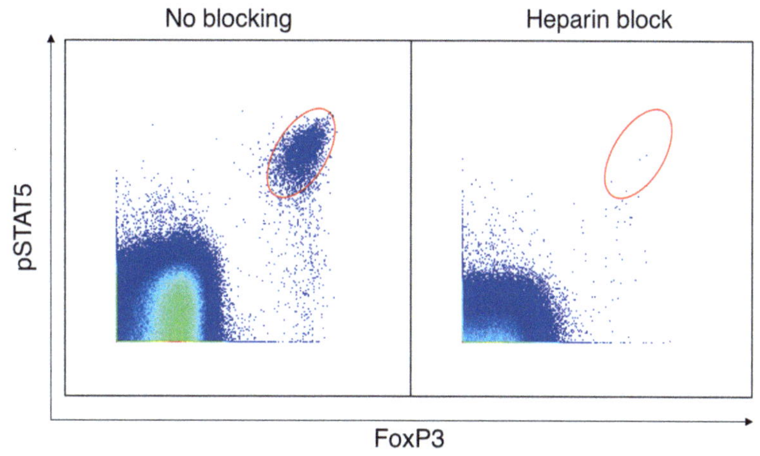

Fig. 3 Nonspecific staining of pSTAT5 and FoxP3 in STB-fixed whole blood. Blocking with heparin dramatically reduces nonspecific intracellular staining of eosinophils in smart tube buffer-fixed whole blood [2]

reduce the amount of antibody needed to achieve saturation. Nonspecific binding of intracellular markers in eosinophils can be greatly reduced by blocking with Heparin (Fig. 3) [2].

9. Another method of removing possible antibody aggregates is to centrifuge the antibody vial for 10 min at 16,000 × *g* at 4 °C and aliquoting the desired volume of reagent from the top of the solution.
10. Exceptions occur as we found that human CD194 (CCR4) staining works best at 4 °C.
11. Temperature: Staining of certain antigens, especially chemokine receptors, can vary depending on the temperature at which staining occurs. Ensure that titrations are performed at the temperature that staining will occur at in the future.
12. Method of fixation: Staining of various intracellular or nuclear antibodies works optimally under different fixation conditions, i.e., methanol vs. 4% paraformaldehyde (PFA) which could affect titration of some reagents.
13. Note that it is important to include a heparin block step after fixation when titrating intracellular reagents on STB-fixed cells.
14. Do not leave cells resuspended in MilliQ water longer than is necessary. Even if cells are fixed, lysis and cell loss can still occur. Cells are most stable during the day of acquisition when left pelleted at 4 °C in a small volume with most of the supernatant removed.
15. Determine the minimum number of events to acquire based on the frequency antigen will be present on cells being titrated.

If the reagent being titrated occurs at low frequencies in the cell population, more events may need to be acquired. It is important to acquire the same number of events for each titration.

References

1. Fernandez R, Maecker H (2015) Cytokine-stimulated phosphoflow of whole blood using CyTOF mass cytometry. Bio Protoc 5(11):e1495
2. Rahman AH, Tordesillas L, Berin MC (2016) Heparin reduces nonspecific eosinophil staining artifacts in mass cytometry experiments. Cytometry A 89(6):601–607. https://doi.org/10.1002/cyto.a.22826
3. Berhanu D, Mortari F, De Rosa SC et al (2003) Optimized lymphocyte isolation methods for analysis of chemokine receptor expression. J Immunol Methods 279(1–2):199–207
4. Smith SG, Smits K, Joosten SA et al (2015) Intracellular cytokine staining and flow cytometry: considerations for application in clinical trials of novel tuberculosis vaccines. PLoS One 10(9):e0138042. https://doi.org/10.1371/journal.pone.0138042

Chapter 7

Surface Barcoding of Live PBMC for Multiplexed Mass Cytometry

Axel Ronald Schulz and Henrik E. Mei

Abstract

Sample barcoding is a powerful method for harmonizing mass cytometry data. By assigning a unique combination of barcode labels to each cell sample, a set of individual samples can be pooled and further processed and acquired as a large, single sample. For assays that require uncompromised profiling of cell-surface markers on live cells, barcoding by metal-labeled antibodies targeting cell-surface epitopes is the barcoding approach of choice. Here we provide an optimized and validated protocol for cell-surface barcoding of ten PBMC samples with palladium-labeled β2-microglobulin (B2M) antibodies used in a 5-choose-2 barcoding scheme, for subsequent immune phenotyping by mass cytometry. We further provide details on the generation of palladium-labeled antibodies utilizing amine-reactive isothiocyanobenzyl-EDTA (ITCB-EDTA) that permits the implementation of antibody-based barcoding not interfering with lanthanide channels typically used for analyte detection in mass cytometry assays.

Key words Mass cytometry, CyTOF, Cell-surface sample barcoding, Antibody-based sample barcoding, Palladium, Immune monitoring, β2-Microglobulin

1 Introduction

Mass cytometry is increasingly applied in biomedical research and enables systematic in-depth characterization of immune cells in healthy state, cancer and autoimmunity [1–4], and for biomarker discovery [5–7]. Typically, such research requires the mass cytometric measurement and analysis of many individual cell samples for which sample barcoding, a method that allows the processing and acquisition of multiple individual cell samples as a single, multiplexed convolute, has been proven to be particularly useful. To apply sample barcoding, cell samples are first labeled with unique metal tags. This allows their subsequent pooling and further processing as a convolute in a single vial, thereby greatly reducing wet work, technical variability (e.g., resulting from pipetting inaccuracies), and reagent consumption. Originally established for conventional flow cytometry [8], the concept of sample barcoding has

Helen M. McGuire and Thomas M. Ashhurst (eds.), *Mass Cytometry: Methods and Protocols*, Methods in Molecular Biology, vol. 1989,
https://doi.org/10.1007/978-1-4939-9454-0_7,

proven to be a very convenient and easy-to-integrate extension of the sample preparation workflow in mass cytometry, substantially improving sample-to-sample data consistency, reducing sample carryover and instrument running time. Considering that mass cytometry lacks light scatter parameters routinely applied for cell doublet removal in conventional flow cytometry data, the application of restricted combinatorial barcoding schemes importantly represents the most rational and efficient way to detect and exclude doublet/multiplet events, facilitating reliable phenotyping of single cells by mass cytometry.

In essence, there are two ways to perform sample barcoding in mass cytometry, by methods requiring or not requiring the fixation and permeabilization of cells prior to barcode labeling of cell samples. For 'intracellular' barcode labeling, small protein-reactive and metal-loaded molecules, such as thiol-reactive mDOTA loaded with lanthanide isotopes [9, 10], or thiol-reactive bromoacetamidobenzyl-EDTA (BABE), or amine-reactive isothiocyanobenzyl-EDTA (ITCB-EDTA) loaded with palladium isotopes [11], mark the total protein content of fixed and permeabilized cells. The commercially available palladium-based sample barcoding kit (Fluidigm) can be used for intracellular barcoding of up to 20 samples. Additional compounds applicable for intracellular barcode labeling are cisplatins containing highly isotopically enriched platinum isotopes [12], and lipid-reactive ruthenium and osmium tetroxides [13]. Importantly, intracellular barcoding requires at least "mild" permeabilization of cells with 0.02% saponin buffer prior to barcode labeling [14], limiting the benefits of barcoding to postfixation/permeabilization steps of the protocol. Furthermore, the detection of fixation-sensitive cell-surface markers, for example, many chemokine receptors, is compromised after performing the intracellular barcoding protocol [15]. The intracellular barcoding approach has proven particularly useful in in vitro cell signaling studies, in which cell activation is stopped by fixation and all cytometric stainings are performed after fixation [9, 16].

In contrast, cell-surface barcoding by metal-labeled antibodies [17] covered in this chapter in detail, can be applied to live cells, as it does not require any fixation or permeabilization prior to barcode labeling and can be applied directly to cells at the very beginning of sample processing. Thus, all relevant protocol steps, including cell-surface staining, cell fixation, permeabilization, and all washing steps, and data acquisition, are performed on the barcoded sample convolute, facilitating the detection of fixation-sensitive markers in barcoded samples. For live-cell barcoding, differently labeled antibody conjugates are combinatorially applied to cells of a given sample to create a unique sample identifier. For barcoding PBMC, combinations of CD45 antibody conjugates have been used [17–20], while for other cell types, different antibody targets might be more suitable as discussed below. Barcode antibodies can be labeled with lanthanides using commercial MAXPAR antibody labeling kits [21], with palladium isotopes using ITCB-EDTA [17] or with platinum isotopes using cisplatins [18]. The use of labels in the Pd and Pt channels is particularly advantageous, as they do not interfere with lanthanide channels commonly used for antibodies detecting the markers of interest serving for assay readout. Cell-surface barcoding with antibodies is especially suitable for complex immunophenotyping studies in which cell fixation interferes with the detectability of cell phenotypes defined by fixation-sensitive markers [19, 20, 22].

Using CD45 as an antibody target for live-cell barcoding restricts the approach to CD45-expressing cells such as PBMC. Since many cell types of interest such as neutrophils, plasma cells, and tumors show low, no, or variable CD45 expression, we investigated other cell-surface antigens with nearly ubiquitous expression on human cells, such as β2-microglobulin (B2M), HLA-ABC, and CD298 [23, 24] as antibody targets for live-cell barcoding. As shown in Fig. 1a, B2M and HLA-ABC were expressed at high levels by all tested subsets of fresh human peripheral blood leukocytes, whereas CD298 expression was high in most subsets too, but was strongly reduced on CD56^{bright} NK cells, limiting its use as a general barcoding marker. Especially on neutrophils, relative signal intensities of B2M and HLA-ABC stainings were higher than that of CD45. We therefore tested Pd-labeled anti-B2M antibody conjugates for their suitability to barcode cell suspensions that are not fully covered by CD45-based live-cell barcoding due to heterogeneous expression of CD45. Our data confirm that B2M antibodies can be used to barcode human blood leukocyte samples (Fig. 1b), and that cell population frequencies obtained from barcoded/pooled samples match those of individually run samples (Fig. 1c). Similarly, B2M-based live-cell barcoding enabled joint processing of healthy and diseased human bone marrow samples, preserving cellular sample composition including CD45^{neg} plasma cells [25] and CD45$^{low/var}$ leukemic cells (Fig. 1d, e). Overall, the B2M-based barcoding approach promises a broader applicability than CD45-based barcoding and facilitates co-processing of human cell samples other than PBMC.

In mass cytometry the use of restricted combinatorial barcoding schemes prevails, in which each barcode is defined by a combination of equal numbers of individual labels, and a digital signal intensity readout for each channel, i.e., absence or presence of signal [11, 17]. Thus, only defined combinations of barcode markers, fulfilling these criteria, are used in restricted barcoding schemes. The theoretical capacity of restricted barcoding schemes equals $n!/(k!(n\text{-}k)!)$, with n being the number of barcode channels and k being the number of labels combined on a given sample, and is illustrated by Pascal's triangle in Fig. 2 [26]. In practice, upscaling the barcoding scheme capacity may be limited by the competition

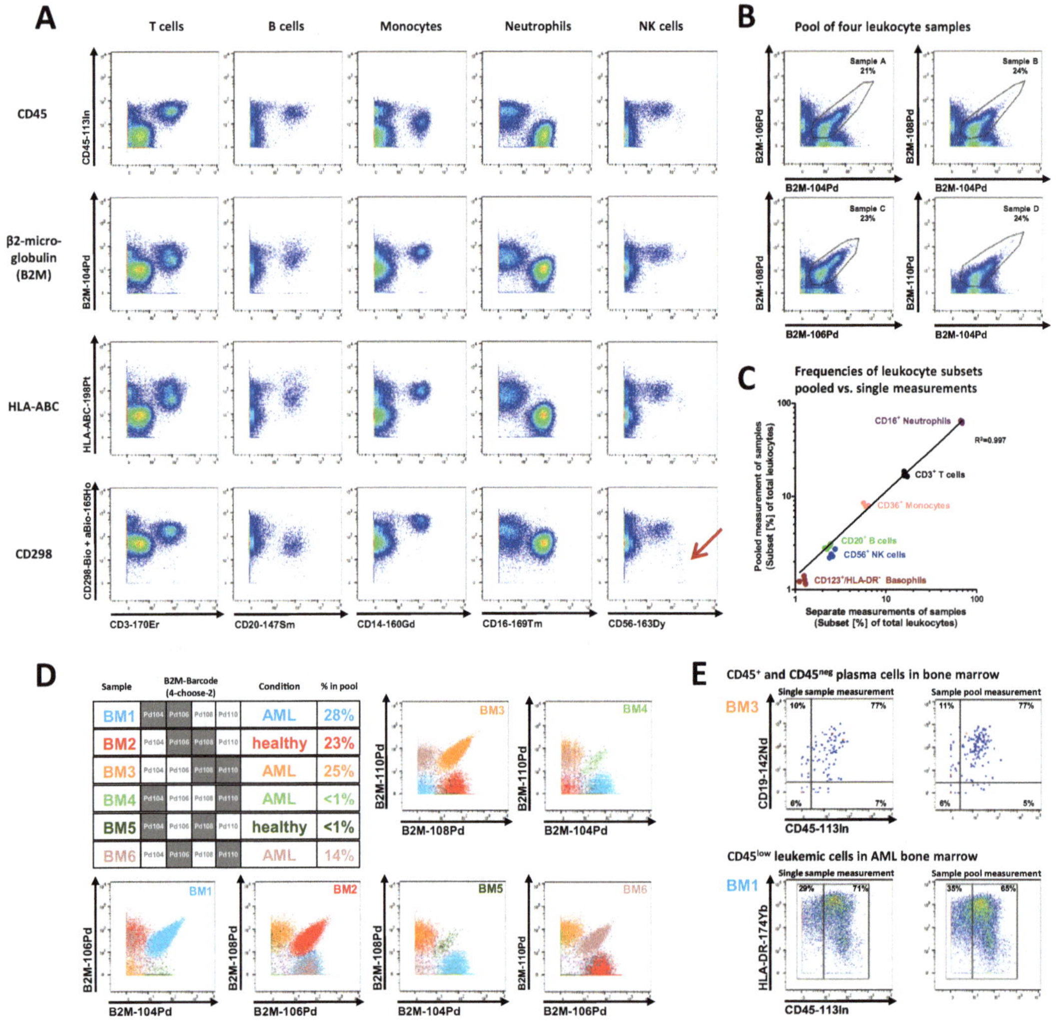

Fig. 1 Cell-surface markers applicable for live-cell barcoding. (**a**) Expression of CD45, β2-microglobulin (B2M), HLA-ABC, and CD298 on major leukocyte populations in fresh peripheral blood. The relative staining intensity of CD45-^{113}In on neutrophils is critically low and expected to interfere with proper deconvolution of a barcoded and pooled leukocyte sample, while B2M-^{104}Pd and HLA-ABC-^{198}Pt achieve higher relative signal intensities on neutrophils and monocytes than CD45-^{113}In, and are strongly expressed by all other major leukocyte subsets. CD298 staining generates high signal intensities on all leukocyte subsets (median, 52-271), except for CD56^{bright} NK cells (median, 8, arrow), possibly complicating accurate barcoding/deconvolution of samples containing NK cells. All stainings were performed together on a single sample of fresh leukocytes isolated from whole blood after RBC lysis. Data was normalized, and live cells were identified by gating according to event length, DNA and 103Rh^{103}-based live/dead staining. (**b**) Application of B2M-based live-cell barcoding to leukocyte samples. Four individual B2M-Pd conjugates were used in a 4-choose-2 scheme to barcode four fresh leukocyte samples (^{104}Pd/^{106}Pd, ^{104}Pd/^{108}Pd, ^{106}Pd/^{108}Pd, and ^{104}Pd/^{110}Pd). Manually drawn gates identify the individual sample populations after pre-gating on live, single cells within data of the pooled sample and are used for the deconvolution of the four individual leukocyte samples. (**c**) Comparison of major leukocyte subset frequencies of leukocyte samples run individually (*x*-axis) or as a B2M-based barcoded sample convolute (*y*-axis). Data indicate the correct reporting of subset frequencies, notably of CD45^{low} neutrophils and basophils in the B2M-based barcoded, deconvoluted samples. (**d**) Application of B2M-based live-cell barcoding on six

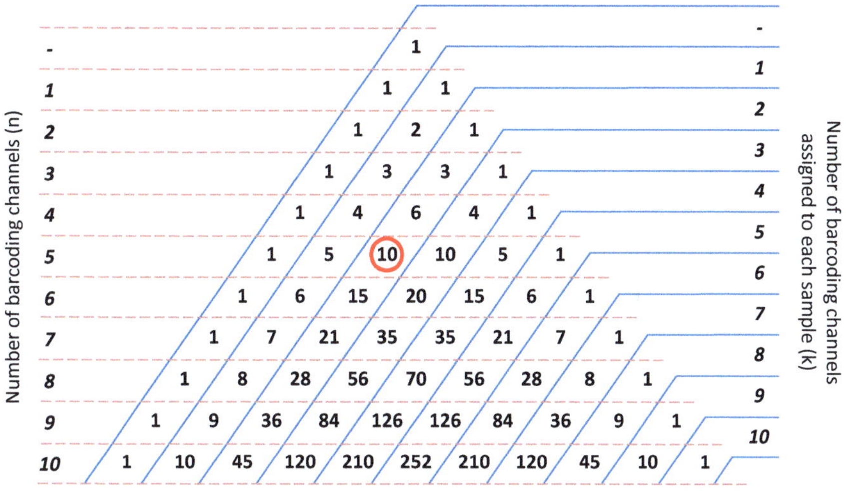

Fig. 2 Pascal's triangle visualizes the capacities of restricted *n*-choose-*k* sample barcoding schemes, in which *n* denotes the number of barcoding channels (left side) and *k* the number of barcoding channels assigned to each sample (right side). For example, a 5-choose-2 barcoding scheme uses combinatorial staining of exactly two out of five differently labeled barcoding antibodies to encode each single sample. The maximum capacity of this barcoding scheme is ten samples (indicated by the red circle)

of barcoding reagents for identical binding sites that can lead to critically low barcode label signal intensities. Employing restricted barcoding schemes effectively filters out doublet events that comprise events from different samples (inter-sample doublets), as true single-cell events carry combination of, k (and only k) labels. Event doublets, resulting either from cell aggregation or poorly separated ion clouds, bear invalid barcode label combinations (e.g., $k + 1$, $k + 2$, or $k + 3$ labels) [11, 17] and can be excluded from data analysis accordingly. After acquisition, the original sample data is extracted from the barcoded, pooled sample by deconvolution by either manual gating and Boolean operations using standard flow cytometry software [17], or using a script developed for that purpose [11]. The better the cytometric separation of individual samples from each other in the barcode channels, the more original

Fig. 1 (continued) frozen bone marrow (BM) leukocyte samples from healthy donors and acute myeloid leukemia (AML) patients. The table shows the barcoding scheme along with the assigned sample and their proportions in the sample pool. Two samples (BM4 and BM5) contained critically low amounts of live cells after thawing due to poor sample quality, as confirmed by independent analyses. Data were deconvoluted by manual gating after pre-gating on live, single cells. Data of a given sample were individually colored after Boolean deconvolution. (**e**) Successful recovery of $CD45^{low/-}$ cell types from B2M-based cell-surface barcoding of human BM leukocytes. Expression of CD19 and CD45 on plasma cells and HLA-DR and CD45 on leukemic cell samples from BM3 and BM1 (from data shown in **d**) are compared between individually run and barcoded/deconvoluted BM samples sample cells will be accurately assigned during deconvolution. In our lab, live-cell barcoding of ten PBMC samples using palladium- and platinum-tagged CD45 antibodies with a 5-choose-2 scheme typically returns about 75% of total input events. Notably, unassigned events contain unwanted doublets events and cell debris [11, 17], so that with sufficient cytometric separation in barcode channels, only insignificant amounts of cells of interest are not recovered from the convolute.

In this chapter, we provide a state-of-the-art procedure for ITCB-EDTA-based palladium labeling of live-cell barcoding antibodies, followed by a protocol for B2M antibody-based sample barcoding for subsequent immune phenotyping by mass cytometry using a 5-choose-2 scheme for ten individual PBMC samples.

2 Materials

2.1 For ITCB-EDTA Aliquotation

1. ITCB-EDTA (Dojindo); M030, 10 mg.
2. Acetonitrile (Sigma/Fluka) TraceSelect; 01324.
3. 1.5 mL safe-lock reaction tubes (Eppendorf).
4. Vacuum concentrator (e.g., Concentrator 5301 (Eppendorf)).
5. Sonication bath (e.g., from JSP, 50 W, 42 kHz).
6. Desiccant (e.g., Silica gel blue (Roth)).
7. Parafilm (Bemis Company).
8. Fume hood.

2.2 For Loading ITCB-EDTA with Pd Isotopes

1. 1 mg aliquots of ITCB-EDTA (*see* Part 1).
2. Palladium nitrate ($Pd(NO_3)_2$), maximum available isotopic purity >95%, 10 mg elemental Pd weight (TRACE Sciences).
3. HCl (30%).
4. L-Buffer (Fluidigm).
5. DMSO.
6. Heat block, preheated to 65 °C for 1.5 mL reaction tubes (e.g., LS1 (VLM)).
7. Liquid nitrogen in a small dewar for snap freezing.
8. Lyophilizer (e.g., Alpha 2-4 LSC basic (Christ), condenser temperature set to −70 °C).

2.3 For Antibody Conjugation

1. 50 μg of antibody (IgG), carrier protein-free (e.g., anti-human B2M, clone 2M2; or anti-human CD45, clone HI30).
2. 20 mM ITCB-EDTA-Pd solution(s) (from Subheading 3.2).
3. 50 kDa Amicon Ultra spin column (Millipore).
4. Water bath or heat block, preheated to 37 °C.

5. Tabletop microcentrifuge, cooled to 4 °C for 1.5 mL reaction tubes (e.g., Heraeus Fresco 21 (Thermo Fisher)).
6. Antibody Stabilizer (supplemented with 0.05% sodium azide); (Candor).
7. Double-concentrated HEPES-buffered saline solution with EDTA (2× HEPES/E). For 100 mL, weigh 1.3 g NaCl, 27 mg $CaCl_2 \cdot 2H_2O$, 23 mg $MgCl_2$ and 83.6 mg KH_2PO_4 into a new (Milli-Q water-rinsed) glass beaker. Add 4 mL 1 M HEPES (final concentration 40 mM) and 2 mL 0.5 M EDTA (final concentration 10 mM). Add 80 mL Milli-Q water and adjust to pH 7.3 using NaOH. Make up to 100 mL with water. Store at 4 °C.
8. PBS made from 10× stock (Rockland). Adjust to pH 7.3 using NaOH and sterile filter through 0.22 μm (Millipore Steritop filtration unit). Store in single-use bottles (e.g., Millipore Stericup flasks) at 4 °C.
9. UV-Vis spectrometer (e.g., NanoDrop 2000c (Thermo Fisher)).

2.4 Barcoding

1. Cell samples (e.g., human PBMC or leukocytes).
2. PBS/BSA/azide buffer. Supplement 1× PBS (*see* Subheading 2.3) with 0.5% BSA (PAN Biotech) and 0.02% sodium azide (Sigma).
3. Tabletop microcentrifuge, cooled to 4 °C for 1.5 mL reaction tubes (e.g., Heraeus Fresco 21 (Thermo Fisher)).
4. Centrifuge, cooled to 4 °C, suitable for deep well plates (e.g., Multifuge 1S-R (Thermo Fisher)).
5. Deep well plates, 2 mL (Corning).
6. Four individual anti-B2M palladium conjugates and one anti-B2M platinum isotope conjugate.
7. FcR Block reagent (Miltenyi Biotec).
8. Filter cap tubes (BD).
9. PCR tubes 0.2 mL (stripes).
10. Multichannel pipettes; 1200 μL and 100 μL.
11. FlowJo software (FlowJo LLC).

3 Methods

3.1 Preparation of ITCB-EDTA Aliquots (Optional—See Note 1)

1. Take a new vial of ITCB-EDTA powder (10 mg) and let it assume room temperature.
2. Spin down ITCB-EDTA vial in microcentrifuge (30 s, maximum speed).
3. Add 10 mL acetonitrile stepwise to the ITCB-EDTA vial and transfer volumes to a 15 mL centrifuge tube. As ITCB-EDTA

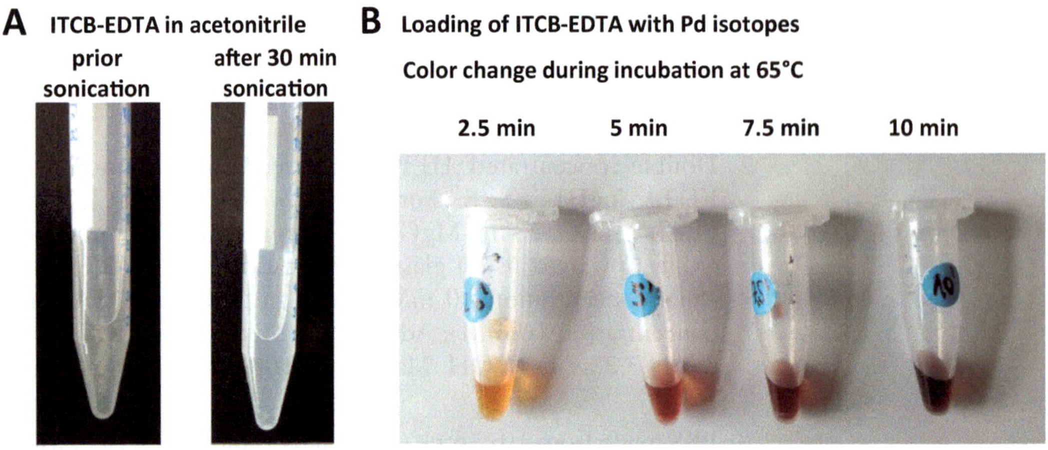

Fig. 3 Expected appearance of ITCB-EDTA in acetonitrile and of ITCB-EDTA during loading with Pd. (**a**) ITCB-EDTA powder does not dissolve immediately in acetonitrile at ambient conditions (left, note white crystals). After 30 min of sonication, crystals have disappeared. The mixture now presents as a homogeneous dispersion (right) that can be accurately aliquoted by volume, ensuring precise proportioning of ITCB-EDTA. (**b**) Color change over time of the ITCB-EDTA loading reaction with 10 mM Pd nitrate at 65 °C

powder does not fully dissolve (Fig. 3a), clumps remain visible that may block regular pipette tips. Cut away approx. 5 mm of the thin end of a 1000 μL pipette tip using clean scissors or a scalpel to achieve a larger orifice for pipetting in this step.

4. Use a water bath sonicator for the dispersing ITCB-EDTA in acetonitrile. After approximately 30 min of sonication ITCB-EDTA crystals have disappeared and the suspension appears as slightly cloudy (Fig. 3a) (*see* **Note 2**).
5. Quickly aliquot the ITCB-EDTA dispersion to fresh 1.5 mL safe-lock reaction tubes (Eppendorf), by dispensing, e.g., 1 mL to achieve 1 mg aliquots.
6. Dry aliquots in the vacuum concentrator (room temperature, takes approx. 30–45 min), place outlet in an active fume hood.
7. When dry, close tubes, seal the tubes (e.g., with Parafilm) and freeze in an airtight bag supplied with desiccant at −20 °C for further use.

3.2 Loading ITCB-EDTA with Pd Isotopes

1. Prepare a 5 N HCl solution. For 1 mL add 0.506 mL 30% HCl to 0.494 mL H_2O (*see* **Note 3**). For 10 mg Pd (metal weight, not total compound weight) approx. 1 mL of 5 N HCl is required.
2. Dissolve isotopically enriched $Pd(NO_3)_2$ in 5 N HCl to generate a 100 mM stock solution (*see* **Notes 4** and **5**).
3. For each Pd isotope, get a 1 mg aliquot of ITCB-EDTA from −20 °C freezer and let it assume room temperature.

4. For each Pd isotope, prepare a 10 mM Pd nitrate solution in L-buffer, by mixing 12 μL of 100 mM $Pd(NO_3)_2$-in-HCl solution with 108 μL L-buffer.
5. Add 100 μL of this 10 mM Pd nitrate solution to a 1 mg ITCB-EDTA aliquot, vortex intensely for 10 s. The molar ratio of $Pd(NO_3)_2$: ITCB-EDTA is 1:2.
6. Directly incubate the mixture in a 65 °C heat block for (exactly) 10 min. Shortly vortex the tube after 2.5, 5, and 7.5 min, then briefly spin down vial content in a microcentrifuge (*see* **Note 6**). The color of the solution turns from yellow to amber/red-brownish (Fig. 3b).
7. Snap freeze ITCB-EDTA/Pd tubes in liquid nitrogen and lyophilize tube content overnight, using perforated lids (*see* **Note 7**).
8. Redissolve tube content in 113.8 μL DMSO to achieve a 20 mM ITCB-EDTA-Pd solution.
9. Store dissolved ITCB-EDTA-Pd at −20 °C.

3.3 Antibody Labeling with ITCB-EDTA-Pd

1. For each conjugate, equilibrate a 50 kDa spin column by adding 300 μL 2× HEPES/E (*see* **Note 8**) and spinning at 12,000 × *g* for 8 min. Discard flow-through.
2. Add 50 μg of antibody (*see* **Note 9**), fill up to 400 μL using 2× HEPES/E and centrifuge at 12,000 × *g* for 8 min. Discard flow-through.
3. Pipette 300 μL 2× HEPES/E to the column and stir in 3.2 μL of 20 mM ITCB-EDTA-Pd solution using a 10 μL pipette. Immediately and thoroughly mix with, e.g., a 100 μL pipette to homogenize and incubate columns for 1 h at 37 °C in a water bath (*see* **Note 10**).
4. Wash antibody conjugates over spin columns by adding 200 μL 2× HEPES/E and centrifuge at 12,000 × *g* for 8 min, discard flow-through.
5. Repeat washes, once with 400 μL 2× HEPES/E and four times with 400 μL PBS. Discard flow-through.
6. Transfer concentrated conjugate solution (approx. 20–30 μL) from filter reservoirs to a 1.5 mL safe-seal reaction tube. Rinse filters twice with 20 μL PBS and transfer liquid to the same 1.5 mL tube.
7. Confirm recovery of antibody conjugate, e.g., by UV-Vis spectroscopy at 280 nm (*see* **Note 11**).
8. Dilute antibody conjugate solution 1:1 v/v with Candor Antibody Stabilization Solution.
9. Label and store at 4 °C.
10. Determine optimal working concentration of conjugates on appropriate target cells (*see* **Note 12**).

3.4 Barcoding of up to Ten PBMC Samples Using Pd and Pt-Labeled B2M Conjugates in a 5-Choose-2 Barcoding Scheme

1. For an optimal barcode staining, all B2M antibody conjugates should be used at similar concentrations to achieve equal competition for cell-surface binding sites (*see* **Note 13**).
2. Freshly prepare mixes of B2M antibody conjugates in 0.2 mL PCR tubes (or stripes) according to the "5-choose-2" barcoding scheme on ice, as outlined in Fig. 4a. For a final barcode antibody staining reaction volume of 50 μL, sufficient to barcode up to 2.5 × 10^6 PBMC, add 10 μL PBS/BSA/azide, then add Fc Block (1 μL), then respective B2M antibody conjugates (e.g., all 1:25 v/v, i.e., 2 μL, resulting in a total volume of 15 μL, *see* **Note 14**). Keep all tubes on ice.
3. Prepare single-cell suspensions, e.g., PBMC. Filter cells through a 30 μm cell strainer and count cells. Adjust the cell density with PBS/BSA/azide to 2.5 × 10^6 cells in 35 μL (*see* **Note 14**).
4. Add each 35 μL (*see* **Note 14**) PBMC sample to a PCR tube containing the individual barcode antibody mix, resuspend them thoroughly, and incubate them for 30 min at 4 °C.
5. Stop barcode staining by adding 100 μL PBS/BSA/azide preferably with a multichannel pipette. Then transfer samples to a new 2 mL V-bottom deep well plate. Wash PCR tubes three times with 200 μL PBS/BSA/azide and transfer all volumes to the deep well plate, to maximize cell recovery. Finally add 1200 μL PBS/BSA/azide preferably with a multichannel pipette to each well and centrifuge at 500 × *g*, 5 min, 4 °C. Carefully aspirate the supernatant.
6. Repeat washes with 1200 μL PBS/BSA/azide three times (*see* **Note 15**).
7. Pool samples by adding 1 mL PBS/BSA/azide into each well and transfer all samples into a new 15 mL centrifuge tube. Using a single pipet tip will minimize cell loss. To recover maximum cell amounts, wash wells of the deep- well plate twice with 500 μL PBS/BSA/azide and transfer all volumes to the 15 mL tube. Optionally, pass the sample pool over a 30 μm cell strainer. Count cells to adjust cell concentration for the subsequent steps such as cell-surface staining.
8. Continue with mass cytometry sample preparation workflow, e.g., live-dead staining, cell-surface staining, sample fixation, and Iridium-intercalator staining.
9. Acquire the cell sample pool on a mass cytometer (*see* **Note 16**).
10. Normalize the FCS files using the inbuilt normalization algorithm of the Helios software (Fluidigm) or other normalization tools e.g. [27].
11. Perform barcode deconvolution either using manual gating and Boolean operators in a flow cytometry analysis software

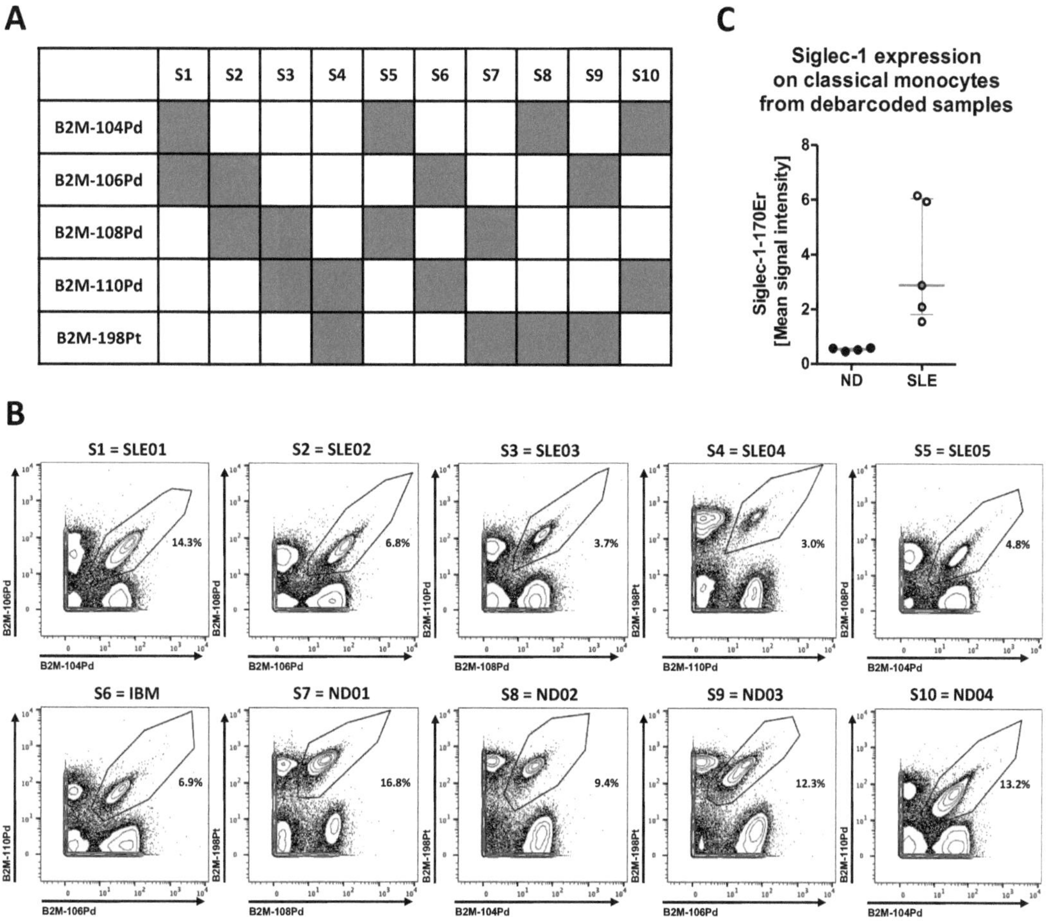

Fig. 4 Barcoding of ten PBMC samples from autoimmune disease patients and healthy controls (normal donors, ND) using B2M-based live-cell barcoding and a 5-choose-2 barcoding scheme. (**a**) A 5-choose-2 sample barcoding scheme was used to barcode ten samples. Gray indicates staining of a sample (top) with a B2M antibody (left). Each sample is encoded by a unique combination of two B2M antibody conjugates. The B2M-^{198}Pt antibody conjugate was produced in-house as described before [18]. (**b**) Deconvolution of all samples by manually drawn gates in bivariate dot plots. After normalizing the sample pool FCS file, exclusion of EQ beads, gating of DNA^{+} cellular events and pre-gating on single-cell events, cells of each original sample were gated as the "double-positive" population in the respective combination of two barcode channels. For example, sample 1 cells exhibit a co-staining of B2M-^{104}Pd and B2M-^{106}Pd and thereby can be identified as a ^{104}Pd^{+}/^{10}6Pd^{+} population. After manual gating, an exclusion of cell events carrying invalid barcode combinations was performed by using the Boolean operators AND and NOT. Cryopreserved PBMC from five patients with systemic lupus erythematosus (SLE), and one with inclusion body myositis (IBM) and four matched healthy controls were used. (**c**) Siglec-1 expression on classical CD14^{+}/CD16^{neg} monocytes in SLE patients and healthy controls determined from debarcoded samples from (**d**). Mean signal intensity of Siglec-1 staining (Clone 7-239, Miltenyi Biotec, coupled to ^{170}Er) was determined in each sample on CD36^{+}CD14^{+}CD16^{neg} classical monocytes after pre-gating on live single CD45^{+} events. Monocytic Siglec-1 expression is a biomarker for disease activity in SLE [28, 29]

such as FlowJo as outlined in Fig. 4b, or a deconvolution algorithm [11] or inbuilt in the Helios software. Typically, about 75% of all cellular events can be assigned to a valid barcode combination. Deconvolution will yield a new set of FCS files, each one reflecting an individual original sample, which can be subjected to further manual or computational analysis. Figure 4c depicts an example of a result from the analysis of PBMC samples from five systemic lupus erythematosus (SLE) patients and four healthy controls that were jointly processed and acquired as a live-barcoded sample convolute using the procedure described in this protocol.

4 Notes

1. Aliquoting ITCB-EDTA powder is optional, but highly recommended. ITCB-EDTA is commonly delivered in 10 mg vials that can be used directly for Pd loading, or aliquoted. Aliquoting using acetonitrile ensures that exact quantities of ITCB-EDTA are used in subsequent steps. Using a lab balance has turned out impractical for this purpose.
2. We have used a water bath sonicator without cooling. Due to the sonication energy the water bath heats up to about 50 °C during the 30 min incubation. We have not observed an adverse effect on the functionality of ITCB-EDTA in that regard.
3. Note that the lot-specific Certificate of Analysis for HCl may show a different acid content, e.g., 32% w/w. An online acid calculator, e.g., https://tinyurl.com/acidcalc may be used to determine the exact amount of HCl required.
4. The volume of 5 N HCl to achieve a 100 mM $Pd(NO_3)_2$ solution differs for the different Pd isotopes. Based on 10 mg palladium (metal weight), the following volumes of 5 N HCl are needed: 102Pd: 981 μL; 104Pd: 962 μL; 105Pd: 953 μL; 106Pd: 944 μL; 108Pd: 927 μL; 110Pd: 910 μL.
5. For troubleshooting, the protocol can be exercised using natural abundance palladium nitrate that is much cheaper than isotopically purified $Pd(NO_3)_2$.
6. We have tested different ITCB-EDTA loading conditions. Best results for CD45 antibody staining were achieved with a Pd loading time of 10 min, returning a solution of typically red-brownish color. Incubation at 65 °C shorter than 5 min returned a yellowish solution that, if used to label CD45 antibodies, provided inferior CD45 staining intensities in the barcode labeling reaction. Incubation times of more than 10 min returned a deep brown solution, and only marginally

increased specific signal of antibody conjugates in immunoassays, but led to increased unspecific background, as exemplified in Fig. 5a using CD8-Pd conjugates. We furthermore tested if a doubling of the molar Pd: ITCB-EDTA loading ratio results in higher signal intensities. As shown in Fig. 5b, loading of ITCB-EDTA with 20 mM Pd nitrate instead of 10 mM did not generate higher signal intensities of a CD8-Pd conjugate.

7. Lyophilization of the ITCB-EDTA-Pd solution is essential to ensure complete removal of water that could otherwise lead to hydrolysis of the amine-reactive moiety of ITCB-EDTA.
8. Double-concentrated HEPES-buffered saline with EDTA (HEPES/E) has a higher buffering capacity than PBS, and is used to maintain stable pH conditions during antibody conjugation after adding ITCB-EDTA-Pd solution. EDTA was added to scavenge possibly remaining unchelated Pd ions.

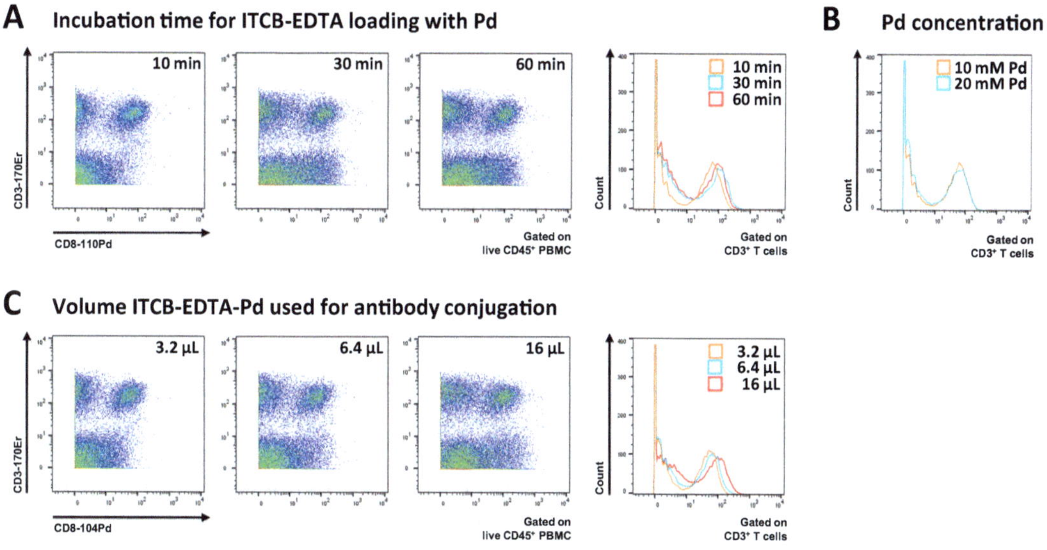

Fig. 5 Impact of ITCB-EDTA loading and antibody conjugation conditions on CD8-Pd staining patterns on PBMC. (**a**) Increasing the incubation time of the ITCB-EDTA loading with ^{110}Pd during the production of ITCB-EDTA-^{110}Pd to 30 or 60 min leads to only marginal increase of specific staining and a significant increase of unexpected, background staining when conjugated to CD8 antibody. (**b**) Doubling the amount of 110Pd during ITCB-EDTA loading to 20 mM does not result in a higher CD8 signal intensity (histogram overlays of pre-gated CD3^{+} T cells). (**c**) Increasing the volume of ITCB-EDTA-^{104}Pd solution from 3.2 µL to 6.4 µL or 16 µL in the antibody conjugation reaction (paralleled by an increase in DMSO content from 2% to approx. 8%) marginally increases CD8-^{104}Pd signal intensity (histogram overlay of pre-gated CD3^{+} T cells), but also increases nonspecific binding in the CD3^{+}CD8^{neg} T cells and CD3^{neg} non-T cells (pre-gated live CD45^{+} cells). In all experiments conjugates of anti-human CD8 clone GN11/134D7 (DRFZ) were used to stain cryopreserved PBMC processed as described before [30]. For analysis, samples were normalized, and live cells were identified by gating according to event length, DNA staining, and ^{103}Rh-based live/dead staining

9. It is possible to upscale the conjugation reaction to at least 100 μg of antibody. Adjust the conjugation reaction volume on the column to 500 μL and add 6.4 μL ITCB-EDTA-Pd. Wash twice with 400 μL HEPES/E and four times with 400 μL PBS.
10. We have tested whether increasing the ITCB-EDTA-Pd volume in the antibody conjugation reaction generates antibody conjugates with higher signal intensities. Using 6.4 μL or 16 μL ITCB-EDTA-Pd instead of 3.2 μL, however, only slightly increased signal intensities of the CD8 antibody conjugates at the expense of antibody signal specificity (Fig. 5c), which might be result of too much DMSO in the conjugation reaction.
11. ITCB-EDTA-Pd has an intrinsic absorption at 280 nm wavelength. Spectroscopic measurements of Pd antibody conjugates are hence inaccurate, but minimally report the successful recovery of antibody from the conjugation reaction.
12. If the antibody conjugate will be used for sample barcoding, it is advisable to experimentally confirm sufficient barcode marker expression on target cells or sample.
13. Since Pd and Pt antibody conjugates underwent different conjugation procedures via thiols vs. amines [18], that may differently interfere with antigen binding of the conjugated antibody, the amount of each conjugate in each barcode labeling mixture should be experimentally confirmed or optimized.
14. The exact volumes depend on the required volumes of barcode antibodies for a given barcode labeling determined by dilutions series, and of the FcR blocking reagent used.
15. Thorough washing of cells after barcode antibody incubation is essential for achieving sufficient signal separation, which has an important impact on cell recovery during data deconvolution. Insufficient washing leads to barcode signal cross-contamination between barcoded samples, entailing increased background signal and poor separation. When cytometric separation in a barcode channel is insufficient, consider additional washing steps and/or increasing wash buffer volumes.
16. The sample size of the barcoded sample convolute may require long sample acquisition times on the mass cytometer. It is recommended to use a sample introduction system for handling large samples, e.g., the Super Sampler (Victorian Airship & Scientific Apparatus LLC).

Acknowledgments

The authors would like to thank all members of the DRFZ mass cytometry group, Michael D. Leipold, PhD, for discussing antibody labeling strategies, Dr. Lars Fransecky for providing leukemia samples, and Dr. Petra Henklein for providing access to the lyophilizer.

References

1. Bendall SC, Simonds EF, Qiu P et al (2011) Single-cell mass cytometry of differential immune and drug responses across a human hematopoietic continuum. Science 332(6030):687–696. https://doi.org/10.1126/science.1198704
2. Levine JH, Simonds EF, Bendall SC et al (2015) Data-driven phenotypic dissection of AML reveals progenitor-like cells that correlate with prognosis. Cell 162(1):184–197. https://doi.org/10.1016/j.cell.2015.05.047
3. Rao DA, Gurish MF, Marshall JL et al (2017) Pathologically expanded peripheral T helper cell subset drives B cells in rheumatoid arthritis. Nature 542(7639):110–114. https://doi.org/10.1038/nature20810
4. O'Gorman WE, Hsieh EW, Savig ES et al (2015) Single-cell systems-level analysis of human Toll-like receptor activation defines a chemokine signature in patients with systemic lupus erythematosus. J Allergy Clin Immunol 136(5):1326–1336. https://doi.org/10.1016/j.jaci.2015.04.008
5. Gaudilliere B, Fragiadakis GK, Bruggner RV et al (2014) Clinical recovery from surgery correlates with single-cell immune signatures. Sci Transl Med 6(255):255ra131. https://doi.org/10.1126/scitranslmed.3009701
6. Nair N, Mei HE, Chen SY et al (2015) Mass cytometry as a platform for the discovery of cellular biomarkers to guide effective rheumatic disease therapy. Arthritis Res Ther 17:127. https://doi.org/10.1186/s13075-015-0644-z
7. Ermann J, Rao DA, Teslovich NC et al (2015) Immune cell profiling to guide therapeutic decisions in rheumatic diseases. Nat Rev Rheumatol 11(9):541–551. https://doi.org/10.1038/nrrheum.2015.71
8. Krutzik PO, Nolan GP (2006) Fluorescent cell barcoding in flow cytometry allows high-throughput drug screening and signaling profiling. Nat Methods 3(5):361–368. https://doi.org/10.1038/nmeth872
9. Bodenmiller B, Zunder ER, Finck R et al (2012) Multiplexed mass cytometry profiling of cellular states perturbed by small-molecule regulators. Nat Biotechnol 30(9):858–867. https://doi.org/10.1038/nbt.2317
10. Zivanovic N, Jacobs A, Bodenmiller B (2014) A practical guide to multiplexed mass cytometry. Curr Top Microbiol Immunol 377:95–109. https://doi.org/10.1007/82_2013_335
11. Zunder ER, Finck R, Behbehani GK et al (2015) Palladium-based mass tag cell barcoding with a doublet-filtering scheme and single-cell deconvolution algorithm. Nat Protoc 10(2):316–333. https://doi.org/10.1038/nprot.2015.020
12. McCarthy RL, Mak DH, Burks JK et al (2017) Rapid monoisotopic cisplatin based barcoding for multiplexed mass cytometry. Sci Rep 7(1):3779. https://doi.org/10.1038/s41598-017-03610-2
13. Catena R, Ozcan A, Zivanovic N et al (2016) Enhanced multiplexing in mass cytometry using osmium and ruthenium tetroxide species. Cytometry A 89(5):491–497. https://doi.org/10.1002/cyto.a.22848
14. Behbehani GK, Thom C, Zunder ER et al (2014) Transient partial permeabilization with saponin enables cellular barcoding prior to surface marker staining. Cytometry A 85(12):1011–1019. https://doi.org/10.1002/cyto.a.22573
15. Dzangue-Tchoupou G, Corneau A, Blanc C et al (2018) Analysis of cell surface and intranuclear markers on non-stimulated human PBMC using mass cytometry. PLoS One 13(3):e0194593. https://doi.org/10.1371/journal.pone.0194593
16. Gaudilliere B, Ganio EA, Tingle M et al (2015) Implementing mass cytometry at the bedside to study the immunological basis of human diseases: distinctive immune features in patients with a history of term or preterm birth. Cytometry A 87(9):817–829. https://doi.org/10.1002/cyto.a.22720

17. Mei HE, Leipold MD, Schulz AR et al (2015) Barcoding of live human peripheral blood mononuclear cells for multiplexed mass cytometry. J Immunol 194(4):2022–2031. https://doi.org/10.4049/jimmunol.1402661
18. Mei HE, Leipold MD, Maecker HT (2016) Platinum-conjugated antibodies for application in mass cytometry. Cytometry A 89(3):292–300. https://doi.org/10.1002/cyto.a.22778
19. Lavin Y, Kobayashi S, Leader A et al (2017) Innate immune landscape in early lung adenocarcinoma by paired single-cell analyses. Cell 169(4):750–765.e717. https://doi.org/10.1016/j.cell.2017.04.014
20. Mrdjen D, Pavlovic A, Hartmann FJ et al (2018) High-dimensional single-cell mapping of central nervous system immune cells reveals distinct myeloid subsets in health, aging, and disease. Immunity 48(2):380–395.e386. https://doi.org/10.1016/j.immuni.2018.01.011
21. Lai L, Ong R, Li J et al (2015) A CD45-based barcoding approach to multiplex mass-cytometry (CyTOF). Cytometry A 87(4):369–374. https://doi.org/10.1002/cyto.a.22640
22. Hartmann FJ, Bernard-Valnet R, Queriault C et al (2016) High-dimensional single-cell analysis reveals the immune signature of narcolepsy. J Exp Med 213(12):2621–2633. https://doi.org/10.1084/jem.20160897
23. Lawson DA, Bhakta NR, Kessenbrock K et al (2015) Single-cell analysis reveals a stem-cell program in human metastatic breast cancer cells. Nature 526(7571):131–135. https://doi.org/10.1038/nature15260
24. Hartmann FJ, Simonds EF, Bendall SC A Universal Live Cell Barcoding-Platform for Multiplexed Human Single Cell Analysis. PMID 30018331
25. Medina F, Segundo C, Campos-Caro A et al (2002) The heterogeneity shown by human plasma cells from tonsil, blood, and bone marrow reveals graded stages of increasing maturity, but local profiles of adhesion molecule expression. Blood 99(6):2154–2161
26. Cossarizza A, Chang HD, Radbruch A et al (2017) Guidelines for the use of flow cytometry and cell sorting in immunological studies. Eur J Immunol 47(10):1584–1797. https://doi.org/10.1002/eji.201646632
27. Finck R, Simonds EF, Jager A et al (2013) Normalization of mass cytometry data with bead standards. Cytometry A 83(5):483–494. https://doi.org/10.1002/cyto.a.22271
28. Biesen R, Demir C, Barkhudarova F et al (2008) Sialic acid-binding Ig-like lectin 1 expression in inflammatory and resident monocytes is a potential biomarker for monitoring disease activity and success of therapy in systemic lupus erythematosus. Arthritis Rheum 58(4):1136–1145. https://doi.org/10.1002/art.23404
29. Rose T, Grutzkau A, Hirseland H et al (2013) IFNalpha and its response proteins, IP-10 and SIGLEC-1, are biomarkers of disease activity in systemic lupus erythematosus. Ann Rheum Dis 72(10):1639–1645. https://doi.org/10.1136/annrheumdis-2012-201586
30. Schulz AR, Stanislawiak S, Baumgart S et al (2017) Silver nanoparticles for the detection of cell surface antigens in mass cytometry. Cytometry A 91(1):25–33. https://doi.org/10.1002/cyto.a.22904

Part III

Sample Preparation

Chapter 8

Automated Cell Processing for Mass Cytometry Experiments

Jaromir Mikes, Axel Olin, Tadepally Lakshmikanth, Yang Chen, and Petter Brodin

Abstract

Mass cytometry is a powerful technology for high-dimensional single-cell measurements in millions of individual cells. Antibodies and other detection probes are coupled to elemental tags, each with a unique mass and detectable at single-cell resolution using an ICP-MS type of instrument. Given the sensitivity of the detection system, any free metal ions must be carefully removed through multiple rounds of washing in order to prevent background signal. This results in significant loss of cells. Together with cells lost during acquisition, the final data can represent as little as 10% of the starting material, seriously limiting the amount of information that can be extracted from small samples. Furthermore, complex staining protocols introduce experimental variations that limit comparisons across experiments. Here we present a cell processing and staining procedure for mass cytometry fully automated using a liquid handling robotic system and we present measures taken to optimize all steps of the protocol. These advances are applicable to both manual and automated protocols and provide a six-fold higher cell yield as compared to a standard protocol. With this increased yield and improved reproducibility this protocol now allows us to perform mass cytometry analysis using as little as 100 μL of whole blood as starting material.

Key words Mass cytometry, Flow cytometry, Lab automation, Liquid handling robotics, Protocol, Cell processing

1 Introduction

Mass cytometry is a powerful method for high-dimensional single-cell analysis [1]. These measurements are enabling important analyses of human immune system variation [2] and are making human immune systems more predictable [3] and informing clinical outcome [4]. Sample preparation for mass cytometry is a laborious and time-consuming multistep process. It requires numerous washing steps to avoid contamination of free metals and poorly

Jaromir Mikes and Axel Olin contributed equally to this work.

Helen M. McGuire and Thomas M. Ashhurst (eds.), *Mass Cytometry: Methods and Protocols*, Methods in Molecular Biology, vol. 1989, https://doi.org/10.1007/978-1-4939-9454-0_8, © Springer Science+Business Media, LLC, part of Springer Nature 2019

bound antibodies, which can otherwise interfere with the specific signal of interest. In addition, sample preparation involves manual addition of antibodies, often in multiple rounds, as well as fixation and permeabilization steps required for analysis of intracellular targets. As a consequence, cell yield is severely reduced. Moreover, only about 30–50% of the stained and injected sample eventually reaches the detector due to an imperfect sample introduction system and ion transmission. Together this leads to a final information content representing as little as 10% of the starting material in some cases. Also, as different cell types and epitopes will vary in their sensitivity to wash, fixation, and permeabilization, each such step will introduce a bias to the cell composition of the sample. Another issue of importance is that with an increase in the number of samples, the total time spent on sample processing and staining is typically very high, often spanning multiple days. Long protocols with many different steps also introduce technical variation between samples and between experiments, impacting the ability for comparisons across experiments. All of these factors generate a strong incentive for carefully controlling time and temperature conditions during the course of such protocols. Here we present an optimized cell processing and staining protocol for improved cell yield and increased throughput.

2 Materials

2.1 Reagents

1. RPMI complete medium: RPMI-1640, 10% FBS, 1% Penicillin-Streptomycin, L-Glutamine (from 100× supplement).
2. Thawing medium: RPMI-1640, 10% FBS, 1% Penicillin-Streptomycin, L-Glutamine, 25 U/mL Benzonase® Nuclease (from 2.5×10^5 U/mL stock).
3. CyFACS: 1× PBS, 0.1% BSA (from 30% stock), 2 mM EDTA (from 0.5 M stock), 0.05% Na-azide (from 10% stock).
4. Freezing buffer: FBS supplemented with 10% DMSO.
5. Metal-free, Ca^{2+} and Mg^{2+} ion free PBS (Rockland Immunochemicals): make 1× final concentration by diluting 10× concentrate with MilliQ grade water and store in LDPE bottles.
6. Lymphoprep™ (STEMCELL Technologies).
7. Maxpar metal-conjugated antibodies (Fluidigm), Table 1.
8. Maxpar X8 polymer multimetal reaction kit (Fluidigm).
9. In-house coupled antibodies: 100 μg of a purified antibody is coupled per reaction (as per manufacturer's recommendations).
10. Antibody stabilizer (Candor Bioscience).

Table 1
Antibody panel

Metal	Marker	Source	Clone
89Y	CD45	Fluidigm	HI30
113In	CD235a/b	Coupled	HIR2
115In	CD57	Coupled	HCD57
141Pr	CCR6 (CD196)	Fluidigm	11A9
142Nd	CD19	Fluidigm	HIB19
143Nd	CD5	Fluidigm	UCHT2
144Nd	CD16	Coupled	3G8
145Nd	CD4	Fluidigm	RPA-T4
146Nd	CD8a	Fluidigm	RPA-T8
147Sm	CD11c	Fluidigm	Bu15
148Nd	CD31	Coupled	WM59
149Sm	p4E-BP1 (Thr37/46)	Fluidigm	236B4
150Nd	pSTAT5 (Tyr694)	Fluidigm	C71E5
151Eu	CD123	Coupled	6H6
152Sm	gdTCR	Fluidigm	11F2
153Eu	pSTAT1 (Tyr701)	Fluidigm	58D6
154Sm	CD3e	Fluidigm	UCHT1
156Gd	p-p38 (Thr180/Tyr182)	Fluidigm	D3F9
157Gd	CXCR3	Coupled	G025H7
158Gd	pSTAT3 (Tyr705)	Fluidigm	D3A7
159Tb	pMAPKAP2 (Thr334)	Fluidigm	27B7
160Gd	CD14	Coupled	M5E2
161Dy	CD161	Coupled	HP-3G10
162Dy	Ki-67	Fluidigm	B56
163Dy	HLA-DR	Coupled	L243
164Dy	CD44	Coupled	BJ18
165Ho	CD127	Fluidigm	A019D5
166Er	pNFkBp65 [pS529]	Fluidigm	K10-895.12.50
167Er	CD27	Fluidigm	L128
168Er	CD38	Fluidigm	HIT2

(continued)

Table 1
(continued)

Metal	Marker	Source	Clone
169Tm	CD45RA	Fluidigm	HI100
170Er	CD20	Coupled	2H7
171Yb	pERK1/2 (Thr202/Tyr204)	Fluidigm	D13.14.4E
172Yb	IgD	Coupled	IA6-2
173Yb	CD56	Coupled	NCAM16.2
174Yb	CD185 (CXCR5)	Coupled	51505
175Lu	p-S6 (Ser235/Ser236)	Fluidigm	D57.2.2.E
176Yb	pCREB	Fluidigm	87G3

11. Live-dead stain: Maleimide-DOTA loaded with 103Rh, 5 mg/mL (Macrocyclics).
12. Fc Receptor (FcR) blocking buffer (Cytodelics).
13. Cell-ID 20-Plex Pd Barcoding Kit (Fluidigm).
14. Sodium Orthovanadate.
15. Intracellular fixation and permeabilization buffer set (eBioscience): Freshly make this buffer by mixing diluent and fix concentrate using 1:4 dilution factor.
16. 10× Saponin-based permeabilization buffer (eBioscience): 10× buffer diluted to 1× in MilliQ water.
17. 16% Paraformaldehyde.
18. EQ™ Four Element Calibration Beads (Fluidigm).
19. 125 μM Cell-ID™ Intercalator-Ir (Fluidigm).
20. Thawed cryopreserved human peripheral blood mononuclear cells (PBMCs) or freshly isolated PBMCs.
21. 0.4% Trypan blue.

2.2 Equipment and Supplies

1. A Bravo liquid handling platform (Agilent technologies, Santa Clara, CA, USA): 9 stages (2 stages equipped with temperature control), 96ST disposable tip heads for volumes up to 250 μL/tip.
2. Vspin microplate centrifuge (Agilent): spinning within washing steps.
3. BenchBot robot (Agilent): microplates handling platform.
4. EL406 Washer Dispenser (BioTek, Winooski, VT, USA): aspiration and dispensing of buffers during washing steps.

5. TC20™ automated cell counter (BioRad).
6. CyTOF® 2 mass cytometer (Fluidigm Inc., South San Francisco, CA) with software version 6.0.626. Noise reduction: a lower convolution threshold of 200, event length 10–150 and a sigma value of 3.
7. SepMate™-15 and -50 tubes (STEMCELL Technologies) for PBMCs gradient separation.
8. Water bath at 37 °C.
9. Centrifuge with adaptors for 5-mL, 15-Ml, and 50-mL tubes.
10. Pipettors: calibrated single channel and multichannel.
11. 35 μm nylon mesh.

3 Methods

3.1 Blood Sampling

1. Blood was collected from anonymous blood bank donors requiring no ethical review according to the Stockholm regional ethics board.
2. Peripheral blood mononuclear cells (PBMC) were isolated through Lymphoprep™ density gradient centrifugation in SepMate™ tubes.
3. Cells were washed and cryopreserved in a freezing buffer at −80 °C until use.

3.2 Staining Protocol

1. Frozen PBMCs were thawed in a water bath at 37 °C followed by slow addition of thawing medium.
2. Cells were washed twice in warm thawing medium by centrifugation at 300 × *g* for 5 min.
3. Cells were resuspended in RPMI complete medium and counted.
4. Two tubes of 2.5×10^7 cells were prepared by taking required volume of counted cells and topping up to 5 mL with RPMI complete medium.
5. To induce phosphorylation of signaling intermediates, 125 μM Sodium Orthovanadate was added to one of the tubes and both tubes were incubated for 15 min at 37 °C.
6. The cells were then fixed by adding 5 mL 2% paraformaldehyde in PBS for a final concentration of 1% and incubated for 10 min at room temperature.
7. All staining was done in CyFACS. Manual and automated staining procedures were performed in parallel as described in detail (Table 2) (*see* also **Notes 1–3**).

Table 2
Detailed manual and automated staining protocols

	Manual protocol	Automated protocol
Day 1	Wash cells twice by centrifugation (800 × *g*/300 s) with CyFACS	Wash cells twice by centrifugation (1000 × *g*/120 s) with CyFACS
	Add 35 μL of surface markers specific antibodies cocktail, incubate at 4 °C/30 min	Add 30 μL of surface markers specific antibodies cocktail, incubate at 4 °C/30 min. Resuspend six times each 10 min
	Wash twice by centrifugation with CyFACS (800 × *g*/300 s)	Wash twice by centrifugation with CyFACS (1000 × *g*/120 s)
	Add 500 μL 1% paraformaldehyde in CyFACS and store overnight at 4 °C	Add paraformaldehyde to reach 1% and store overnight at 4 °C
Day 2	Wash twice by centrifugation with CyFACS (800 × *g*/300 s)	Wash twice by centrifugation with CyFACS (1000 × *g*/120 s)
	Permeabilize with 200 μL 0.3% Saponin at 4 °C/45 min	Permeabilize with 200 μL 0.3% Saponin at 4 °C/45 min
	Wash twice by centrifugation with CyFACS (800 × *g*/300 s); in first step, top up with 1 mL of CyFACS and wait 2 min	Wash twice by centrifugation with CyFACS (1000 × *g*/120 s); in first step, add 50 μL of CyFACS and wait 2 min
	Add 35 μL of intracellular markers specific antibodies cocktail #1, incubate at room temp for 60 min	Add 30 μL of intracellular markers specific antibodies cocktail #1, incubate at RT/60 min. Resuspend six times each 10 min
	Wash twice by centrifugation with CyFACS (800 × *g*/300 s)	Wash twice by centrifugation with CyFACS (1000 × *g*/120 s)
	Permeabilize with 500 μL 100% Methanol at 4 °C/10 min	Permeabilize with 100% methanol (170 μL) at 4 °C/10 min
	Top 1 mL of CyFACS over methanol, incubate for 5 min and centrifuge (800 × *g*/300 s)	Top 70 μL CyFACS over methanol, incubate for 5 min and centrifuge (1000 × *g*/180 s)
	Wash once by centrifugation with CyFACS (800 × *g*/300 s)	Wash once by centrifugation with CyFACS (1000 × *g*/120 s)
	Add 35 μL of intracellular markers specific antibodies cocktail #2, incubate at RT/60 min	Add 30 μL of intracellular markers specific antibodies cocktail #2, incubate at RT/60 min. Resuspend six times each 10 min
	Wash twice by centrifugation with CyFACS (800 × *g*/300 s)	Wash twice by centrifugation with CyFACS (1000 × *g*/120 s)
	Add 500 μL 4% paraformaldehyde and store overnight at 4 °C	Add paraformaldehyde to reach 4% and store overnight at 4 °C

(continued)

Table 2 (continued)

	Manual protocol	Automated protocol
Day 3	Add DNA intercalator 1:2000 and incubate at RT/20 min	Add DNA intercalator 1:2000 and incubate at RT/20 min
	Wash twice by centrifugation with CyFACS (800 × *g*/300 s)	Wash twice by centrifugation with CyFACS (1000 × *g*/120 s)
	Wash once by centrifugation with PBS (800 × *g*/300 s)	Wash once by centrifugation with PBS (1000 × *g*/120 s)
	Wash once by centrifugation with MilliQ diH_2O (800 × *g*/300 s) and resuspend cells in MilliQ diH_2O	Wash once by centrifugation with MilliQ diH_2O (1000 × *g*/180 s) and resuspend cells in MilliQ diH_2O

8. Samples were stained with a combination of 27 antibodies to surface antigens and 11 antibodies to intracellular antigens (Table 1).
9. Following staining and final washes, cells were filtered through a 35 μm nylon mesh, diluted to 5.0×10^5 cells/mL in EQ™ Four Element Calibration Beads, diluted 1:10 in MilliQ water and acquired on a CyTOF2 instrument.

3.3 Data Analysis

1. All files were normalized as previously described [5].
2. Each data point was arcsinh-transformed with a cofactor of 5 and thereafter randomized using a Gaussian randomization function with a standard deviation of 0.3.
3. To allow a stringent comparison of staining indices and population frequencies between manually and automatically processed samples, the openCyto R package [6] was used and automated gating performed using predefined marker combinations (*see* **Note 4**).
4. For each cell population, the population frequency as a percentage of the parent population was calculated and the coefficient of variance (standard deviation/mean) for each population was calculated (*see* **Note 5**).

4 Notes

1. *Optimizing cell yield in a manual cell processing protocol*

   We started with a standard Maxpar (Fluidigm) protocol for phospho-protein staining, with an initial surface staining post fixation followed by intracellular staining performed in two

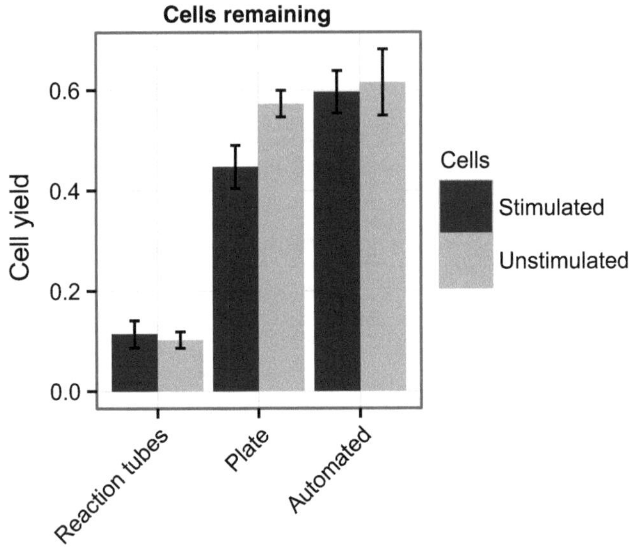

Fig. 1 Fraction of cells remaining at the end of the indicated cell processing protocol. Cells were either processed manually in 1.5 mL reaction tubes, in a 96-deep well plate after adoption of aspiration conditions learned from automation optimization or by automated platform itself. The rim over the bottom of the deep well where the straight walls slope to V-bottom in these plates allows a better control of the aspiration conditions, mimicking conditions and reproducibility achievable by automated platform. This proves that high cell yields are achievable even by manual or semi-automated processing (*see* **Notes 1** and **5**). Means ± StdDev of the same sample processed in triplicate for each condition, stained and acquired in the same batch, are presented

steps that includes one mild permeabilization with saponin and one harsh permeabilization with methanol. The manual sample preparation protocol was performed in 1.5 mL reaction tubes. Using these tubes, staining of resting and stimulated cells (Sodium Orthovanadate) resulted in cell yields of 10.2% and 11.4%, respectively, at the end of the protocol (Fig. 1). Next, we evaluated each step of this protocol to determine where cells were lost and how to prevent it. We discovered that loss of cells occurred at almost every step, but especially in the aspiration step during each round of washing. Optimization of this procedure was done using our liquid handling robotic platform, in which many different conditions can be carefully compared. We found that slowing down the aspiration movement and using at least a 4 mm distance between the pipette tip and the cell pellet led to a significantly increased yield after each wash step, with only a few percent lost even after nine consecutive wash/spin cycles. This was easily reproduced in a standard manual protocol by less thorough aspiration of the supernatant after centrifugation. Furthermore, centrifugation at high speeds (1000 × *g*, 2 min) for a short duration also improved yields as a whole compared to centrifugation at lower speeds (400 × *g*, 5 min). For manual staining, we achieved an increased overall cell yield from 11% to 50% (Fig. 1) by switching from 1.5 mL reaction tubes to V-bottomed square-shaped well 2 mL plates. One reason for this improvement was that it was easier to achieve reproducible vacuum aspiration thanks to the rim over the bottom of the well where the straight walls slope to V-bottom in these plates. The implementation of all of these improvements minimized the cell loss during wash/spin cycles of non-permeabilized cells almost completely.

2. *Intracellular permeabilization*

   Permeabilization of cells using either saponin or methanol is known to increase cell loss but is essential for successful staining of intracellular antigens. Of all the strategies we tried to optimize, the most successful approach was permeabilization and subsequent washing, addition of protein in the form of BSA (even as little as 0.02% final concentration) followed by a short incubation at room temperature (2–5 min) before washing away the permeabilization reagent. In our hands, simple addition of CyFACS reaching 1/5 of the total volume prior to the first washing step after saponin permeabilization improved cell yield from 84% to 93%. Methanol permeabilization proved to be harsher, with yields as low as 42% after first spin and less than 16% after an additional five wash/spin cycles using CyFACS. Simple addition of a small volume of concentrated BSA (10%) just prior to the centrifugation step, reaching a final concentration of 0.1% BSA in methanol, improved yield to 82% after a first washing step, 73% after an additional three rounds of washing, and 67% yield after nine washes in CyFACS. Higher BSA concentration (>0.5%) did not improve this further, but instead caused aggregates and a reduced cell yield. In addition, prolonged spin cycles during the washes immediately following permeabilization steps helped to improve reproducibility of staining profiles.

3. *Remarks on semi-automation*

   As from the above-presented improvements, the washing steps are the most prone to subjective errors as well as time-consuming when executed manually. Semi-automation of washing steps, using cell washers and multichannel pipettes, is the most economical approach in labs not equipped with similar robotic platform. As cell washers might differ in their performance (e.g., vacuum pumps with different suction performance), simple testing of different setups using multiple wash cycles with given buffers and cell yield evaluation will help to set the optimal conditions for any given protocol.

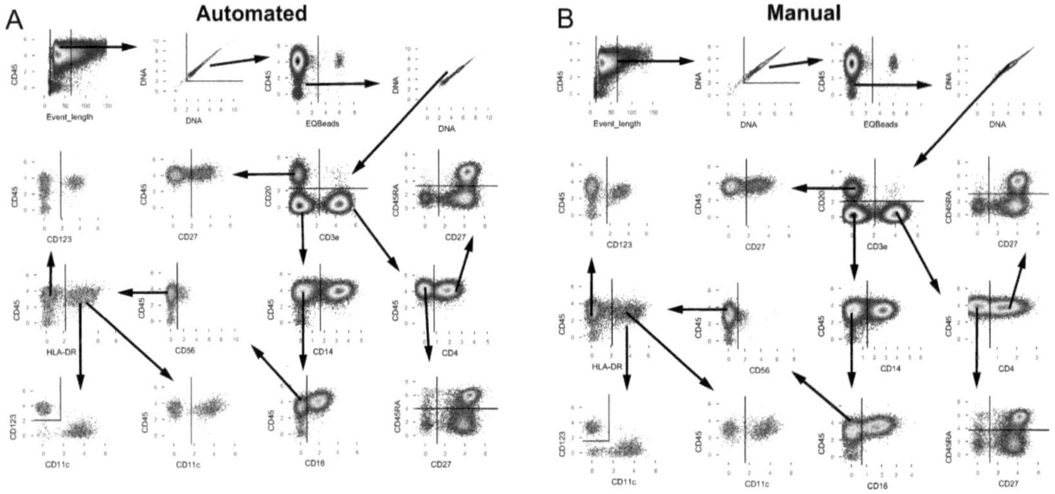

Fig. 2 Staining patterns and gating scheme for automatically (**a**) and manually (**b**) processed cells from the same sample. An example demonstrating similar staining profiles and the principle of the gating scheme in a sample processed both manually (optimized protocol) or by automated platform is presented. To avoid subjective error, an automated gating pipeline programmed in openCyto R package was used

4. *Staining indices*

To compare staining resolutions between manual and automated sample preparation for mass cytometry, we calculated normalized signal to background values, i.e., staining indices for 14 surface markers in unstimulated samples. The staining index was defined as the $(MFI_{pos} - MFI_{neg})/SD_{neg}$ where pos and neg are the positive and negative peaks, respectively. Only markers with a bimodal distribution were used for these calculations. The average staining index for each marker varied minimally between manual and automated gating and in all cases the separation between the positive and negative population was large enough to allow unambiguous gating (Fig. 2). The standard deviation within replicates was low for most markers with a few notable exceptions. Interestingly, the standard deviations of staining indices were higher for manually processed samples as compared to automatically processed samples for all markers except one (Fig. 3). This further suggests the improved reproducibility of staining procedures on a liquid handling robotic platform. Apart from the improvements mentioned above, the fully automated staining procedure described herein allows for one individual operator to simultaneously monitor the acquisition of samples on two or more mass cytometry instruments, while monitoring the automated preparation of additional samples for coming experiments.

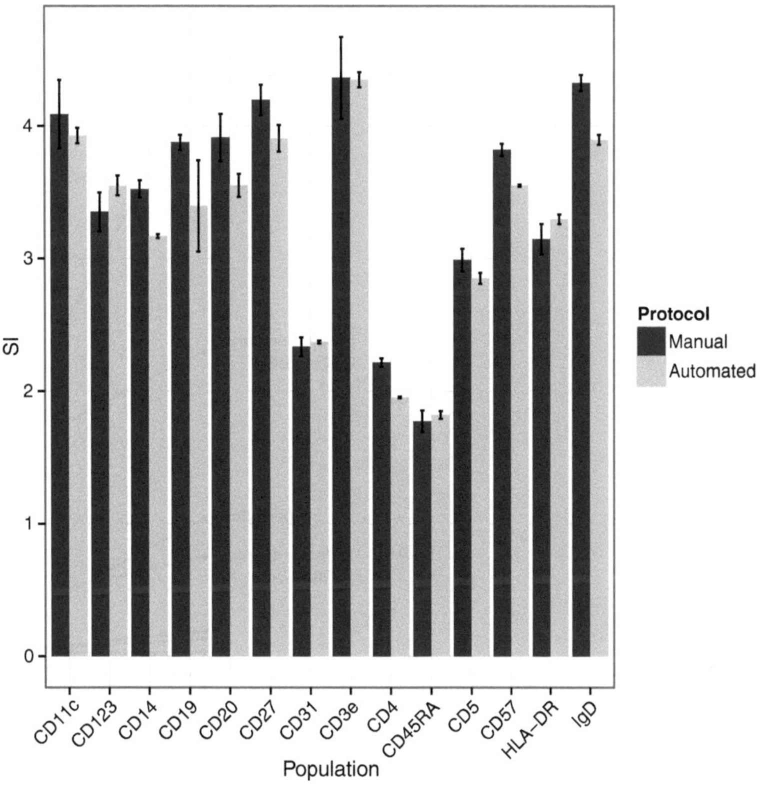

Fig. 3 Staining indices for the indicated markers after manual and automated sample processing, respectively. Only markers with a bimodal distribution in unstimulated samples were chosen to compare staining resolutions between manual and automated sample preparation for mass cytometry. The staining index was defined as the $(MFI_{pos} - MFI_{neg})/SD_{neg}$ where pos and neg are the positive and negative peaks, respectively

5. *Population frequencies*

To be able to accurately determine cell frequencies with mass cytometry, it is important that sample manipulation prior to acquisition is reproducible and does not affect sample composition or staining patterns. To determine whether our automated approach was comparable to manual preparation, we devised an automated gating strategy for the most common cell types among PBMCs. Overall, staining patterns were comparable between manual and automated staining protocols (Fig. 2). The subset frequencies as percentages of parent populations were calculated for each cell type and were comparable. To quantify variability, the coefficient of variance for each cell population frequency was calculated for samples processed in our automated and manual protocols, respectively (Fig. 4). For most cell types, cell frequencies could be reproducibly determined

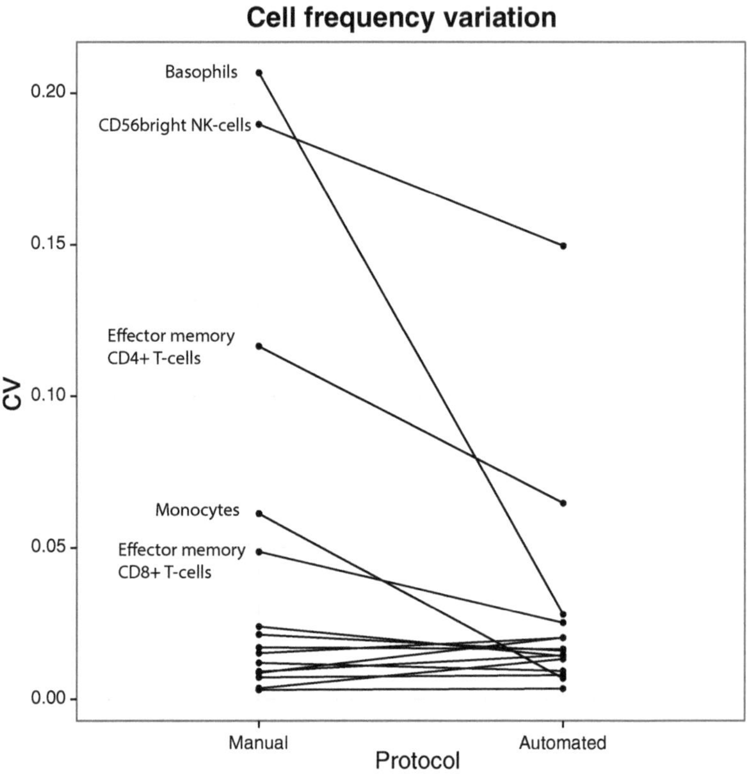

Fig. 4 Average coefficients of variance (CV) for cell frequencies. Samples were processed manually or by automated platform in triplicate and CVs were calculated for each gated population to assess reproducibility in sample composition

using either protocol, but smaller cell populations such as basophils and CD56bright NK cells had more reproducible frequencies when stained using the automated liquid handling robotic protocol (Fig. 4). We conclude that manual and automated protocols give comparable staining patterns, but that the automated protocol reduces the variability, especially for smaller populations of cells.

References

1. Bandura DR, Baranov VI, Ornatsky OI et al (2009) Mass cytometry: technique for real time single cell multitarget immunoassay based on inductively coupled plasma time-of-flight mass spectrometry. Anal Chem 81(16):6813–6822. https://doi.org/10.1021/ac901049w
2. Brodin P, Jojic V, Gao T et al (2015) Variation in the human immune system is largely driven by non-heritable influences. Cell 160(1–2):37–47. https://doi.org/10.1016/j.cell.2014.12.020
3. Kaczorowski KJ, Shekhar K, Nkulikiyimfura D et al (2017) Continuous immunotypes describe human immune variation and predict diverse responses. Proc Natl Acad Sci U S A 114(30):E6097–E6106. https://doi.org/10.1073/pnas.1705065114
4. Lakshmikanth T, Olin A, Chen Y et al (2017) Mass cytometry and topological data analysis reveal immune parameters associated with complications after allogeneic stem cell transplantation.

Cell Rep 20(9):2238–2250. https://doi.org/10.1016/j.celrep.2017.08.021

5. Finck R, Simonds EF, Jager A et al (2013) Normalization of mass cytometry data with bead standards. Cytometry A 83(5):483–494. https://doi.org/10.1002/cyto.a.22271

6. Finak G, Frelinger J, Jiang W et al (2014) OpenCyto: an open source infrastructure for scalable, robust, reproducible, and automated, end-to-end flow cytometry data analysis. PLoS Comput Biol 10(8):e1003806. https://doi.org/10.1371/journal.pcbi.1003806

Chapter 9

Live Cell Barcoding for Efficient Analysis of Small Samples by Mass Cytometry

Lisa E. Wagar

Abstract

In mass cytometry, sample loss is of considerable concern due to the relative inefficiency of cell event collection compared to similar techniques such as flow cytometry. Cell stimulation and the harsh conditions required in the later stages of certain sample preparations also contribute to cell loss. Low starting cell numbers are especially susceptible to these effects, potentially limiting the ability to use mass cytometry. Here is presented a live cell barcoding scheme and additional efficiency methods to improve recovery and achieve consistent staining for small samples.

Key words Mass cytometry, CyTOF, Barcoding, Cell recovery

1 Introduction

Mass cytometry (CyTOF) provides the benefit of high-parameter data collection compared to traditional flow cytometry. However, a major limitation of CyTOF technology is the substantial cell loss during sample preparation and data collection. Depending on cell stimulation and staining conditions, up to 60% of starting material can be lost during sample processing. Compounding this problem, approximately half of all cell events are lost during data acquisition.

These inherent technical limitations should be considered during experimental design. As with other single-cell assays, high viability starting material greatly improves the quality of downstream data. A viability cutoff should be established based on starting cell numbers and stimulation conditions, wherein samples are discarded if they do not meet the minimum threshold. The cutoff value should be selected using a preliminary test of the specific user-defined stimulation and staining protocol, though starting values are recommended here. The choice of stimulation conditions (if any) and length of stimulation should also be selected

Helen M. McGuire and Thomas M. Ashhurst (eds.), *Mass Cytometry: Methods and Protocols*, Methods in Molecular Biology, vol. 1989, https://doi.org/10.1007/978-1-4939-9454-0_9,

based on the quality of starting material and desired marker readouts. In the protocol presented here, strong mitogen stimulation was used for 6 h to detect immune cell functional capacity. This will necessarily result in downregulation of certain markers (e.g., CD4 expression on CD4 T cells) and potentially some cell death.

Nonetheless, there are solutions to limit cell loss during the various phases of sample processing and acquisition. Combining multiple small samples into larger pools of barcoded samples can help improve recovery. Various barcoding techniques, originally developed for flow cytometry, have been adapted for CyTOF. Combinatorial staining of fixed and methanol-permeabilized cells has been quite successful using metal-loaded mDOTA reagents [1] and palladium-loaded isothiocyanobenzyl-EDTA reagents [2]. A 20-plex barcoding kit from Fluidigm is available based on similar technology. A gentler saponin-based permeabilization approach has permitted barcoding of paraformaldehyde-fixed cells without affecting surface staining quality [3]. However, some surface marker staining can be affected by cell fixation. For optimal staining quality and flexibility of antibody clones, live cell barcoding is ideal. Therefore, an anti-CD45 antibody-based live cell barcoding [4, 5] approach is included in this protocol. Debarcoding tools are available, including a Matlab implementation used in Zunder et al. [2], an R package called *premessa*, and an application from Fluidigm, though not all tools are currently compatible with all barcoding schemes. Sample deconvolution can also be performed using relatively straightforward Boolean gating schemes with conventional cytometry gating software.

Another method to improve recovery is to include irrelevant or "carrier" cells during the staining process. They have been used with varying degrees of success [6, 7] depending on experimental setup. For the protocol below, peripheral blood mononuclear cells (PBMCs) from a healthy control donor were previously aliquoted into several tubes and cryopreserved. Each time a new CyTOF run is performed, an aliquot of the control donor is thawed and treated under the same conditions as the test samples. A portion of the control donor cells is used in each barcoded pool and serves the dual purpose of contributing carrier cells and indicating staining quality for the pool.

Combining these approaches provides several benefits in addition to improved cell recovery. The effective volume of staining reagents (and thus cost) is significantly reduced by using a few barcoded pools rather than several small samples. Barcoded samples are particularly amenable to continuous injector systems, meaning more hands-off time during sample acquisition and also less time spent cleaning between samples. In downstream analysis, doublets are easily removed from samples that are barcoded in certain configurations (e.g., five possible barcode channels, with each sample labeled by only two). Carrier cells provide the added benefit of controlling for technical and batch variation during analysis.

2 Materials

All supplies must be free of contact with glass or metal, as detergents and other materials are common sources of heavy metal contamination. It is recommended to use disposable plastic materials and to test any liquid reagents (e.g., MilliQ water source, buffers) using solution mode to identify possible contamination sources. Dilutions of cell resting medium, cell stimulation cocktail, protein transport inhibitor, cisplatin, fix buffer, perm wash buffer, DNA intercalator, and all antibody staining cocktails (barcoding, surface, and intracellular) should all be freshly prepared. An overview of the workflow is shown in Fig. 1.

2.1 Cell Stimulation

1. Frozen PBMC samples, including a control donor aliquot.
2. Complete medium: 450 mL RPMI-1640 with glutamax, 50 mL fetal bovine serum, 5 mL 100× penicillin/streptomycin, 5 mL 100× non-essential amino acids, 5 mL 100× sodium pyruvate.
3. Cell resting medium: 1 μL Benzonase (Sigma) per 10 mL complete medium (10,000× dilution); prepare 2 mL for each PBMC sample.
4. Cell stimulation cocktail: Cell Stimulation Cocktail concentrate with phorbol 12-myristate 13-acetate (PMA) and ionomycin, 500× (eBioscience), diluted to 1× with complete medium.
5. Protein transport inhibitor: Protein Transport Inhibitor concentrate with Brefeldin A and Monensin, 500× (eBioscience), diluted to 50× with complete medium.
6. Cell enumeration materials, including trypan blue.
7. Sterile plastic-ware appropriate for cell culture.
8. Deep well block with U or V bottom, 2 mL capacity per well.
9. 37 °C water bath.
10. 37 °C incubator with 5% CO_2 and humidity.

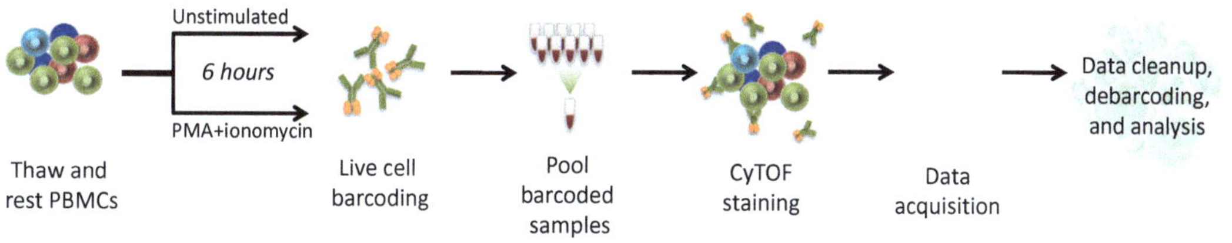

Fig. 1 Workflow. Schematic depiction of the steps detailed in this protocol

11. Temperature-controlled centrifuges equipped with rotors for 15 mL tubes, 1.5 mL tubes, and 96-well plates.
12. Aspirator with vacuum trap.

2.2 CyTOF Staining

1. Deep well block with U or V bottom, 2 mL capacity per well.
2. Previously tested and titrated metal-conjugated antibodies for surface and intracellular staining, including anti-human CD45 antibodies labeled with five different metals (*see* **Note 1**).
3. 0.1 μm spin filters.
4. MilliQ H_2O.
5. CyPBS: 100 mL 10× PBS, Calcium and Magnesium free (Rockland) with 900 mL MilliQ H_2O; 0.2 μm filtered.
6. CyFACS buffer: CyPBS with 0.1% w/v bovine serum albumin, 2 mM EDTA, 0.05% v/v sodium azide; 0.2 μm filtered.
7. Cisplatin (Fluidigm).
8. Fix buffer: Dilute fresh 16% stock paraformaldehyde (Electron Microscopy Sciences) to 2% using CyPBS.
9. Perm wash buffer: Dilute 10× permwash (eBioscience) to 1× with MilliQ H_2O.
10. DNA intercalator: Iridium 191/193 intercalator (Fluidigm) diluted according to manufacturer's recommendation (typically 1 μM) with fix buffer.

2.3 Running CyTOF Prepared Samples

1. EQ 4-element calibration beads (Fluidigm).
2. Filter-top 5 mL tubes.
3. Continuous injector system (multiple models available).
4. Ice.

2.4 Debarcoding Samples in fcs Files

1. Flowjo (Treestar) software.

3 Methods

All procedures should be performed at room temperature unless otherwise indicated.

3.1 Thaw and Rest PBMC Samples

1. Collect samples from liquid nitrogen and keep on dry ice until ready to thaw.
2. For each sample, aliquot 10 mL complete medium into a 15 mL tube. Warm tubes to 37 °C in a water bath.
3. Working quickly, thaw PBMC samples (recommended no more than four at a time) in a 37 °C water bath until mostly liquid, about 1 min.

4. Quickly transfer thawed cells into the pre-warmed tubes of complete medium. Carefully rinse the cryopreservation tubes with extra pre-warmed complete media to efficiently collect all cells.
5. Once all samples are thawed, centrifuge at 500 *g*, 10 min.
6. Aspirate supernatant. Resuspend cell pellets thoroughly to a final volume of 2 mL in cell resting medium. Aim for an estimated cell concentration between 2×10^6 and 1×10^7 cells/mL.
7. Loosen the caps on the 15 mL tubes and place them in a 37 °C incubator for 1 h.
8. Approximately 45 min into the 1 h rest, gently mix the cell solution and collect 10 μL of resuspended cells from each sample for enumeration. Determine the total live cell count and percent viability of each sample using trypan blue exclusion (*see* **Note 2**).

3.2 Cell Stimulation and Transport Inhibitor Addition

1. After the 1 h rest is complete, add 10 mL complete medium to each 15 mL tube and gently mix the samples to resuspend the settled cell pellets.
2. Centrifuge the samples at 500 *g*, 7 min.
3. Aspirate supernatants. Resuspend samples with 2 mL complete medium.
4. Determine the volume required to get 5×10^6 cells (per condition). If there are less than 1×10^7 cells in a given sample, it can be split in half (yielding 1 mL for each condition).
5. For each sample, transfer the corresponding volume into a deep well block (one well for unstimulated cells, one well for stimulated cells) (*see* **Note 3**).
6. Centrifuge the block at 500 *g*, 7 min.
7. Meanwhile, prepare the cell stimulation cocktail (500 μL per stimulated sample) (*see* **Note 4**).
8. Discard supernatants by tipping the block upside down and gently patting the surface on paper towel to remove residual liquid. Add 500 μL of cell stimulation cocktail to one of the two cell pellets for each sample and 500 μL of complete medium to the other pellet to serve as an unstimulated control.
9. Gently resuspend the cell pellets and transfer each into the wells of a 24-well tissue culture plate (*see* **Note 5**).
10. Incubate samples for 1 h in a 37 °C incubator.
11. Add 10 μL of diluted protein transport inhibitor to all samples, including controls (10 μL per well using the 50× preparation).

Metal 2 \ Metal 1	104Pd	106Pd	108Pd	113In
89Y	Donor A (ns) time point 1	Donor A (pi) time point 1	Donor A (ns) time point 2	Donor A (pi) time point 2
104Pd		Donor B (ns) time point 1	Donor B (pi) time point 1	Donor B (ns) time point 2
106Pd			Donor B (pi) time point 2	Control (ns)
108Pd				Control (pi)

Fig. 2 Sample barcode staining scheme. Staining scheme for a sample barcoded pool. Each sample is labeled by two of five possible metal-conjugated anti-CD45 antibodies. Both unstimulated (ns) and PMA-ionomycin stimulated (pi) samples from the same donor are incorporated into the same pool. If more than one time point or other conditions are tested, they should be pooled into the same tube if possible. A sample of a control donor (contributes carrier cells to the pool) is also included

12. Gently mix each well to thoroughly distribute the inhibitor.
13. Return the samples to a 37 °C incubator for 5 h (*see* **Note 6**).

3.3 Live Cell Barcoding

1. Transfer all samples into a deep well block for washing and staining. Rinse each well of the 24-well plate with 500 μL CyFACS buffer to harvest as many cells as possible.
2. Top up each well to 1.8 mL with CyFACS buffer.
3. Centrifuge at 500 *g* for 7 min at 4 °C.
4. Aspirate supernatants. Wash cells with 1.8 mL CyFACS buffer.
5. Centrifuge at 500 *g* for 7 min at 4 °C.
6. In the meantime, prepare the barcoding reagents (*see* sample barcoding scheme, Fig. 2) in CyFACS buffer. For each sample, 50 μL of barcode mixture is required (*see* **Note 7**).
7. Aspirate supernatants. Add anti-human CD45 barcodes, ensuring a unique barcode is used for each sample to be pooled later (*see* **Note 8**).
8. Incubate samples for 30 min at 4 °C.
9. Top up samples to 1.8 mL with CyFACS buffer.
10. Centrifuge at 500 *g* for 7 min at 4 °C.
11. Aspirate supernatants. Wash cells with 1.8 mL CyFACS buffer.
12. Centrifuge at 500 *g* for 7 min at 4 °C.
13. Resuspend each cell pellet with 100 μL CyFACS buffer.

14. Pool compatible barcodes (up to ten; *see* sample barcode scheme for an example) into 1.5 mL tubes and place on ice. Collect 10 μL of each barcoded pool and enumerate the total number of live cells and percent viability.

3.4 Surface Staining

1. Based on the cell counts, prepare the surface antibody cocktail (*see* **Note 9**).
2. Filter the surface antibody cocktail through a 0.1 μm spin filter by centrifugation at 12,000 *g* for 1 min.
3. Centrifuge the pooled samples at 500 *g* for 5 min at 4 °C.
4. Aspirate supernatants. Add the appropriate volume of surface antibody cocktail to each pooled sample (50 μL per 1 × 10^7 cells to be stained) and gently resuspend cells.
5. Incubate at 4 °C for 30 min.
6. Top up each sample with CyFACS (up to 1.2 mL).
7. Centrifuge at 500 *g* for 5 min at 4 °C.
8. Aspirate supernatants. Wash cells with 1.2 mL CyFACS buffer and place on ice.

3.5 Live/Dead Staining and Sample Fixation

1. Prepare the cisplatin live/dead stain by diluting the stock 1/1000 with CyPBS. For each pooled sample, 200 μL is needed.
2. Centrifuge the pooled samples at 500 *g* for 5 min at 4 °C.
3. Aspirate supernatants. Add 200 μL of diluted cisplatin to each sample and thoroughly resuspend the cell pellets.
4. Incubate for 5 min at room temperature.
5. Top up each sample with CyFACS (up to 1.2 mL).
6. Centrifuge at 500 *g* for 5 min at 4 °C.
7. Aspirate supernatants. Wash cells with 1.2 mL CyFACS buffer.
8. Centrifuge at 500 *g* for 5 min at 4 °C.
9. Aspirate supernatants as completely as possible. Add 1 mL fix buffer to each sample.
10. Incubate for at least 30 min at 4 °C (*see* **Note 10**).

3.6 Intracellular Staining

1. Top up each sample with Permwash buffer (up to 1.2 mL).
2. Centrifuge at 800 *g* for 5 min at 4 °C.
3. Aspirate supernatants. Repeat Permwash washes two more times for a total of three washes.
4. Prepare intracellular staining cocktail, using Permwash buffer as the diluent (*see* **Note 11**).

5. Filter the intracellular antibody cocktail through a 0.1 μm spin filter by centrifugation at 12,000 *g* for 1 min.
6. To cell pellets, add intracellular staining cocktail (50 μL per 1×10^7 cells to be stained).
7. Incubate at 4 °C for 30 min.
8. Top up each sample with Permwash buffer (up to 1.2 mL).
9. Centrifuge at 800 *g* for 5 min at 4 °C.
10. Top up each sample with Permwash buffer (up to 1.2 mL).

3.7 DNA Intercalator Staining and Preparation for Acquisition

1. Prepare the DNA intercalator by diluting the stock intercalator with fix buffer.
2. Centrifuge the samples at 800 *g* for 5 min at 4 °C.
3. Aspirate supernatants as completely as possible. Add DNA intercalator stain (50 μL per 1×10^7 cells).
4. Incubate at 4 °C for 30 min.
5. Top up each sample with CyPBS (up to 1.2 mL).
6. Centrifuge at 800 *g* for 5 min at 4 °C.
7. Aspirate supernatants. Repeat CyPBS wash.
8. Aspirate supernatants. Repeat washes three times using MilliQ H_2O (*see* **Note 12**). Following washes, keep pelleted samples on ice.
9. Prepare a 10% solution of EQ 4-element calibration beads by diluting the stock with MilliQ H_2O and keep on ice.
10. Just prior to running the sample, dilute the cells to 7×10^5 cells/mL using the 10% calibration beads and pass through a filter-top 5 mL tube. Keep on ice until immediately before collection.
11. Collect the CyTOF data. Use a continuous injector system for optimal cell recovery.
12. Use the internal calibration beads to normalize the sample. Debarcoding can be done using multiple methods as discussed in the introduction. A sample gating scheme to be applied prior to Boolean debarcoding is shown in Fig. 3, using FlowJo software.

4 Notes

1. The channels for anti-CD45 antibodies should be selected carefully to ensure no cross-talk of signal between barcodes. For five metals, 89Y, 104Pd, 106Pd, 108Pd, and 113In are suggested. *See* Mei et al. [5] for a protocol to couple antibodies to Pd-labeled polymers. Barcodes should be titrated to

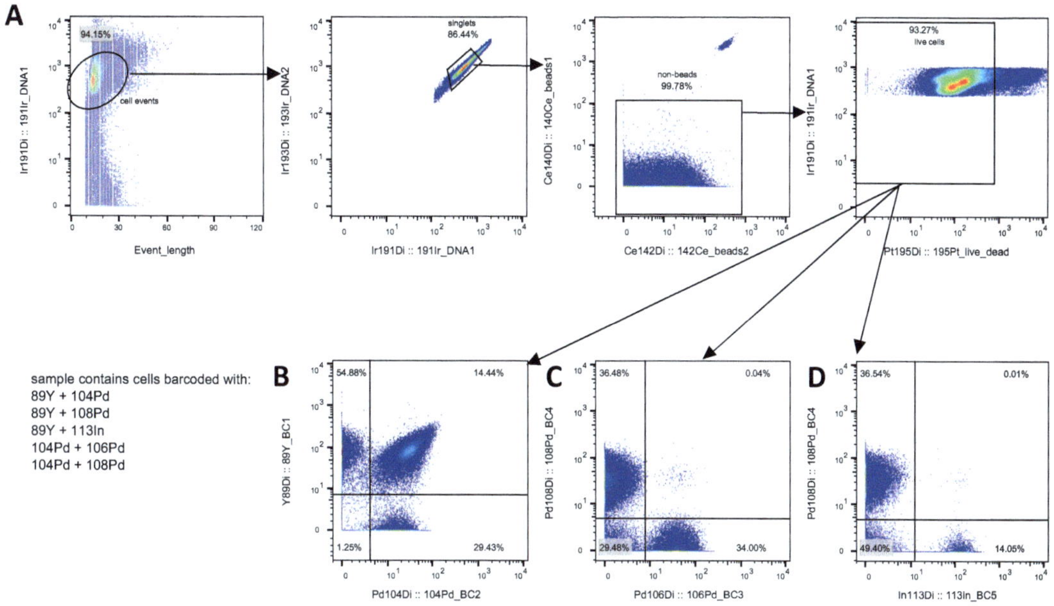

Fig. 3 Deconvolution example. (**a**) Gating strategy shows in top row the steps to identify live single cells. The dot plots on the second row demonstrate how differential staining of CD45 with different metal tags can be achieved with barcoded samples. (**b**) Cells barcoded simultaneously with 89Y + 104Pd are evident as double positive events. Because the barcode combinations of (**c**) 108Pd + 106Pd, or (**d**) 108Pd + 113In were not used, very few cells are found expressing these patterns. The few cells that are found with these markers likely represent doublets

ensure adequate separation of positives and negatives for each channel. Concentrations will vary based on the channel, chelating agent, and number of barcoding channels used to define a sample. Starting titration values should be in the 2–10 μg/mL range.

2. In small samples (PBMCs collected from less than 3 mL of blood), it is recommended to discard samples with less than 70% viability if possible (or run only an unstimulated condition). Cell event yields after stimulation and staining tend to be insufficient for reliable analysis from samples with low initial viability.
3. Smaller tubes or a deep well block are best for fully aspirating supernatants to achieve accurate stimulation volume.
4. Metal-labeled antibodies against the degranulation markers CD107a (clone H4A3, BD) and CD107b (clone H4B4, BD) can be incorporated during the stimulation if desired. Although the antibodies should be titrated depending on use, the concentration should be around 0.5 μg/mL. Be sure to add the antibodies to both the unstimulated and stimulated cultures.
5. For cell counts in the 1–5 × 10^6 range, stimulation in 24-well plates is recommended.

6. A total of 6 h of stimulation is recommended for optimal cytokine detection with PMA + ionomycin. Incubation time will require optimization for other stimuli. Cells should not be treated with protein transport inhibitor for more than 6 h.
7. A five-choose-two system is recommended to yield ten unique barcodes: (1) 89+104, (2) 89+106, (3) 89+108, (4) 89+113, (5) 104+106, (6) 104+108, (7) 104+113, (8) 106+108, (9) 106+113, and (10) 108+113.
8. When barcoding samples, combine stimulation conditions from the same donor into one barcoded pool to reduce intra-donor variation in staining. A control donor should be added into each barcoded pool (using two available barcodes). *See* sample barcode scheme in Fig. 2 for details.
9. For CyTOF panels that will be used consistently for barcoded and pooled samples, it is recommended to titrate antibodies using 1×10^7 cells. Per 1×10^7 cells, 50 μL of antibody cocktail should be used, with Fc blocking agent of choice. Antibody concentrations typically range from 2–10 μg/mL and must be validated before use. When determining the amount of surface antibody cocktail to prepare, round the number of cells counted to the nearest five million for each barcoded pool. In pools with a large number of dead cells (>30% of total cells), consider increasing the volume of staining cocktail by up to 50 μL.
10. Samples can be kept overnight (and up to 2 days) in Fix buffer if desired, though sample loss may be increased for fixation longer than overnight.
11. Similar to the surface staining cocktail procedure, the concentration of intracellular antibodies should be titrated using 1×10^7 cells. To determine the volume of cocktail, use the cell counts previously calculated after pooling the barcoded samples (50 μL per 1×10^7 cells).
12. Samples should be kept pelleted until immediately before sample acquisition. If cells will not be run immediately (more than 4 h), they should be kept pelleted in CyPBS and washed three times with MilliQ H_2O just before use. This should improve sample recovery.

References

1. Bodenmiller B, Zunder ER, Finck R et al (2012) Multiplexed mass cytometry profiling of cellular states perturbed by small-molecule regulators. Nat Biotechnol 30(9):858–867. https://doi.org/10.1038/nbt.2317
2. Zunder ER, Finck R, Behbehani GK et al (2015) Palladium-based mass tag cell barcoding with a doublet-filtering scheme and single-cell deconvolution algorithm. Nat Protoc 10(2):316–333. https://doi.org/10.1038/nprot.2015.020
3. Behbehani GK, Thom C, Zunder ER et al (2014) Transient partial permeabilization with saponin enables cellular barcoding prior to surface

marker staining. Cytometry A 85(12):1011–1019. https://doi.org/10.1002/cyto.a.22573
4. Lai L, Ong R, Li J et al (2015) A CD45-based barcoding approach to multiplex mass-cytometry (CyTOF). Cytometry A 87(4):369–374. https://doi.org/10.1002/cyto.a.22640
5. Mei HE, Leipold MD, Schulz AR et al (2015) Barcoding of live human peripheral blood mononuclear cells for multiplexed mass cytometry. J Immunol 194(4):2022–2031. https://doi.org/10.4049/jimmunol.1402661
6. Yao Y, Liu R, Shin MS et al (2014) CyTOF supports efficient detection of immune cell subsets from small samples. J Immunol Methods 415:1–5. https://doi.org/10.1016/j.jim.2014.10.010
7. Newell EW, Sigal N, Bendall SC et al (2012) Cytometry by time-of-flight shows combinatorial cytokine expression and virus-specific cell niches within a continuum of CD8+ T cell phenotypes. Immunity 36(1):142–152. https://doi.org/10.1016/j.immuni.2012.01.002

Part IV

Specific Applications

Chapter 10

Staining of Phosphorylated Signalling Markers Protocol for Mass Cytometry

Diana Shinko, Thomas M. Ashhurst, Helen M. McGuire, and Kellie A. Charles

Abstract

Mass cytometry is a multi-parametric technique that offers insight into functional and biological systems at a single cell level (Tanner et al., Cancer Immunol Immunother 62:955–965, 2013). One of the major advantages of mass cytometry is the ability to measure multiple intracellular markers, including phosphorylated proteins that are part of major signaling pathways, such as NF-κB, JAK/STAT, and ERK/MAPK. Here we describe an optimized mass cytometry protocol for staining human clinical blood samples with panels that include phosphorylated antibodies.

Key words Mass cytometry, Phosphorylation, Signaling pathway, Inflammatory pathways, Signaling proteins

1 Introduction

Mass cytometry, a multi-parametric single cell technique, has allowed the comprehensive investigation of immune profiles in various fields of research, including immunology, oncology, and virology [1–3]. One of the major advantages of mass cytometry is the extensive number of intracellular proteins that can be measured in a single panel [4]. This allows the investigation of functional responses or regulatory pathways of inflammatory responses by the measurement of phosphorylated signaling proteins involved in major inflammatory markers such as NF-κB, JAK/STAT, and MAPK-ERK pathways. The changes in phosphorylation of these proteins can indicate the activation status of the signaling pathways. This ability to investigate the inflammatory signaling pathways increases the depth of understanding of immune regulation, particularly in human clinical samples.

In this chapter, we describe an optimized protocol for staining human clinical blood samples using a panel that includes antibodies

Helen M. McGuire and Thomas M. Ashhurst (eds.), *Mass Cytometry: Methods and Protocols*, Methods in Molecular Biology, vol. 1989, https://doi.org/10.1007/978-1-4939-9454-0_10,

targeting phosphorylated signaling proteins. For optimal intracellular staining, cells must be fixed to stabilize the cell membrane and permeabilized to allow the antibody to enter the cell. This protocol will describe the fixation and permeabilization necessary for the optimal staining of the phospho-specific antibodies. Furthermore, the protocol describes the steps to include additional intracellular markers such as FoxP3 or transcription factors in the same panel.

2 Materials

1. Prepare all solutions using ultrapure water (18 MΩ cm).
2. Store all reagents at room temperature (RT) unless otherwise specified.

2.1 Reagents

1. Staining buffer: 0.5% Bovine Serum Albumin (BSA), 0.02% sodium azide, 2 mM Ethylenediaminetetraacetic acid (EDTA), in phosphate-buffered saline (PBS) (*see* **Note 1**). Store at 4 °C.
2. FoxP3/Transcription Factor Staining Buffer Set (eBioscience™, USA) (*see* **Note 2**):
   - 1× fixation/permeabilization buffer: Comes in 4× concentrate. Dilute the fixation/permeabilization buffer (one part) with FoxP3 fixation/permeabilization diluent (three parts). 1 mL is required for each sample.
   - 1× Permeabilization buffer: Comes in 10× concentrate. Dilute the perm buffer with ultrapure water (1:10). 4 mL is required for each sample.
3. Metal conjugated-antibodies master mixtures:
   - Surface antibodies: Prepare a master mix of surface antibodies in staining buffer using a Centrifugal Filter Unit (0.1 μm pore size) (*see* **Note 3**). Centrifuge at 12,000 *g* for 4 min at 4 °C. 50 μL per sample is required (*see* **Note 4**).
   - FoxP3/transcription factors antibodies: Prepare a master mix of the FoxP3 or other transcription factor antibodies in 1× permeabilization buffer using the Centrifugal Filter Unit (0.1 μm pore size). Centrifuge at 12,000 *g* for 4 min at 4 °C. 50 μL per sample will be required (*see* **Note 4**).
   - Phospho-specific antibodies: Prepare a master mix of the phospho-specific antibodies in staining buffer using a Centrifugal Filter Unit (0.1 μm pore size). Centrifuge at 12,000 *g* for 4 min at 4 °C. 50 μL per sample will be required (*see* **Notes 4–6**).
   - Store all the master mixtures on ice or at 4 °C.
4. 100% ice-cold methanol.
5. 4% Paraformaldehyde (PFA), diluted in PBS.

6. Cell ID™ DNA intercalator-Ir mix: stock DNA iridium intercalator (Fluidigm, USA) diluted in 1:4000 v/v in PBS to a final concentration of 125 nM (*see* **Note** 7).
7. 1× EQ™ four element calibration beads mixture (Fluidigm, USA): 1:10 dilution of the EQ™ four element calibration beads in ultrapure water.
8. Thawed cryopreserved human peripheral blood mononuclear cells (PBMCs) (*see* **Note 8**).

2.2 Equipment

1. 5 mL polystyrene round bottom test tube with snap cap (*see* **Note 9**).
2. 5 mL round bottom polystyrene test tube, with Cell Strainer Snap Cap (*see* **Note 10**).
3. Refrigerated centrifuge for 5 mL tubes.
4. Tabletop refrigerated centrifuge, for 1.5 mL tubes.
5. Centrifugal Filter Unit (0.1 μm pore size).

3 Methods

Prior to fixation, centrifuge all samples at 500 *g* for 5 min at 4 °C. After fixation, centrifuge at 800 *g* for 8 min at RT.

3.1 Surface Staining

1. Suspend the samples in staining buffer and transfer to the 5 mL polystyrene round bottom test tubes (*see* **Note 9**). Centrifuge the samples at 500 *g* for 5 min at 4 °C. Discard supernatant.
2. Resuspend the samples in 50 μL of surface antibodies master mix and incubate for 30 min at 4 °C.
3. Top up with 1 mL staining buffer and centrifuge at 500 *g* for 5 min. Discard the supernatant.
4. Repeat the wash in **step 3** twice for a total of three washes.

3.2 Intracellular Staining: FoxP3/Transcription Factors (See Notes 11 and 12)

1. Pulse vortex the samples to completely dissociate the pellet.
2. Add 1 mL of the 1× FoxP3 fixation/permeabilization buffer and pulse vortex.
3. Incubate in the dark for 30–60 min at RT.
4. Without washing, add 2 mL of 1× permeabilization buffer to each tube. Centrifuge at 800 *g* for 8 min at RT. Discard supernatant.
5. Add 50 μL of the FoxP3/transcription factors antibody master mix to the cells and incubate at RT for at least 30 min.
6. Add 2 mL of 1× permeabilization buffer to each tube. Centrifuge samples at 800 *g* for 8 min. Discard supernatant.

7. Add 1 mL staining buffer to each tube. Centrifuge at 800 *g* for 8 min. Discard supernatant.
8. Repeat **step 7** twice for a total of three washes.

3.3 Methanol Permeabilization

1. Add 1 mL of 100% cold methanol slowly to the samples (*see* **Note 13**). Incubate for 30 min on ice.
2. Add 2 mL staining buffer. Centrifuge at 800 *g* for 5 min. Discard supernatant.

3.4 Intracellular Staining: Phospho-specific Antibodies

1. Resuspend the samples in 50 μL of the phospho-specific antibody master mix and incubate at RT for 45–60 min.
2. Top up with 1 mL staining buffer. Centrifuge at 800 *g* for 8 min. Discard supernatant.
3. Repeat **step 2** twice for a total of three washes.

3.5 Fixation

1. Resuspend in 100 μL of 4% PFA. Incubate overnight at 4 °C (*see* **Note 7**).
2. Top up with 1 mL staining buffer. Centrifuge at 800 *g* for 5 min. Discard supernatant.

3.6 DNA Intercalation and Final Wash

1. Resuspend in 100 μL of the DNA intercalator mix (*see* **Note 7**). Incubate for 20 min at RT.
2. Add 1 mL of ultrapure water. Centrifuge at 800 *g* for 8 min. Discard supernatant.
3. Repeat **step 2** twice for a total of three washes.
4. Resuspend the sample in 1 mL of ultrapure water and count the cells.
5. Centrifuge at 800 *g* for 8 min. Discard supernatant and leave the sample as a pellet at 4 °C until ready to acquire. Do not resuspend in the beads mixtures until ready to acquire the sample.

3.7 Addition of Calibration Beads

1. Add the required volume of 1× EQ™ four element calibration beads to the sample just before acquiring the sample and filter it by transferring the sample to a 5 mL Round Bottom Polystyrene Test Tube with Cell Strainer Snap Cap (*see* **Note 9**). Cells need to be resuspended at $<8 \times 10^5$ cells/mL.
2. Run the samples on the mass cytometer.

4 Notes

1. It is recommended to filter the staining buffer through a 0.22 μM pore size filter prior to use to sterilize the buffer. Additionally, ensure that buffers are not stored in glass bottles because common dish soaps used in the cleaning of glass bottles

contain barium. The barium contaminants can persist on the glass bottles even with multiple rinses, thus leading to interference with the results, detector aging, as well as possible oxidization signals in the M + 16 channel [5].

2. Any FoxP3/transcription factors staining kit can be used for the staining of FoxP3 or transcription factors. This protocol specifically integrates the steps from the manufacturer's instructions for the FoxP3/Transcription Factor Staining Buffer Set kit (eBioscience™, USA). If using a different kit, ensure you follow the instructions from the manufacturer regarding dilution of the reagents and the specific fixation and permeabilization steps that are needed for Subheading 3.2 of this protocol.
3. The master mix should be prepared and filtered relatively close to when the mixture is planned to be added to the cells. Do not spin-filter too far in advance. Use a 0.5 mL, 0.1 μm pore size centrifugal filter unit to filter the antibody master mixtures prior to use to remove any precipitates from the mixture.
4. A staining volume of either 50 μL or 100 μL per sample can be used. We found that for less than three million cells per sample (we typically use one to two million cells per sample), 50 μL per sample is sufficient for staining. If using more than three million cells, we recommend increasing the staining volume accordingly.
5. Other non-phospho-specific intracellular antibodies can also be added to this master mixture if the staining buffer used is the appropriate buffer for the specific intracellular antibodies. However, ensure the fixation and permeabilization reagents and steps in this protocol are also appropriate for all the additional intracellular antibodies.
6. The sensitivity and specificity of the phospho-specific antibodies need to be validated, particularly if conjugating antibodies in house. To achieve this, the PBMCs must be stimulated with a combination of ligands known to induce the phosphorylation of the specific intracellular proteins. For instance, in a panel that includes phospho-specific antibodies targeting signaling protein in the JAK/STAT, MAPK/ERK, and NF-κB pathways, PBMCs were stimulated with either a combination of phorbol 12-myristate 13-actetate (PMA; 80 nM) and ionomycin (1.3 μM) or a mixture of cytokines (100 ng/mL IL-6, 100 ng/mL IL-2, 20 ng/mL IFNα, 2 ng/mL GM-CSF) for 15 min at 37 °C, fixed, and stained with the phospho-specific antibodies. A sample of PBMCs were left unstimulated as a negative control. The median intensity of the staining (arcsinh ratio) represented in the heatmap (Fig. 1) demonstrates the phosphorylation changes between the unstimulated and stimulated samples. The phosphorylation occurred in the expected stimulation condition for the specific signaling protein in the immune cell type. For example, in myeloid cells, PMA with

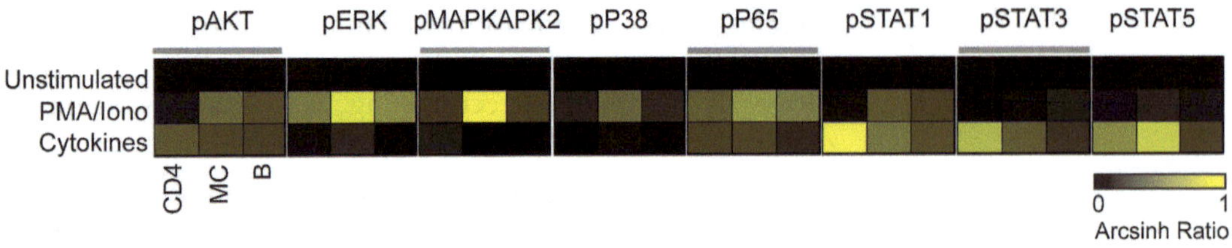

Fig. 1 Phospho-specific antibody validation. PBMCs from healthy controls were stimulated with either a combination of phorbol 12-myristate 13-actetate (PMA; 80 nM) and ionomycin (1.3 μM) or a mixture of cytokines (100 ng/mL IL-6, 100 ng/mL IL-2, 20 ng/mL IFNα, 2 ng/mL GM-CSF). A sample was left unstimulated as a negative control. The median intensity of the staining (arcsinh ratio) represents the phosphorylation changes of the signaling proteins between the unstimulated and stimulated samples. *CD4* $CD4^+$ T cells, *MC* myeloid cells, *B* B cells

ionomycin induced the phosphorylation of pAKT, pERK, and pMAPKAPK2 but not pSTAT3 or pSTAT5, which were in turn induced by the cytokine mixture.

7. This protocol advises fixing the samples in 4% PFA overnight and staining with the DNA iridium intercalator the following day (or on the day of the acquisition). However, it is also possible to combine the two steps by diluting the DNA iridium intercalator directly in 4% PFA. You can dilute the DNA iridium intercalator in 4% PFA (1:4000), add 100 μL to the samples, and incubate the samples at 4 °C overnight or until the day of acquisition. We recommend running the samples through the mass cytometer within a week of the fixation.
8. This protocol was optimized using cryopreserved human peripheral blood mononuclear cells (PBMCs). It can, however, also be used for any human cell samples. Isolate PBMCs from human blood samples using Ficoll-hypaque density gradient centrifugation (density 1.077 ± 0.003 g/dL). Gently add the blood on a layer of Ficoll and centrifuge at 550 *g* for 20 min at RT (deceleration OFF) and remove the layer of PBMCs.
9. This protocol utilizes 5 mL polystyrene round bottom test tubes with snap caps for staining the samples. However, it is also possible to use 1.5 mL microcentrifuge tubes for the staining and decreasing the washing volumes accordingly. The samples can be transferred to the 5 mL round bottom tubes prior to running the samples on the mass cytometer, at the filtration step of the protocol (Subheading 3.7, **step 1**).
10. Filter the samples using the 5 mL round bottom polystyrene test tubes with cell strainer snap cap. Alternatively, you can run the samples through a filter (35 μm pore size filter) into a new tube.
11. If you are not using an anti-FoxP3 antibody or other transcription factor antibodies, you can omit this section and move onto the staining of phospho-specific antibodies directly. However, you will need to fix the samples with 4% PFA prior to the methanol permeabilization. To achieve this, add 100 μL of 4% PFA

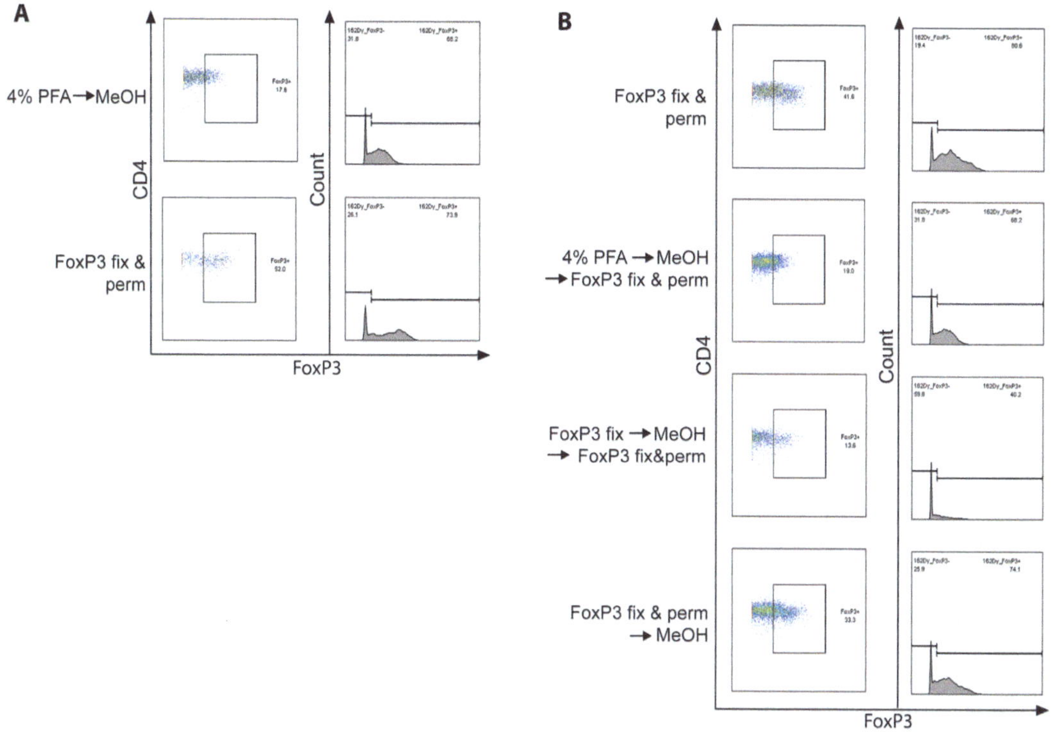

Fig. 2 Fixation and permeabilization protocol optimization for FoxP3 staining. (**a**) PBMCs were either fixed with 4% PFA and permeabilized using methanol (PFA → MeOH) or fixed and permeabilized using the Foxp3/Transcription Factor staining Buffer Set (FoxP3 fix&perm; eBioscience™, USA). (**b**) PBMCs were fixed and permeabilized using various combination: Fixed and permeabilized with FoxP3 fix&perm; Fixed with 4% PFA and permeabilized with methanol first followed by the FoxP3 fix&perm then stain for FoxP3 (4% PFA → MeOH → FoxP3 fix&perm, FoxP3 stain); Fixed with the fixation buffer from the Foxp3/Transcription Factor staining Buffer Set first, permeabilized by methanol followed by the FoxP3 fix&perm then stain for FoxP3 (FoxP3 fix → MeOH → FoxP3 fix&perm, FoxP3 stain); Fixed and permeabilized with the FoxP3 fix&perm, stain for FoxP3 followed by methanol permeabilization (FoxP3 fix&perm then FoxP3 stain → MeOH)

to the samples and incubate for 10–20 min at RT. wash the samples three times with 1 mL of staining buffer before moving on to the methanol permeabilization section (Subheading 3.3).

12. It is important to follow the steps on this protocol in the correct order. In a panel containing FoxP3, we optimized the order of the fixation and permeabilization steps to obtain the optimal FoxP3 staining without interfering with the phospho-specific staining. Fixation with 4% PFA followed by methanol permeabilization (4%PFA → MeOH) was insufficient in obtaining an optimal staining for FoxP3 when compared to the FoxP3 fixation and permeabilization buffers (FoxP3 fix&perm) obtained from a FoxP3/Transcription Factor staining Buffer Set (eBioscience™; Fig. 2a). A combination of fixation and permeabilization reagents were then tested in various combinations

and the results indicated that using the FoxP3 fixation and permeabilization from the FoxP3/Transcription Factor staining Buffer Set and staining for FoxP3 first followed by methanol permeabilization (FoxP3 fix&perm → MeOH) produced the optimal staining for FoxP3 (Fig. 2b).

13. Ensure the cells are well suspended in the methanol without vortexing as this may cause cell loss. If the cells are clumping when you acquire them, you can optimize this step by resuspending the cells in 100 μL of staining buffer before adding 900 μL of methanol (90% methanol is just as effective at permeabilizing the cells).

References

1. Tanner SD, Baranov VI, Ornatsky OI et al (2013) An introduction to mass cytometry: fundamentals and applications. Cancer Immunol 62(5):955–965. https://doi.org/10.1007/s00262-013-1416-8
2. Gaudilliere B, Fragiadakis GK, Bruggner RV et al (2014) Clinical recovery from surgery correlates with single-cell immune signatures. Sci Transl Med 6(255):255ra131. https://doi.org/10.1126/scitranslmed.3009701
3. Lavin Y, Kobayashi S, Leader A et al (2017) Innate immune landscape in early lung adenocarcinoma by paired single-cell analyses. Cell 169(4):750–765.e717. https://doi.org/10.1016/j.cell.2017.04.014
4. Bendall SC, Nolan GP (2012) From single cells to deep phenotypes in cancer. Nat Biotechnol 30(7):639–647. https://doi.org/10.1038/nbt.2283
5. Leipold MD, Newell EW, Maecker HT (2015) Multiparameter phenotyping of human PBMCs using mass cytometry. Methods Mol Biol 1343:81–95. https://doi.org/10.1007/978-1-4939-2963-4_7

Chapter 11

Multiplex MHC Class I Tetramer Combined with Intranuclear Staining by Mass Cytometry

Yannick Simoni, Michael Fehlings, and Evan W. Newell

Abstract

Antigen-specific $CD8^+$ T cells play a crucial role in the host protective immune response against viruses, tumors, and other diseases. Major histocompatibility complex (MHC) class I tetramers allow for a direct detection of such antigen-specific $CD8^+$ T cells. Using mass cytometry together with multiplex MHC class I tetramer staining, we are able to screen more than 1000 different antigen candidates simultaneously across tissues in health and disease, while retaining the possibility to deliver an in-depth characterization of antigen-specific $CD8^+$ T cells and associated phenotypes. Here we describe the method for a MHC class I tetramer multiplexing approach together with intracellular antibody staining for a parallel phenotypic cell characterization using mass cytometry in human specimens.

Key words Mass cytometry, MHC class I tetramer, Antigen-specific T cell, Multiplex tetramer staining

1 Introduction

$CD8^+$ T lymphocytes represent an important actor of the immune system, implicated in cell-mediated immunity. A unique feature of $CD8^+$ T cells is their ability to discriminate between healthy and abnormal (e.g., infected or cancerous) cells. Each $CD8^+$ T lymphocyte expresses on their surface a unique T cell receptor (TCR) that recognizes a specific antigen. $CD8^+$ T cells specifically target and kill cancer cells, or infected cells (viral or bacterial infection). When infected or dysfunctional, target cells express on their surface viral or tumor antigens presented in the context of major histocompatibility complex (MHC) class I. After recognition of target cells, through the interaction between TCR and MHC class I, $CD8^+$ T cell induce apoptosis by production of cytotoxic proteins (i.e., Granzyme, Perforin) or through expression of Fas ligand on their surface. Furthermore, they can produce various cytokines that activate adaptive and innate immune responses (e.g., IFNγ, TNFα, IL-2) [1, 2].

Helen M. McGuire and Thomas M. Ashhurst (eds.), *Mass Cytometry: Methods and Protocols*, Methods in Molecular Biology, vol. 1989,
https://doi.org/10.1007/978-1-4939-9454-0_11,

However, in chronic viral infection or cancer, $CD8^+$ T cells can lose their function and become ineffective at killing target cells [3, 4]. A better characterization of these cells by studying their antigen specificity and phenotype is required for the development of new treatments such as immunotherapy or vaccination. Although several methods have proven to be useful for profiling the heterogeneity of CD8+ T cells, identification of antigen-specific $CD8^+$ T cells and deep phenotyping in the context of these pathologies remains a challenge [5].

Antigen-specific $CD8^+$ T cells can be identified indirectly with functional assays such as ELISPOT. However, this method does not provide phenotypic information about antigen-specific cells, and requires a large number of cells (minimum 20,000 $CD8^+$ T cells for each peptide screened) [6]. Alternatively, antigen-specific cells can be directly identified through the use of recombinant antigen-MHC class I proteins linked together as multimers, such as tetramers [7, 8] (Fig. 1). Direct identification of antigen-specific $CD8^+$ T cells using MHC class I tetramers has the advantage of permitting the simultaneous identification of cells specific for different antigens together with obtaining phenotypic information, even in samples with relatively small numbers of cells (minimum 5000 $CD8^+$ T cells for several peptides). During the past two decades, MHC class I tetramers have been used extensively in flow cytometry. However, the limited number of fluorochromes available limits the screening of a large number of antigen-MHC class I tetramers (i.e., <20 MHC class I tetramers are currently possible). Mass cytometry represents a powerful alternative for the study of immune cells, with more than 40 different channels available [9–14]. Coupled to the multiplex MHC class I tetramer approach, this technique allows us to screen up to several hundred epitopes at the same time [15–18]. Recently, we have applied this method to screen, identify, and characterize tumor neoantigen-specific $CD8^+$ T cells in cancer [19, 20], and during Dengue and Hepatitis B virus infections (unpublished data). Multiplex MHC class I tetramers work by tagging each peptide-MHC class I tetramer with a unique combination of usually three heavy-metal labels. For example, using nine different heavy-metal-labeled streptavidins, we can screen for 84 different MHC class I tetramers (Fig. 2). In addition, the cells can be simultaneously stained with more than 20 probes to analyze T cell surface markers, functional markers, and transcription factor markers (Fig. 3).

Here, we describe the technical details of this approach for the study of virus-specific $CD8^+$ T cells in humans.

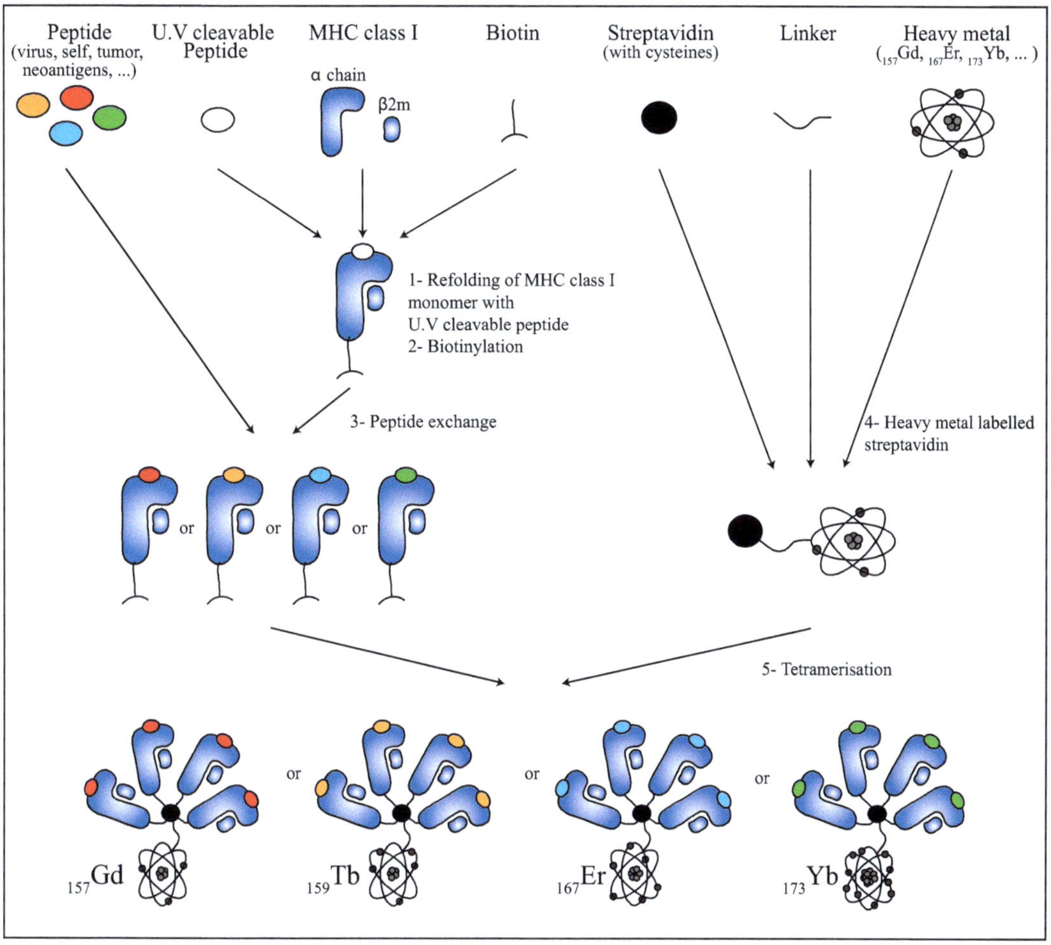

Fig. 1 MHC class I tetramer for mass cytometry. Schematic for MHC class I tetramer preparation for mass cytometry

2 Materials

2.1 MHC Class I Monomer Reagents and Components

1. UV cleavable peptides:
   - YLLEMLWJK for HLA-A*02:01 allele [21].
   - RVFAJSFIK for HLA-A*11:01 allele [22].
   - VYGJVRACL for HLA-A*24:02 allele
   - (**J**: 3-amino-3-(2-nitro)phenyl-propionic acid-linker—UV sensitive).
2. Peptides of interest are dissolved in DMSO to a stock concentration of 1 mM.
3. Biotinylated MHC class I monomer refolded with appropriate UV cleavable peptides were prepared as described [17, 23].

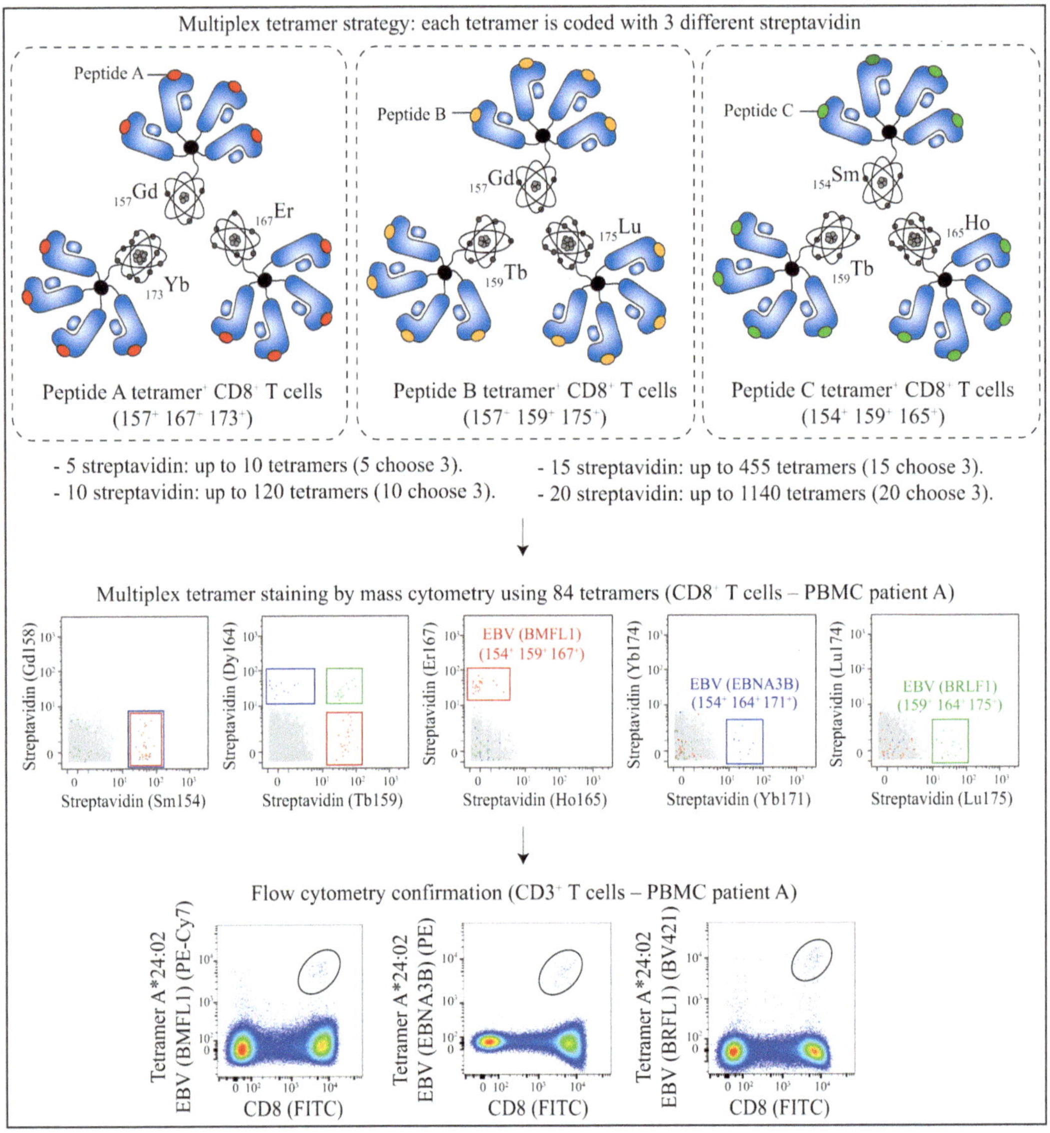

Fig. 2 Multiplex MHC class I tetramer staining for mass cytometry. (Upper panel) Schematic for triple coded MHC class I tetramer. (Middle panel) Example of multiplex MHC class I tetramer staining by mass cytometry using 84 simultaneous tetramers, each one coded with three different streptavidin out of nine streptavidin (9 choose 3 = 84). Cells were gated on CD8+ T cells from PBMC of patient A, HLA-A*24:02 positive. (Lower panel) Flow cytometry confirmation of 3 tetramer+ CD8+ T cells identified by mass cytometry. Cells were gated on CD8+ T cells from PBMC of patient A. Peptide sequence: EBV (BMFL1): DYNFVKQLF, EBV (EBNA3B): TYSAGIVQI, EBV (BRFL1): TYPVLEEMF

4. UV light chamber (A143-002F UV Chamber 365 nm—epak) or equivalent.
5. 96 well plate, round bottom.
6. PBS: phosphate buffered saline (PBS), pH 7.2. Stored at 4 °C.

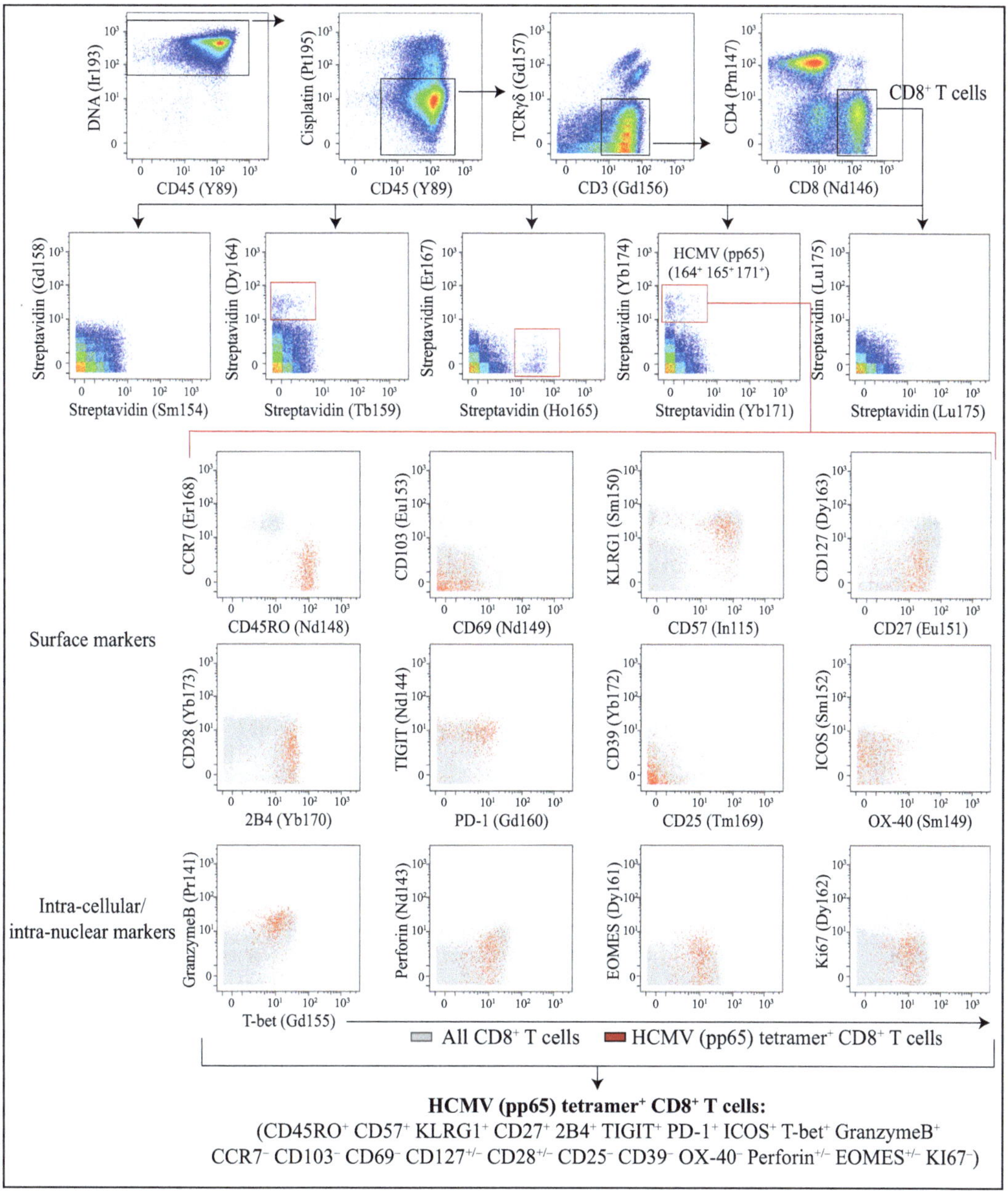

Fig. 3 Multiplex MHC class I tetramer and in-depth phenotyping of tetramer$^+$ CD8$^+$ T cells. Example of multiplex MHC class I tetramer staining by mass cytometry. Cells were gated on CD8$^+$ T cells from PBMC of patient B, HLA-A*02:01 positive. HCMV (pp65)-specific CD8$^+$ T cells were analyses for surface marker and transcription factor. Peptide sequence: HCMV (pp65): NLVPMVATV

2.2 Streptavidin, Antibodies, Linker, and Heavy Metals

1. Streptavidin with six free cysteine is produced as described [17].
2. Streptavidin is coupled to heavy metal using DN3 polymer linker as reported previously [24].
3. Antibodies are selected for their compatibility with the heavy-metal-labeling protocol.

2.3 MHC Class I Tetramerization and Cell Staining

1. PBS, pH 7.2. Stored at 4 °C.
2. PBS-BSA: PBS + 0.5% bovine serum albumin (BSA) + 0.02% sodium azide. Stored at 4 °C.
3. Centrifugal filter 50 kDa.
4. Centrifugal filter 0.1 μm.
5. Cisplatin: 100 mM cisplatin in DMSO. Stored at −80 °C.
6. 16% Paraformaldehyde (PFA).
7. 96 well plate, V bottom.
8. DNA intercalator (Fluidigm).
9. H_2O: MilliQ water or equivalent.
10. Intracellular staining buffer kit.

2.4 Tissue Preparation

1. DNAse, stock concentration 1 mg/mL.
2. Collagenase IV, stock concentration 50 mg/mL.
3. Digestion media: DMEM + Collagenase IV (1 mg/mL).
4. Washing media: RPMI +5% FBS + DNAse (15 μg/mL).
5. Water bath, 37 °C.
6. 70 μm nylon cell strainer (BD Falcon).
7. Freezing media: 90% FBS + 10% DMSO.

3 Methods

3.1 Preparation of a Single-Cell Suspension from Tissues

Multiplex MHC class I tetramer staining can be performed in peripheral blood mononuclear cells (PBMC) or human tissues [16, 20]. The following protocol is optimized for the analysis of human tumor samples.

1. Under sterile conditions, all areas of tissue necrosis are trimmed away. The tissue is sliced into 1–2 mm^3 pieces using scissors.
2. For enzymatic digestion, the pieces are incubated in a water bath for 10–20 min at 37 °C in 10 mL of digestion media (*see* **Note 1**).
3. Stop digestion by adding cold washing media.
4. Centrifuge at 400 *g* for 5 min.
5. Remove supernatant and resuspend the pellet in 20 mL of washing media.
6. Pass solution through a 70 μm nylon cell strainer. Scrape the remaining pieces through the filter using the flat end surface of the syringe insert, with additional wash media used to subsequently rinse the cell strainer.
7. Centrifuge at 400 *g* for 5 min.

8. Remove supernatant and resuspend the pellet in 20 mL of washing media.
9. (optional) If samples remain chunky, the cell suspension can be passed through a 70 μm nylon cell strainer a second time.
10. Centrifuge at 400 *g* for 5 min.
11. Freeze cells in freezing media.

3.2 Peptide Exchange for MHC Class I Monomers

The following protocol is optimized for staining 10 samples, each with two to five million cells (*see* Fig. 1).

1. Synthesize, refold, biotinylate and purify MHC class I monomers according to the established protocols [17, 23].

   IMPORTANT: Each MHC class I monomer allele (e.g., HLA-A*02:01, HLA-A*11:01) must be refolded with the corresponding UV cleavable peptide (refer to Subheading 2.1).
2. Dilute biotinylated MHC class I monomers to 0.1 mg/mL in PBS. 100 μL of diluted MHC class I monomers will be used for exchange with any peptide of interest.

   Example: For 10 peptides, prepare 1 mL (10 × 100 μL) of diluted MHC class I monomers at 0.1 mg/mL. For 200 peptides, prepare 20 mL (200 × 100 μL) of diluted MHC class I monomers at 0.1 mg/mL.
3. Transfer 100 μL of diluted MHC class I monomers to one well of a 96 well plate.
4. Add 5 μL of the peptide of interest per individual well (stock 1 mM).
5. Mix by pipetting.
6. Place the plate on ice and put the plate in a UV chamber. Expose the plate for 5 min (*see* **Note 2**).
7. Rotate the plate by 180° and expose to UV light for another 5 min.
8. Incubate the plate at 4 °C (can be stored up to 3 weeks). (Now reagent is referred to as the Peptide-MHC class I monomer).

3.3 MHC Class I Tetramerization Using a Triple Coding Approach

For multiplex tetramer staining, each MHC class I tetramer is generated with a combination of three different metal-labeled streptavidins. Using nine different metal-labeled streptavidins, 84 possible combinations (9 choose 3) are generated. Each specific combination is linked to a different peptide (Fig. 2).

1. Dilute each metal-labeled streptavidin to 50 μg/mL in PBS and mix using an automated pipetting device in 96 well plates.
2. Prepare three tubes of each biotinylated MHC class I monomer (100 μL in each).

3. For tetramerization, transfer 10 μL of each metal-labeled streptavidin to the relevant peptide-MHC class I monomer tube, according to the coding scheme. Mix by pipetting and incubate for 10 min at room temperature.
4. Add another 10 μL of metal-labeled streptavidin mixture to the relevant peptide-MHC class I monomers tube, mix by pipetting, and incubate for 10 min at room temperature.
5. Repeat **step 4** one more time.
6. Incubate peptide-MHC class I tetramers with free biotin (10 μM) for 10 min at room temperature.
7. Combine all peptide-MHC class I tetramers (as per coding scheme) and concentrate down to a final volume of 500 μL using a 50 kDa concentrator.
8. Add 500 μL of PBS-BSA 0.5% to a final volume of 1 mL.
9. Filter peptide-MHC class I tetramers using a 0.1 μM filter (referred as tetramer cocktail).

3.4 Multiplex MHC Class I Tetramer Staining and Antibody Surface Staining

1. Human samples (e.g., PBMC, tumor) are prepared according to an established procedure. HLA phenotypes are determined by DNA genotyping.
2. Thaw cryopreserved sample into 10 mL of PBS-BSA in a 15 mL conical tube.
3. Centrifuge at 400 *g* for 5 min.
4. Pour off the supernatant and resuspend the cells in 100 μL of PBS-BSA by pipetting.
5. Transfer the cells to one well of a 96 well round bottom plate.
6. Centrifuge at 400 *g* for 2 min.
7. Pour off supernatant by flicking the plate over a waste container and resuspend the cells with 100 μL of peptide-MHC class I tetramer cocktail.
8. Incubate for 1 h at room temperature.
9. Top up to 200 μL with PBS-BSA and centrifuge at 400 *g* for 2 min. Remove supernatant by flicking the plate over a waste container.
10. Repeat **step 9**.
11. Dilute cisplatin stock to 200 μM in PBS and resuspend the cells in 80 μL of cisplatin solution (*see* **Note 3**).
12. Incubate for 5 min at 4 °C.
13. Add 100 μL washing media, centrifuge, and flick plate.
14. Top up to 200 μL with PBS-BSA and centrifuge at 400 *g* for 2 min. Remove supernatant by flicking the plate over a waste container.

15. Resuspend the cells with 50 μL of surface antibodies cocktail.
16. Incubate for 15 min at 4 °C.
17. Fill to 200 μL with PBS-BSA and centrifuge at 400 *g* for 2 min. Remove the supernatant by flicking the plate over a waste container.
18. Repeat **step 17** a further two times.
19. Dilute PFA stock to 2% with PBS and resuspend the cells in 100 μL of 2% PFA (*see* **Note 4**).
20. Seal plate and incubate overnight at 4 °C.
21. Centrifuge the plate at 800 *g* for 2 min and remove the supernatant by flicking the plate over a waste container.
22. Resuspend the cells in 100 μL of DNA intercalator diluted 1:1000 in MilliQ water.
23. Incubate for 15 min at 4 °C.
24. Fill to 200 μL with MilliQ water and centrifuge at 800 *g* for 2 min. Remove the supernatant by flicking the plate over a waste container.
25. Repeat **step 24** another three times.
26. Resuspend in 100 μL of MilliQ water and filter througha 0.2 μm mesh strainer.
27. Dilute the cells to 500,000 cells/mL in MilliQ water.

Optional: For Intracellular and/or Intranuclear staining

19′. Add 100 μL of Fix/perm buffer and incubate for 30 min at 4 °C.

20′. Centrifuge the plate at 800 *g* for 2 min and remove the supernatant by flicking the plate over a waste container.

21′. Resuspend the cells with 50 μL of Intracellular and/or intranuclear staining antibodies cocktail.

22′. Incubate for 30 min at 4 °C.

23′. Top up to 200 μL with permeabilization buffer and centrifuge at 800 *g* for 2 min. Remove the supernatant by flicking the plate over a waste container.

24′. Repeat **step 26** a further two times.

25′. Top up to 200 μL with PBS-BSA and centrifuge at 800 *g* for 2 min. Remove the supernatant by flicking the plate over a waste container.

26′. Dilute PFA stock to 2% with PBS and resuspend the cells in 100 μL of 2% PFA by pipetting (*see* **Note 4**).

27′. Seal plate and incubate overnight at 4 °C.

28′. Centrifuge the plate at 800 *g* for 2 min and remove the supernatant by flicking the plate over a waste container.

29′. Resuspend the cells in 100 μL of DNA intercalator diluted 1:1000 in MilliQ water.

30′. Incubate for 15 min at 4 °C.

31′. Fill to 200 μL with MilliQ water and centrifuge at 800 *g* for 2 min. Remove the supernatant by flicking the plate over a waste container.

32′. Repeat **step 30′** another three times.

33′. Resuspend in 100 μL of MilliQ water and filter through a 0.2 μm mesh strainer.

34′. Dilute the cells to 500,000 cells/mL in MilliQ water.

3.5 Data Analysis

1. Samples are acquired on a CyTOF®, CyTOF2®, or Helios®.
2. Use Fluidigm acquisition software to create .fcs files for subsequent analysis (*see* **Note 5**).
3. The signal of each parameter is then normalized based on the EQ beads (Fluidigm) as previously described [25].
4. Standard cytometry analysis software (e.g., FlowJo) can be used for mass cytometry data analysis.
5. To identify live $CD8^+$ T cells by manual gating, we use cisplatin (live-dead), DNA (debris exclusion), CD45 (immune cell marker), CD3 (T cell marker), and CD8 to identify these cells (Fig. 3). Then, manual gating can be used to deconvolute tetramer positive cell populations (each coded with a unique combination of three heavy metals) (Figs. 2 and 3). Data can be confirmed using conventional flow cytometry (Fig. 2). Phenotypes of $tetramer^+$ $CD8^+$ T cells can further be studied using standard bi-axial dot plots or with high-dimensional analysis tools such as One-SENSE, t-SNE, H-SNE, SPADE, PCA, or UMAP [26–30].

4 Notes

1. Time of incubation depends on several factors (tissue size, quality, donor, etc.). This will need to be determined empirically at each laboratory.
2. If using a different UV source, ensure an equivalent UV exposure of ~120,000 μJ in total. Heat is generated during UV irradiation hence the necessity to keep the wells in contact with wet ice throughout the UV process.
3. Cisplatin stock must be diluted just prior to use. Live cells will stain less brightly than dead cells.
4. Dilute PFA just prior to use.

5. On .fcs files, any zero values are randomized using a uniform distribution of values between zero and minus-one using an R script [16]. Note that all other integer values measured by the mass cytometer are randomized in a similar fashion by default; thus, we think it is reasonable and useful in most cases to apply the same randomization to zero values.

Acknowledgments

The authors thank all members of Evan Newell lab.

Competing interests: E.W.N. is a board director and shareholder of immunoSCAPE Pte.Ltd. M.F. is Director, Scientific Affairs and shareholder of immunoSCAPE Pte. Ltd. All other authors declare no competing financial interests.

References

1. Wong P, Pamer EG (2003) CD8 T cell responses to infectious pathogens. Annu Rev Immunol 21:29–70. https://doi.org/10.1146/annurev.immunol.21.120601.141114
2. Zhang N, Bevan MJ (2011) CD8(+) T cells: foot soldiers of the immune system. Immunity 35(2):161–168. https://doi.org/10.1016/j.immuni.2011.07.010
3. Wherry EJ, Kurachi M (2015) Molecular and cellular insights into T cell exhaustion. Nat Rev Immunol 15(8):486–499. https://doi.org/10.1038/nri3862
4. Wherry EJ (2011) T cell exhaustion. Nat Immunol 12(6):492–499
5. Newell EW, Becht E (2018) High-dimensional profiling of tumor-specific immune responses: asking T cells about what they "see" in cancer. Cancer Immunol Res 6(1):2–9. https://doi.org/10.1158/2326-6066.CIR-17-0519
6. Whiteside TL, Zhao Y, Tsukishiro T et al (2003) Enzyme-linked Immunospot, cytokine flow cytometry, and tetramers in the detection of T-cell responses to a dendritic cell-based multipeptide vaccine in patients with melanoma. Clin Cancer Res 9(2):641–649
7. Altman JD, Moss PA, Goulder PJ et al (1996) Phenotypic analysis of antigen-specific T lymphocytes. Science 274(5284):94–96
8. Altman JD, Davis MM (2003) MHC-peptide tetramers to visualize antigen-specific T cells. Curr Protoc Immunol Chapter 17:Unit 17.3. https://doi.org/10.1002/0471142735.im1703s53
9. Simoni Y, Fehlings M, Kloverpris HN et al (2018) Human innate lymphoid cell subsets possess tissue-type based heterogeneity in phenotype and frequency. Immunity 48(5):1060. https://doi.org/10.1016/j.immuni.2018.04.028
10. Simoni Y, Fehlings M, Kloverpris HN et al (2017) Human innate lymphoid cell subsets possess tissue-type based heterogeneity in phenotype and frequency. Immunity 46(1):148–161. https://doi.org/10.1016/j.immuni.2016.11.005
11. Simoni Y, Newell EW (2017) Toward meaningful definitions of innate-lymphoid-cell subsets. Immunity 46(5):760–761. https://doi.org/10.1016/j.immuni.2017.04.026
12. Simoni Y, Newell EW (2018) Dissecting human ILC heterogeneity: more than just three subsets. Immunology 153(3):297–303. https://doi.org/10.1111/imm.12862
13. Wong MT, Chen J, Narayanan S et al (2015) Mapping the diversity of follicular helper T cells in human blood and tonsils using high-dimensional mass cytometry analysis. Cell Rep 11(11):1822–1833. https://doi.org/10.1016/j.celrep.2015.05.022
14. Wong MT, Ong DE, Lim FS et al (2016) A high-dimensional atlas of human T cell diversity reveals tissue-specific trafficking and cytokine signatures. Immunity 45(2):442–456. https://doi.org/10.1016/j.immuni.2016.07.007
15. Newell EW, Klein LO, Yu W et al (2009) Simultaneous detection of many T-cell specificities using combinatorial tetramer staining. Nat Methods 6(7):497–499. https://doi.org/10.1038/nmeth.1344
16. Newell EW, Sigal N, Bendall SC et al (2012) Cytometry by time-of-flight shows combinatorial cytokine expression and virus-specific cell

niches within a continuum of CD8+ T cell phenotypes. Immunity 36(1):142–152. https://doi.org/10.1016/j.immuni.2012.01.002
17. Newell EW, Sigal N, Nair N et al (2013) Combinatorial tetramer staining and mass cytometry analysis facilitate T-cell epitope mapping and characterization. Nat Biotechnol 31(7):623–629. https://doi.org/10.1038/nbt.2593
18. Fehlings M, Chakarov S, Simoni Y et al (2018) Multiplex peptide-MHC tetramer staining using mass cytometry for deep analysis of the influenza-specific T-cell response in mice. J Immunol Methods 453:30–36. https://doi.org/10.1016/j.jim.2017.09.010
19. Fehlings M, Simoni Y, Penny HL et al (2017) Checkpoint blockade immunotherapy reshapes the high-dimensional phenotypic heterogeneity of murine intratumoural neoantigen-specific CD8(+) T cells. Nat Commun 8(1):562. https://doi.org/10.1038/s41467-017-00627-z
20. Simoni Y, Becht E, Fehlings M et al (2018) Bystander CD8(+) T cells are abundant and phenotypically distinct in human tumour infiltrates. Nature 557(7706):575–579. https://doi.org/10.1038/s41586-018-0130-2
21. Rodenko B, Toebes M, Hadrup SR et al (2006) Generation of peptide-MHC class I complexes through UV-mediated ligand exchange. Nat Protoc 1(3):1120–1132. https://doi.org/10.1038/nprot.2006.121
22. Bakker AH, Hoppes R, Linnemann C et al (2008) Conditional MHC class I ligands and peptide exchange technology for the human MHC gene products HLA-A1, -A3, -A11, and -B7. Proc Natl Acad Sci U S A 105(10):3825–3830. https://doi.org/10.1073/pnas.0709717105
23. Toebes M, Coccoris M, Bins A et al (2006) Design and use of conditional MHC class I ligands. Nat Med 12(2):246–251. https://doi.org/10.1038/nm1360
24. Leong ML, Newell EW (2015) Multiplexed peptide-MHC tetramer staining with mass cytometry. Methods Mol Biol 1346:115–131. https://doi.org/10.1007/978-1-4939-2987-0_9
25. Finck R, Simonds EF, Jager A et al (2013) Normalization of mass cytometry data with bead standards. Cytometry A 83(5):483–494. https://doi.org/10.1002/cyto.a.22271
26. Cheng Y, Wong MT, van der Maaten L et al (2016) Categorical analysis of human T cell heterogeneity with one-dimensional soli-expression by nonlinear stochastic embedding. J Immunol 196(2):924–932. https://doi.org/10.4049/jimmunol.1501928
27. Becht E, Dutertre C-A, Kwok IWH, et al (2018) Evaluation of UMAP as an alternative to t-SNE for single-cell data. bioRxiv. https://doi.org/10.1101/298430
28. Qiu P, Simonds EF, Bendall SC et al (2011) Extracting a cellular hierarchy from high-dimensional cytometry data with SPADE. Nat Biotechnol 29(10):886–891. https://doi.org/10.1038/nbt.1991
29. Amir el AD, Davis KL, Tadmor MD et al (2013) viSNE enables visualization of high dimensional single-cell data and reveals phenotypic heterogeneity of leukemia. Nat Biotechnol 31(6):545–552. https://doi.org/10.1038/nbt.2594
30. van Unen V, Hollt T, Pezzotti N et al (2017) Visual analysis of mass cytometry data by hierarchical stochastic neighbour embedding reveals rare cell types. Nat Commun 8(1):1740. https://doi.org/10.1038/s41467-017-01689-9

Chapter 12

Analysis of the Murine Bone Marrow Hematopoietic System Using Mass and Flow Cytometry

Thomas M. Ashhurst, Darren A. Cox, Adrian L. Smith, and Nicholas J. C. King

Abstract

The hematopoietic system produces erythrocytes (red blood cells), leukocytes (white blood cells), and thrombocytes (platelets) throughout the life of an organism. Long-lived hematopoietic stem cells give rise to early progenitors with multi-lineage potential that progressively differentiate into lineage-specific progenitors. Following lineage commitment, these progenitors proliferate and expand, before eventually differentiating into their mature forms. This process drives the up- and downregulation of a wide variety of surface and intracellular markers throughout differentiation, making cytometric analysis of this interconnected system challenging. Moreover, during inflammation, the hematopoietic system can be mobilized to re-prioritize the production of various lineages, in order to match increased demand, often at the expense of other lineages. As such, the response of the hematopoietic system in the bone marrow (BM) is a critical component of both immunity and disease. Because of the complexity of the hematopoietic system in steady state and disease, high-dimensional cytometry technologies are well suited to the exploration of these complex systems. Here we describe a protocol for the extraction of murine bone marrow, and preparation for examination using high-dimensional flow or mass cytometry. Additionally, we describe methods for performing cell cycle assays using bromodeoxyuridine (BrdU) or iododeoxyuridine (IdU). Finally, we describe an analytical method that allows for a system-level analysis of the hematopoietic system in steady state or inflammatory scenarios.

Key words Bone marrow, Hematopoiesis, High-dimensional cytometry, Mass cytometry, CyTOF, Cell cycle

1 Introduction

1.1 Bone Marrow Hematopoiesis

The murine hematopoietic system consists of a complex progression of developmentally related populations. These populations exist as part of a developmental continuum, starting with long-lived hematopoietic stem cells, progressing through a range of developmental intermediates, and culminating in a wide variety of mature phenotypes. This progression occurs in concert with the up- and downregulation of a range of overlapping surface and

Helen M. McGuire and Thomas M. Ashhurst (eds.), *Mass Cytometry: Methods and Protocols*, Methods in Molecular Biology, vol. 1989, https://doi.org/10.1007/978-1-4939-9454-0_12,

intracellular markers. Developmental intermediates are some of the most difficult cells to investigate by cytometry, due to both the downregulation of robust stem cell markers, but also the lack of expression of maturation markers, with many of these cell types existing at the intersection of various lineages. Because these cells exhibit the most proliferative potential for different hematopoietic lineages, understanding the mobilization of these progenitors, and their downstream lineages, is critical in understanding disease pathogenesis.

1.2 High-Dimensional Cytometry of the Bone Marrow

Because of the complexity of the hematopoietic system, high-dimensional (HD) single cell approaches are necessary to fully appreciate the nuances of such systems, as low-dimensional panels are not able to interrogate the system as a whole. Modern mass cytometry enables characterization of the immune system by analyzing up to 47 parameters simultaneously on single cells, allowing for greater resolution of complex progenitors and developmental intermediates. Here we describe a workflow for mapping of the murine hematopoietic system in the bone marrow using both flow and mass cytometry, similar to existing approaches for the human bone marrow [1, 2].

Critical to our analysis was the ability to profile cell cycle status across the hematopoietic lineage. Bromodeoxyuridine (BrdU) is a thymidine analogue that can be incorporated into replicating DNA during S phase of the cell cycle [3] following ex vivo incubation of BrdU with cells, or intraperitoneal (*i.p.*) injection of BrdU into an animal. In flow cytometry, BrdU can be labeled intracellularly using a fluorophore-conjugated antibody following surface staining, fixation, permeabilization, and deoxyribonuclease (DNase) digestion (which exposes the antibody-binding site on the incorporated BrdU molecule) [4]. Cells can also be treated with acid to denature the DNA for labeling, but DNase digestion is generally considered to be superior [5]. Different permeabilization substances can be used, where paraformaldehyde (PFA) and saponin are used to fix and permeabilize the cell, and other compounds such as Triton X-100 (a detergent) or dimethyl sulfoxide (DMSO) are used to aid nuclear membrane permeabilization [6]. Some additional discussion on permeabilization methods can be found later in this protocol. An alternative to BrdU is Iododeoxyuridine (IdU), which substitutes the bromine (Br) group for an iodine (I) group. This approach is preferred for cell cycle analysis in mass cytometry experiments because the presence of IdU can be detected directly on the mass cytometer at 127 daltons (or atomic mass units, AMU), without the need for permeabilization, DNase digestion, or antibody labeling of the compound [1]. Additional protocols on the use of IdU in mass cytometry can be found in Chapter 13.

To perform mass cytometric analysis of cell cycle beyond S-phase, other intracellular factors can be examined, such as Ki67, phosphorylated retinoblastoma protein (pRb), phosphorylated histone H3 (pHH3), and Cyclin B1. More information on these proteins, and their role in mass cytometry cell cycle assays, can be found in Chapter 13. Additionally, cytokines and transcription factors can also be examined.

1.3 Data Analysis of the Bone Marrow Hematopoietic System

In addition to more advanced cytometry technologies, new methodologies in data analysis must be applied in interrogating the hematopoietic system. Because of the complexity of this system, a system-level approach must be undertaken, as manual analysis may fail to capture this complexity. Two elements of computational data analysis are relevant here: dimensionality reduction and clustering. Dimensionality reduction aims to reduce the complexity of HD data to 2 or 3 dimensions, while maintaining HD relationships. A key tool in cytometry is t-distributed stochastic neighbor embedding (tSNE), a nonlinear dimensionality reduction and clustering tool [7]. This tool uses Barnes-Hut approximations [7, 8], used for visualizing large data sets for hundreds of thousands of data points. The Barnes-Hut implementation of the tSNE algorithm (bh-tSNE) has become a routine tool in interpreting HD mass cytometry data [9]. By clustering on phenotyping markers, a two-dimensional plot is generated where similar subsets of cells are grouped more closely together than dissimilar subsets, representing higher-dimensional relationships. While dimensionality reduction allows for a system-level visualization of a dataset, it does not provide an automated method for segregating data into discrete quantifiable populations. To achieve this, a number of clustering approaches can be taken. In particular, FlowSOM [10] is useful for analyzing large and complex datasets, due to the high-speed and scalable processing of large datasets, and sensitivity to rare subsets among clusters of varying densities [11].

A typical workflow using these approaches would involve taking a bone marrow (BM) dataset, running FlowSOM to generate clusters, and then running tSNE to visualize the populations as single cells on a 2D plot. However, we have found that the use of tSNE is typically limited to datasets of up to approximately 10^5 events. When used on datasets above 10^5 cells, data crowding starts to occur which can skew the visualization of data. A limitation of 10^5 cells in our analysis is suboptimal for two reasons. Firstly, the BM contains many rare cell types, including subsets of stem cells such as the linage$^-$SCA-1$^+$c-kit$^+$ cells (LSK, which contain the majority of long- and short-term stem cells) that are estimated to represent less than 0.05% of total bone marrow cells [12]. As such, an analysis of 10^5 bone marrow cells would contain fewer than 50 LSK cells. Secondly, although a limitation of 10^5 cells may be sufficient (though not optimal) for the analysis of a single sample, when multiple samples are combined for analysis, the limit of 10^5 total events still applies. As such, if 10 samples were merged together, then only 10^4 cells from each sample could be used, which would only contain fewer than 5 LSK cells from each sample.

Because of this limitation, we describe a workflow that could circumvent this inherent limitation in tSNE. Firstly, we use FlowSOM to cluster the full (large) dataset (1–3 × 10^6 cells per sample in this protocol, used on datasets of up to 5 × 10^7 cells in our group to date), then we subsample the full dataset to approx. 10^5 cells for dimensionality reduction by tSNE. The tSNE dataset can be interrogated to determine what known cellular population is represented by each specific cluster. Once the cluster identities are known, the larger dataset can be used for numerical and statistical analysis. Furthermore, clusters representing rare subsets (e.g., LSK stem and progenitor cells) which might appear at low frequencies in the original tSNE analysis can be exported from the large dataset, and interrogated in a new, targeted tSNE analysis in more detail.

1.4 Procedure Overview

In this protocol we detail the steps necessary to perform cellular and cell cycle analysis of the murine bone marrow hematopoietic system using mass cytometry. This protocol can similarly be used for BM and cell cycle analysis using fluorescence flow cytometry, with key differences highlighted in the protocol and accompanying notes. A visual overview of this protocol is provided in Fig. 1a–h. Typically we perform IdU measurement by mass detection (not requiring DNase digestion) and perform intracellular permeabilization using methanol to facilitate the labeling of phosphorylated intracellular cell cycle factors pRb, Ki67, and pHH3. Throughout this protocol we have provided procedures notes on labeling of other cytoplasmic or transcription factors, and suggested approaches to combine these protocols to enable optimal staining of each type of intracellular target.

2 Materials

2.1 Animals

1. Use animals according to ethics approvals. In this protocol, we use C57BL/6 female mice between 8 and 12 weeks old.

2.2 Buffers

1. Ultrapure water.
2. Phosphate buffered saline (PBS): filter solution through a 0.2 μm vacuum filter into sterile bottles. *See* **Note 1** for details about avoiding metal contaminants in solutions prepared for mass cytometry.
3. FACS buffer: PBS with 5 mM EDTA and 5% fetal calf serum (FCS) or fetal bovine serum (FBS). Make mixture and store at 4 °C for up to 3 weeks prior to use. Though we have stored FACS buffer for longer, long-term stability has not been tested.

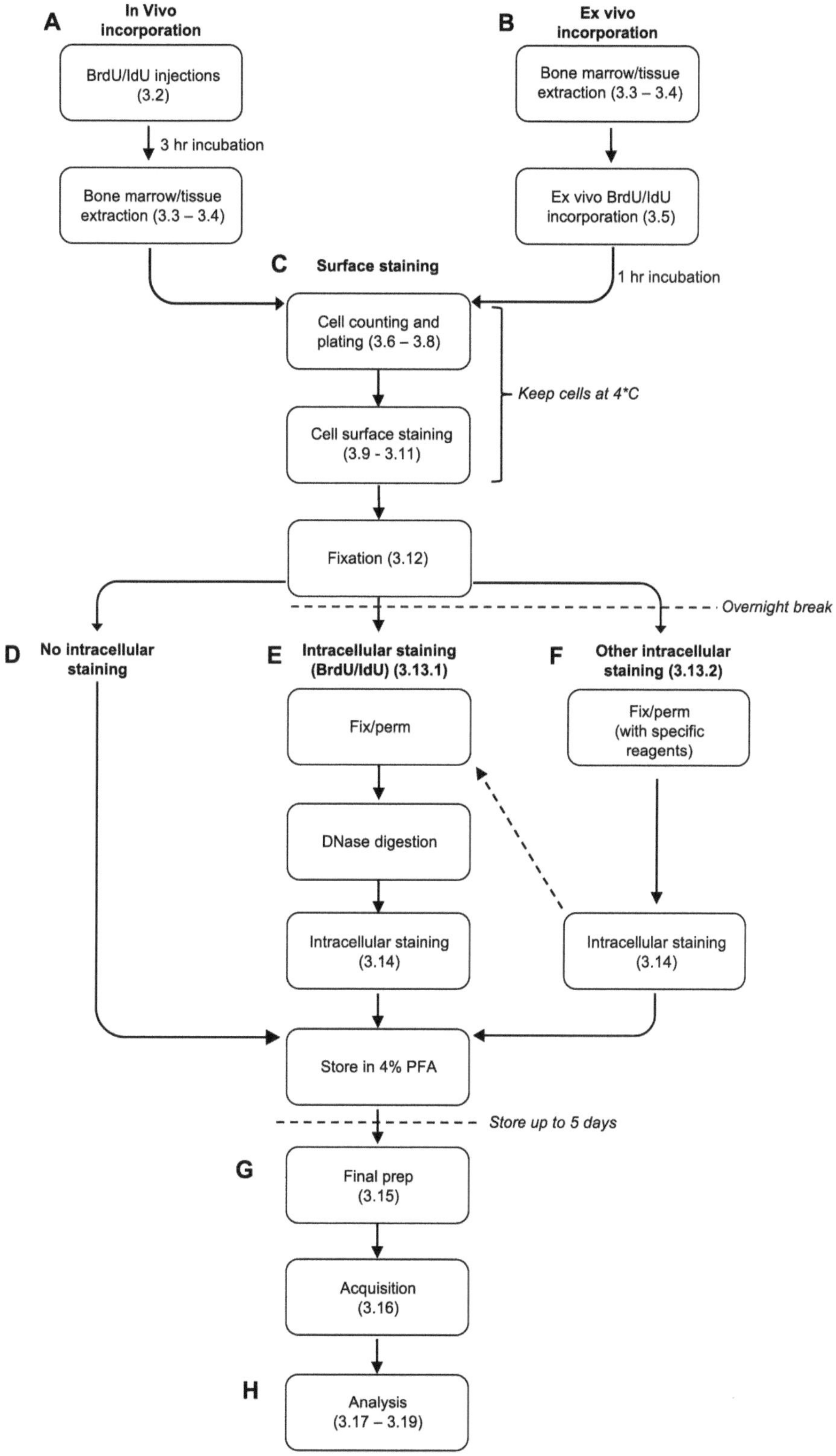

Fig. 1 Procedure overview. An overview of the procedures described in this protocol. (**a**) and (**b**) indicate the pathways for in vivo or ex vivo incubations respectively for BrdU or IdU. This is followed by (**c**) surface staining and fixation, and optional pathways with (**d**) no intracellular staining, (**e**) intracellular antibody labelling of BrdU/IdU (if required), or (**f**) intracellular staining for nuclear factors or other targets. Subsequently, (**g**) cells are prepared and acquired by mass cytometry, and (**h**) analysed

4. Fixative containing formaldehyde: A 4% solution of paraformaldehyde (PFA, referred to as formaldehyde when in solution form) or Fixation Buffer (BioLegend).

2.3 Thymidine Analogues

Two approaches are possible for cell cycle analysis. The first is antibody labeling of bromodeoxyuridine (BrdU), and the second is antibody labeling or mass detection of iododeoxyuridine (IdU).

1. Bromodeoxyuridine (BrdU) powder (Sigma, molecular weight = 307.10 g/mol, store at −30 °C).
2. Iododeoxyuridine (IdU) powder (Sigma, molecular weight = 354.10 g/mol, store at −30 °C).
3. If using IdU, dimethyl sulfoxide (DMSO, store at RT) is also required.

2.4 Reagents for Intracellular Antibody Labeling of BrdU/ IdU

These reagents are required for intracellular antibody labeling of incorporated BrdU or IdU. If measuring IdU by mass detection (as opposed to antibody labeling) in mass cytometry, these reagents are not required. Many of these reagents, including fixative, are provided in "BrdU kits" from various suppliers (*see* **Note 2**).

1. Cytofix/Cytoperm buffer (BD Biosciences).
2. Permeabilization buffer containing saponin: Perm/Wash Buffer (BD Biosciences).
3. Cytoperm Permabilization Buffer Plus (BD Biosciences, 1×, store at 4 °C).
4. Deoxyribonuclease (DNase) I from bovine pancreas. Store powder at −30 °C. Stock solutions can be made up in PBS (*see* below), filtered, and frozen at −80 °C if possible, or at −30 °C.
5. Metal-conjugated (or fluorophore-conjugated, for flow cytometry) anti-BrdU antibodies (clone 3D4, purified or fluorophore-conjugated format from BD Biosciences or BioLegend). Store at 4 °C.

2.5 Other Intracellular Targets

When labeling intracellular targets, the optimal method of fixation and permeabilization must be considered. For example, optimal labeling of intracellular phosphorylated targets (e.g., pRb, pHH3) is achieved following methanol permeabilization. Transcription factors are optimally labeled following permeabilization using special buffers, but some (e.g., Ki67) are sufficiently labeled following methanol permeabilization. The reagents required for different kinds of intracellular targets are detailed in **Note 3**. Various approaches to intracellular staining can be combined to optimally stain each type of intracellular factor, and the approaches are described in Subheading 3.13.

In our standard mass cytometry assays, we measure IdU by mass detection, and perform methanol permeabilization for intracellular labeling of intracellular cell cycle factors.

1. Intracellular antibodies: Anti-pRb (clone J112-906, BD Biosciences, metal conjugated in house), anti-Ki67 (clone B56, BD Biosciences, metal conjugated in house), and anti-pHH3 (clone HTA28, BD Biosciences, metal conjugated in house). We have found the GNS-1 clone of Cyclin B1 only labeled monocytes in mouse BM, and so has been excluded from our typical assays.
2. Methanol (100%, store at 4 °C) for permeabilization. Warning: Methanol will rapidly degrade most fluorophores, although FITC and Alexa-Fluor dyes are resistant to methanol [13]. Metal tags used in mass cytometry are unaffected by methanol.

2.6 Preparation of DNase for Antibody Labeling of IdU or BrdU

A key requirement for antibody labeling of BrdU or IdU is the digestion of cellular DNA to reveal the binding epitope. This is not required for measurement of IdU directly by mass detection in mass cytometry. The activity of DNase is measured in Kunitz units/mL, not simply mg/mL. Although typical concentrations are stated for DNase products (for example, DNase I from Sigma, DN25, has a typical concentration of >400 units/mg of protein, where protein is >85% of the DNase powder), the units/mg and percentage of protein are lot/batch-specific. In this protocol, we used lot #SLBF7798V of DNase (Sigma, DN25) which has 787 units/mg protein, and protein at 87% of the DNase powder, giving 684.69 units/mg of DNase powder. Ideally, DNase should be titrated after the stock solution has been stored frozen, to account for any specific loss of activity due to freezing. As DNase loses activity over time, even when stored frozen, it is important that DNase is titrated for this specific type of assay, with repeated titration at least once a year to ensure consistent performance, or when moving to a new batch of DNase. As an example, we titrated fresh DNase I (Sigma, DN25, lot #SLBF7798V) on cells that had incorporated BrdU to identify an optimal concentration of DNase to digest DNA and expose binding epitopes that could be labeled to provide maximal signal (Fig. 2). We found this to be 1 mg/mL, which equates to 684.69 units/mL, or 68.469 units per 100 μL cell suspension (*see* **Note 4**).

1. DNase: Dilute in PBS to a desired stock concentration (3 mg/mL in our case) and filter into 5 mL tubes through a 0.2 μm syringe filter.
2. Freeze aliquots at −80 °C if possible, or −30 °C.
3. Thaw one aliquot and titrate reagent to determine optimal working concentration (1 mg/mL in our case).
4. On the day of use, thaw aliquots and dilute to working solution in PBS.

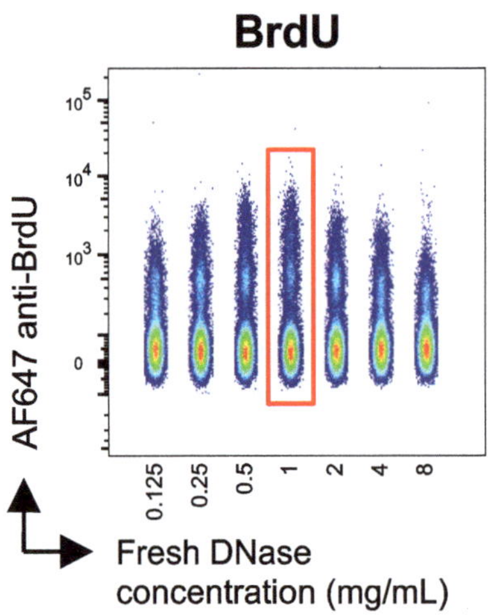

Fig. 2 Titration of DNase for antibody labelling of BrdU and IdU. Titration of DNase concentration showing the resultant antibody labelling of BrdU. Box indicates the maximal labelling by the antibody

2.7 Cell Preparation and Labeling

1. 0.4% Trypan blue solution for cell counting.
2. For red cell lysis: Red blood cell (RBC) lysis buffer (BD Pharm Lyse, BD Biosciences). 10× solution can be stored at 4 °C. 1× solution can be made up by diluting in ultrapure water and stored at 4 °C for up to 1 week prior to use.
3. Cisplatin viability dye (Fluidigm, 5 mM stock). Make up stock solution into 5 μL aliquots, and store at −30 °C. Dilute stock to 5 μM (1/1000) in PBS on the day of use. We have found that stock aliquots will tolerate at least three freeze-thaw cycles.
4. If performing flow cytometry: Viability stain, such as the UV-excitable Live/Dead blue viability stain (Thermo Fischer Scientific).
5. Metal-conjugated antibodies (or fluorophore-conjugated antibodies for flow cytometry).
6. DNA Iridium intercalator (Fluidigm, 500 μM). Make up stock solution into 5 μL aliquots, and store at −30 °C. Dilute stock 1/2000 (0.25 μM for non-permeabilized cells) or 1/4000 (0.125 μM for permeabilized cells) in PBS on the day of use. We have found that stock aliquots will tolerate at least three freeze-thaw cycles, or can be stored at 4 °C for 1 month once defrosted.

2.8 Equipment

1. 50 mL syringes and 0.2 μm syringe filters.
2. Syringe (5 mL) and needle (30 ½ gauge).
3. Forceps.
4. 15 mL tubes or 5 mL tubes.
5. Centrifuge.

6. P10, P50, P200, and P1000 pipettes and corresponding tips.
7. 96-well plate (or Eppendorf tubes).
8. Hemocytometer and/or automated cell counter (Sysmex XP-100 or similar).
9. 40 μm filter paper or cell strainer caps (Falcon 5 mL round bottom polystyrene test tube with cell strainer snap cap).
10. For mass cytometry: 1.5 mL Eppendorf tubes or 5 mL BD FACS tubes or similar.
11. For flow cytometry: 1.2 mL dilution tubes.
12. Mass cytometer (CyTOF, CyTOF 2, or Helios; Fluidigm).
13. EQ calibration beads (Fluidigm).
14. For flow cytometry: Flow cytometer.

2.9 Software

1. FlowJo analysis software (FlowJo LLC) or similar (www.flowjo.com).
2. R software (www.r-project.com).
3. RStudio (https://www.rstudio.com).
4. cytofkit R package (https://github.com/JinmiaoChenLab/cytofkit).

3 Methods: Laboratory

3.1 Preparation of Reagents for Cell Cycle Analysis

We have found that the 3D4 clone of anti-BrdU antibody will label IdU at a similar level to BrdU (Fig. 3a), with a similar antibody titration curve (Fig. 3b). While IdU can be detected using antibody labeling, it is unnecessary in mass cytometry, as IdU is detectable on the mass cytometer with an atomic mass of 127 daltons [1]. This makes IdU ideal for mass cytometry applications, although BrdU (requiring antibody labeling) might be more suitable for some studies, such as when using cryopreserved samples that have already been exposed to BrdU, or when constrained by ethics approval requirements.

A direct comparison of antibody-labeled BrdU and IdU reveals similar staining patterns, with comparable levels of incorporation on a range of cell types (Fig. 3c). As such, antibody labeling of BrdU- or IdU -treated cells can be performed in either a flow cytometry (fluorophore conjugated) or mass cytometry (metal conjugated) context, following fixation, permeabilization, and DNase treatment. Historically, we have used the BrdU Flow Kit (BD Biosciences) to perform this assay, although the components of the kits can be acquired separately. While we have generally found consistency in the incorporation of IdU when detected by mass, and BrdU when detected by antibody labeling, this should

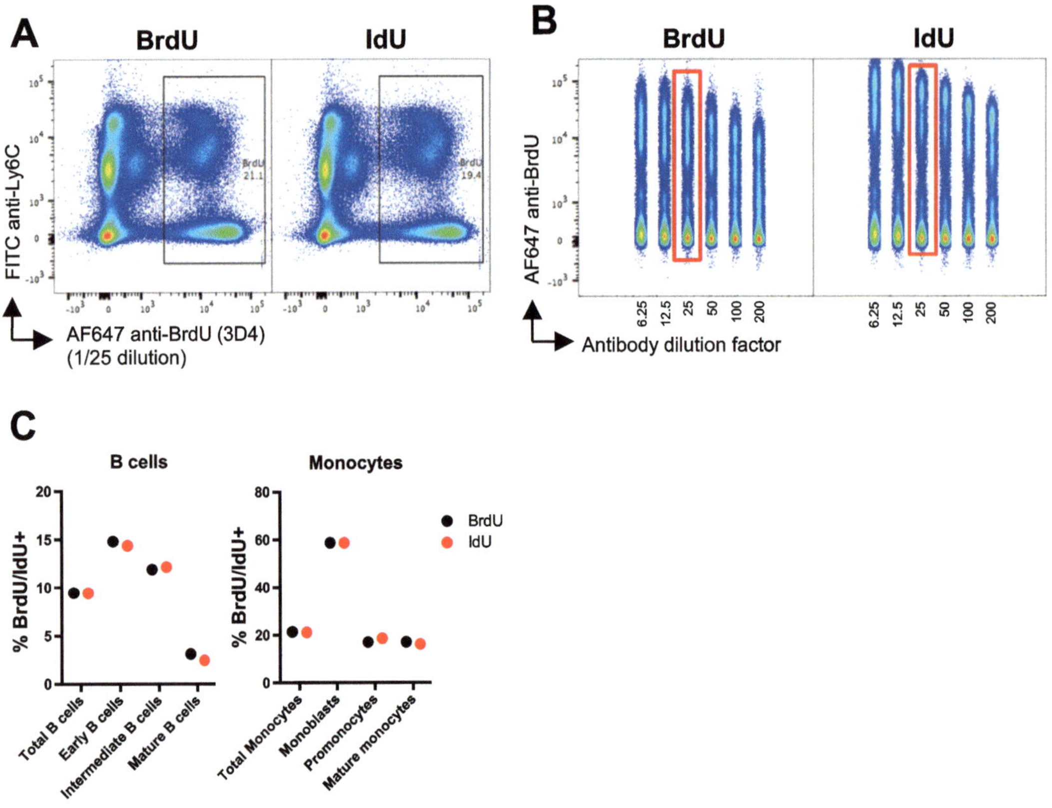

Fig. 3 Antibody labelling of BrdU and IdU. (**a**) Antibody labelling of cells following incubation with BrdU or IdU. Cells were labelled with the 3D4 clone of AF647 anti-BrdU. (**b**) Titration of AF647 anti-BrdU antibody on cells with incorporated BrdU or IdU. Box indicates optimal concentration. (**c**) Comparison of BrdU or IdU incorporation in different cell types

be validated, as there may be differences between antibody- and mass-based detection.

Assays involving BrdU or IdU can be performed in vivo (typically through intraperitoneal (*i.p.*) injection) or ex vivo (incubation of cells in culture) (*see* Fig. 4). While the in vivo approach is preferred, as cells continue through cycle in a normal environment, experimental limitations sometimes prevent this approach. Typically, *i.p.* injections of IdU or BrdU are performed 1–3 h prior to harvest. Ex vivo incubations, on the other hand, are typically performed for up to 1 h after extraction, but longer ex vivo incubations are not recommended (*see* **Note 5**, and further protocols in Chapter 13). Incorporation of BrdU/IdU is fairly internally consistent across the entire hematopoietic landscape (Fig. 4a), where more proliferative progenitors incorporate BrdU/IdU, compared to mature cells. However, we have found that more immature cells incorporate BrdU/IdU in an ex vivo context (10 μM IdU, 1 h), than immature cells in an in vivo context (400 μL injection of 8 mM IdU, 3 h prior to harvest), despite the shorter incubation

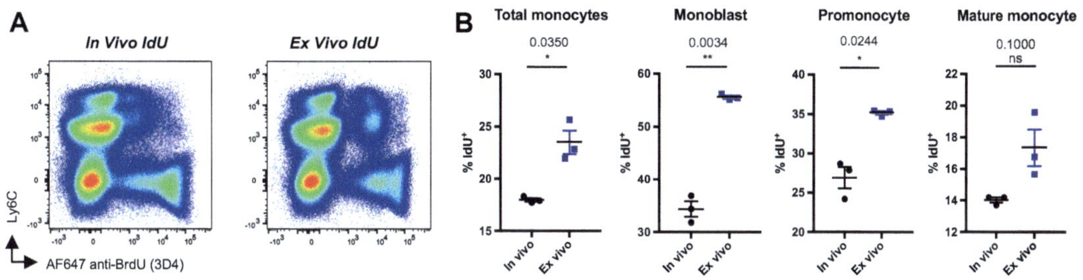

Fig. 4 In vivo vs. ex vivo incorporation of BrdU or IdU. (**a**) Antibody labelling of IdU following in vivo (injection) or ex vivo (incubation in culture) incorporation of IdU. (**b**) Percentage of cells in various monocyte differentiation stages incorporating IdU after 3 h in vivo, or 1 h ex vivo

time (Fig. 4b). This may be due to differences in the local concentration of IdU between the two environments. Alternatively, it is possible that the cells placed in culture become stimulated, leading to increased incorporation.

3.1.1 Option 1: IdU Preparation for Flow or Mass Cytometry

1. Dissolve IdU powder in DMSO at 576.52 mg/mL (1.62 M) (*see* **Note 6**). Agitate to ensure powder fully dissolves in DMSO. Make up aliquots of 12 μL (enough stock for five injections plus extra) and store at −80 °C until use.
2. IdU *i.p.* Injections: thaw stock IdU solution and dilute 1/200 in PBS (final concentration 8.14 mM, 2.88 mg/mL) and filter using a 0.2 um syringe filter. Each mouse (8–12 weeks old) will receive a 400 μL injection of 8.14 mM solution (1.15 mg per dose). If necessary, store any leftover reagent at 4 °C for a short period of time (*see* **Note 7**).
3. IdU ex vivo incubations: Thaw stock IdU solution and dilute 1/1628 in PBS (final concentration 1 mM, or 0.35 mg/mL) and filter using a 0.2 μm syringe filter. Each 3 mL suspension of BM will receive 30 μL of 1 mM solution (final concentration 10 μM). If necessary, store any leftover reagent at 4 °C for a short period of time (*see* **Note 7**).

3.1.2 Option 2: BrdU Preparation for Flow or Mass Cytometry

1. Dissolve BrdU powder in PBS at 10 mg/mL (32.56 mM). Agitate to ensure powder fully dissolves in PBS. Filter through 0.2 μm syringe filter, and make up aliquots of 600 μL (enough stock for five injections plus extra), and store at −80 °C until use.
2. BrdU *i.p.* Injections: thaw stock BrdU solution and dilute 1/2 in PBS (to 5 mg/mL, 16.28 mM). Filter using a 0.2 μm syringe filter prior to use. Each mouse (8–12 weeks old) will receive a 200 μL injection of 16.28 mM solution (1 mg per dose). If necessary, store any leftover reagent at 4 °C for a short period of time (*see* **Note 7**).

3. BrdU ex vivo incubations: Dilute stock solution 1/32.56 in PBS (final concentration 1 mM, or 0.31 mg/mL) and filter through 0.2 μm syringe filter. Each 3 mL suspension of BM will receive 30 μL of 1 mM solution (final concentration 10 uM). If necessary, store any leftover reagent at 4 °C for a short period of time (*see* **Note 7**).

3.2 If Performing Injections (In Vivo) with IdU/ BrdU

If performing injections of BrdU/IdU for in vivo incorporation, follow these steps. Otherwise, proceed to Subheading 3.3.

1. For BrdU, inject each mouse (8–12 weeks old) with 200 μL of 16.28 mM solution (5 mg/mL, 1 mg per dose).
2. For IdU, inject each mouse (8–12 weeks old) with 400 μL of 8.14 mM solution (2.88 mg/mL, 1.15303 mg per dose).

3.3 Terminal Anesthesia

1. Use an appropriate method of terminal anesthesia, such as inhalation of isofluorane, inhalation of CO_2, injection of Ketamine/Xylazine, or injection of Avertin anesthetic [14] according to relevant ethics approvals.
2. If also collecting blood and/or performing a circulatory system perfusion with PBS (*see* **Note 8**), it is preferable to use a compound that results in complete anesthesia, but also maintains a strong heartbeat, such as Avertin anesthetic.

3.4 Bone Marrow Cell Isolation

In our protocol, we flush femurs and/or tibias using PBS. Other protocols describe grinding whole femurs to release cellular contents, but this is not considered here.

1. Isolate single femur.
2. Remove skin and muscle with a scalpel or scissors (Fig. 5a). Using a scalpel, locate the proximal epiphysis and the femoral head, and separate them from the hip by cutting the connecting femoral ligaments. Once detached from the hip, use a scalpel to locate the distal epiphysis and use forceps to overextend the knee joint, effectively separating the femur from the rest of the leg.
3. Remove ends (proximal and distal) of the femur, as close to the epiphyseal ends as possible while exposing the hollow center of the femur (Fig. 5a).
4. Flush marrow out of the femur (into a collection tube) with 3 mL of PBS (Fig. 5a). Use a 6 mL syringe and a 30 ½ gauge needle to flush the femurs, and preferably collect fluid into 5 mL tubes, or 15 mL tubes if necessary. Flush with 1.5 mL of PBS from one end, and then turn the bone around and flush with 1.5 mL of PBS from the other end (3 mL total). Larger volumes can be used for flushing if necessary, which may aid inexperienced users. It helps to move the needle around inside the femur to loosen the marrow.

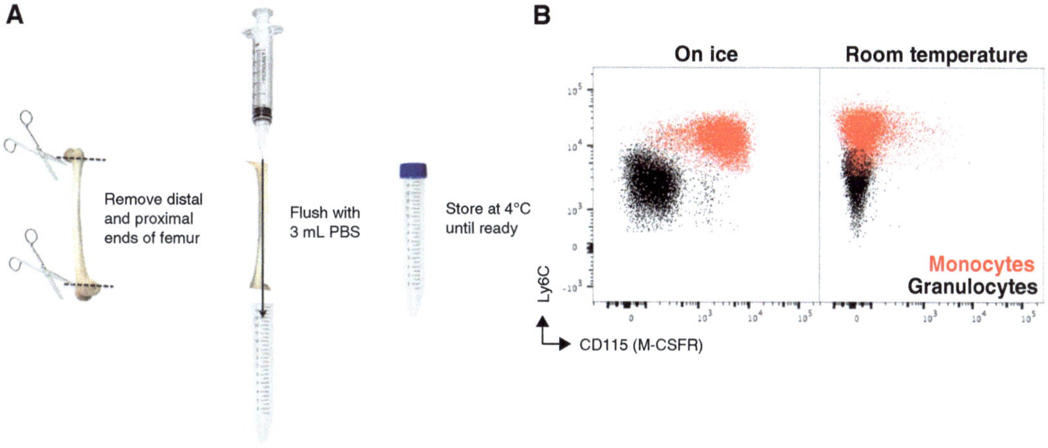

Fig. 5 Bone marrow isolation from femurs. (**a**) Diagrammatic depiction of bone marrow removal from the femur by flushing. (**b**) Effects of sample storage on surface expression of CD115 (M-CSFR). Representative CD115 staining from different experiments where samples were kept on ice (left panel) or at RT (right panel) prior to staining

5. Once all the BM has been flushed, the bone should appear completely white.
6. If performing ex vivo incubation of cells with BrdU/IdU: Proceed to Subheading 3.5 for ex vivo incubation. Do not centrifuge the cells, and do not store them at 4 °C before ex vivo incubation, as this will disturb the cell cycle.
7. If following in vivo BrdU/IdU incorporation, or if not performing BrdU/IdU incorporation: Keep on ice until ready to process, as samples are stable at 4 °C in this state.
8. Centrifuge at 500 × *g* for 5 min at 4 °C, and discard supernatant.
9. Resuspend cells in 1 mL of FACS buffer and keep marrow suspension at 4 °C to prevent surface marker turnover (*see* **Note 9** and Fig. 5b).

3.5 If Performing Ex Vivo Incubation of Cells with BrdU/IdU

If performing ex vivo incubation of cells with BrdU/IdU, follow these steps. Otherwise, proceed to Subheading 3.6.

1. Add 10 μL of 1 mM IdU/BrdU solution per 1 mL suspension of cells (final concentration 10 μM). In this example, add 30 μL of 1 mM solution to the 3 mL suspension of BM cells.
2. Incubate for 1 h at 37 °C.
3. Centrifuge at 300 × *g* for 5 min at 4 °C, and discard supernatant.
4. Resuspend cells in 1 mL of FACS buffer and keep marrow suspension at 4 °C to prevent surface marker turnover (*see* **Note 9** and Fig. 5b).

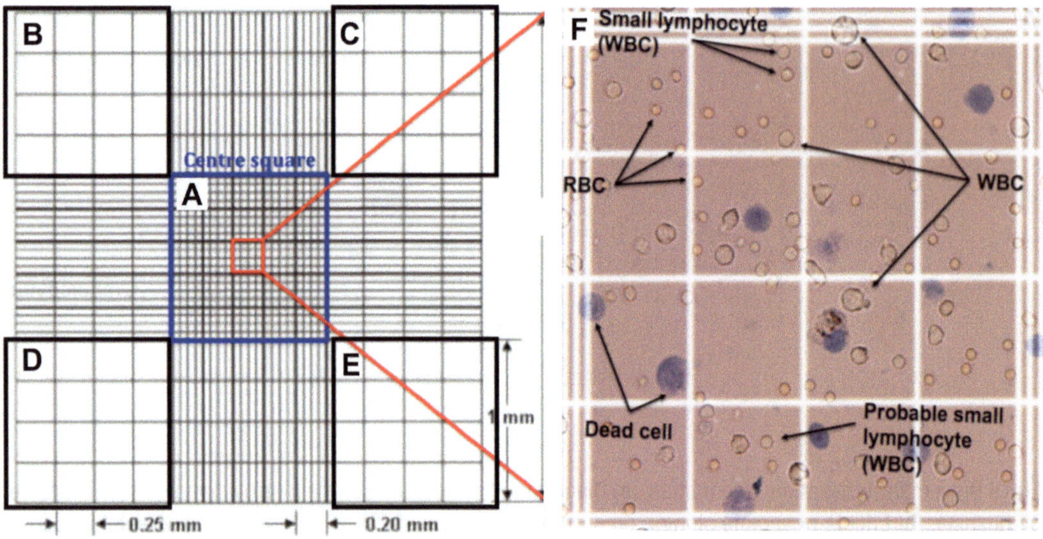

Fig. 6 Cell counting. A microscopic view of a haemocytometer counting grid. (**a**) the large centre square (blue) with additional gridlines, where counting is performed in this protocol. (**b–e**) additional squares for counting to ensure statistical robustness, if required. (**f**) Magnified view of a segment of the centre square showing a variety of cell types including large white blood cells (WBC), small WBCs, red blood cells (RBC), and dead cells, the nuclei of which stain with trypan blue due to the leakiness of the cell membrane, which also makes these cells appear much larger than live cells

3.6 Cell Counting

Cell counting is a critical step to determine the number of live leukocytes (white blood cells, WBC) and erythrocytes (red blood cells, RBC) per femur, and is accurately performed using a microscope and hemocytometer (Fig. 6a–e). This step enables the distribution of a specific number of cells per well in the staining plate. Some researchers prefer to perform an erythrocyte (or red blood cell, RBC) lysis prior to counting, as they are concerned about misidentifying erythrocytes as leukocytes during counting. This RBC lysis removes all erythrocytes prior to counting, resulting in easier counting of leukocytes. However, we have found that differentiating leukocytes from erythrocytes is straightforward at a higher magnification based on size (erythrocytes are smallest), color (erythrocytes are red in color), and birefringence (leukocytes exhibit birefringence, erythrocytes do not) (Fig. 6f). Additionally, RBC lysis prior to cell staining will have a small impact on CD115 staining levels (*see* **Note 9**). To circumvent the need for RBC lysis prior to staining, we count the number of leukocytes (and erythrocytes) in the center square only (Fig. 6a), at a high magnification. Alternatively, an automated cell counter can be used to speed up counting. However, care must be taken to ensure that any automated cell counter is accurately performing counts, as most automated cell counters are calibrated to human blood cells (*see* **Note 10**).

3.6.1 Counting Using a Hemocytometer

1. Dilute some 0.4% Trypan blue stock ½ in PBS (i.e., 1 mL trypan blue with 1 mL PBS). Certain cells (for instance, some cells from the spleen after RBC lysis) appear to be more sensitive to Trypan blue, exhibiting greater levels of cell death when counted in stock solution.
2. Add 10 μL of the 1 mL cell suspension to 40 μL of FACS buffer (1/5 dilution), then add 10 μL of this solution to 10 μL of 0.4% trypan blue (final dilution factor 1/10).
3. Take 10 μL of this mixture and load onto the hemocytometer.
4. Place hemocytometer under the microscope.
5. Find the counting grid.
6. Count the total number of *LIVE* (clear) and *DEAD* (blue) leukocytes in the *large center square* ("A," blue in Fig. 6a) which consists of 5 × 5 smaller squares (edges marked by triple lines in Fig. 6f), each of which contains 4 × 4 smaller squares (edges marked by single lines in Fig. 6f).
7. Count the total number of erythrocytes in the same area, if desired.
8. If fewer than 100 total live leukocytes are counted in the center square, then count the cells in the other large squares progressively, making sure each square is completely counted (Fig. 6b–e) until at least 100 live leukocytes have been counted in total.
9. Calculate cells/sample using the following formula:

$$\frac{\text{Cells}}{\text{sample}} = \frac{\text{Live cells counted}}{\text{Large squares counted}} \times 10^4 \times \text{dil.factor} \times \text{vol.}(\text{mL}) \text{ of sample}$$

3.6.2 Counting Using an Automated Cell Counter, Sysmex XP-100

1. Turn machine ON.
2. While the machine starts up, transfer 100 μL of sample to a 1–1.5 mL tube.
3. Raise the sample tube to the probe so that the probe is submerged in the sample fluid.
4. Press the GREEN start button.
5. Hold sample tube in place (with the probe submerged) during aspiration until you hear beeping, then remove the tube. The machine will aspirate approx. 60 μL of sample. Leukocyte, or white blood cell (WBC) counts are given as $n \times 10^9$ cells/L ($= n \times 10^6$ cells/mL).
6. Calculate cells per sample (*see* **Note 10**):

$$\text{WBC per sample} = \text{WBC count} \times \left(10^6 / \text{mL}\right) \times \text{vol.}(\text{mL}) \text{ of sample}$$

3.7 Required Cell Numbers

1. Calculate volume containing the required cell numbers: $V_{req} = (C_{req}/C_{sample}) \times V_{sample}$.
2. Vol (required, mL) = (cells required/cells in sample, total) × Vol (sample total, mL).
3. Example: If we want 3 × 10⁶ WBCs from a bone marrow sample with 1.4 × 10⁷ WBCs in 1 mL of FACS buffer.

$V_{req} = ?$

$C_{req} = 3 \times 10^6$

$C_{sample} = 1.4 \times 10^7$

$V_{sample} = 1$ mL

$V_{req} = (C_{req}/C_{sample}) \times V_{sample}$

$V_{req} = (3 \times 10^6 \text{ cells}/1.4 \times 10^7 \text{ cells}) \times 1$ mL

$V_{req} = 0.214$ mL

$V_{req} = 214$ μL (i.e., 214 μL of sample contains 3×10^6 WBCs)

3.8 Plating Cells

Using a pipette, transfer the necessary volume for a desired number of cells into relevant wells of a U- or V-bottom 96-well plate.

1. For mass cytometry: Typically stain 3×10^6 cells per sample (up to 6×10^6 if looking at stem/progenitors from the bone marrow).
2. For flow cytometry: Typically stain 1×10^6 cells per sample, due to greater recovery of cells in flow cytometry compared to mass cytometry (up to 3×10^6 if looking at stem/progenitors from the bone marrow).

3.9 Viability Stain and Fc Block for Mass Cytometry

If performing flow cytometry, *see* modifications in **Note 11.**

1. Centrifuge at 300 × *g* for 3 min at 4 °C.
2. Resuspend in 50 μL of 5 μM cisplatin mix (5 μM = 1/1000 dilution of 5 mM stock in PBS).
3. Incubate for 5 min at RT.
4. Top up with 200 μL FACS buffer.
5. Centrifuge at 300 × *g* for 3 min at 4 °C.
6. Resuspend in 50 μL of purified anti-mouse CD16/32 (1/100) (*see* **Note 12**).
7. Incubate for 30 min at 4 °C.
8. Top up with 200 μL FACS buffer.
9. Centrifuge at 300 × *g* for 3 min at 4 °C.

3.10 Surface Staining

1. Resuspend in 50 μL of surface antibody master mix (metal-conjugated antibodies for mass cytometry, fluorophore-conjugated antibodies for flow cytometry) in FACS buffer.

2. Incubate for 30 min at 4 °C.
3. Top up with 200 μL FACS buffer.
4. Centrifuge at 300 × *g* for 3 min at RT.
5. Resuspend in 250 μL of FACS buffer.
6. Centrifuge at 300 × *g* for 3 min at RT.
7. *Repeat* **steps 5** and **6**.

3.11 Optional RBC Lysis

For many applications, leukocytes are the desired target population, and erythrocytes are not relevant to the analysis. In this case, there is the option of performing a RBC lysis prior to fixation. Because the acquisition rate in mass cytometry is only 400 events/sec, any metal signals present on the erythrocytes will result in them being counted by the mass cytometer. We have found that the DNA intercalator will bind to a number of erythrocytes at levels that are sufficient to trigger detection (DNA1^{lo} staining on Ter119^{+} cells, *see* Fig. 7b), and will bind to erythrocyte progenitors at high levels (DNA^{hi} staining on Ter119^{+} cells, *see* Fig. 7b). Because erythrocytes are approximately 50% of the cells present in the BM, the acquisition time required to collect a desired number of leukocytes will be doubled if erythrocytes are retained. Preferably, RBC lysis should be performed after surface staining, to reduce the impact on expression of some cellular markers, such as CD115.

1. Resuspend in 50 μL of 1× RBC lysis buffer.
2. Incubate for 10 min at RT.
3. Add 250 μL FACS buffer and mix.
4. Centrifuge at 300 × *g* for 3 min at 4 °C.
5. *Repeat* **steps 3** and **4**.

3.12 Fixation

If no intracellular permeabilization and staining is required, we recommend that samples be incubated in fixative overnight at a minimum, to improve sample stability when the samples are transferred to ultrapure water. Samples can be stored for up to 5 days suspended in fixative before proceeding to "final cell processing" and "acquisition." If performing any intracellular permeabilization, we recommend that samples be fixed, prior to being incubated in a fix/perm buffer, as the prior fixation improves sample recovery and cellular structure in our experience. If running flow cytometry, samples should only be fixed for 10–20 min where possible (subject to any relevant biosafety approvals), and stored in FACS buffer for up to 3 days before proceeding to "final cell processing."

1. Resuspend samples in 50–100 μL of 4% PFA or fixation buffer (*see* **Note 13**).
2. If no intracellular permeabilization is required: Incubate overnight at 4 °C, and proceed to Subheading 3.15 (final cell processing for mass cytometry).

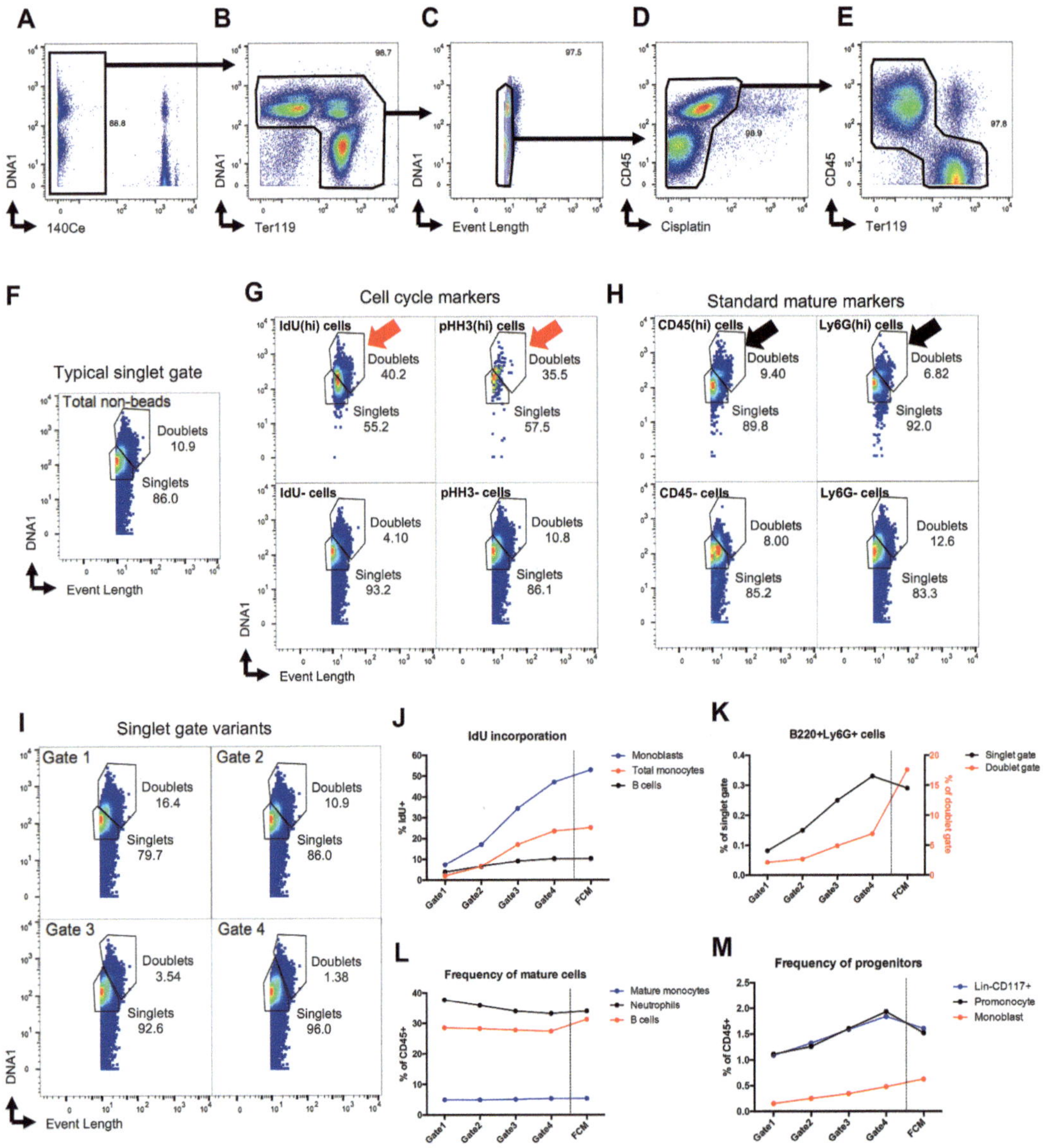

Fig. 7 Initial gating and singlet discrimination of bone marrow mass cytometry data. (**a**) Initial gating of non-beads, followed by (**b**) DNA⁺ and/or Ter119⁺ cells, (**c**) singlets, (**d**) live cells, and (**e**) CD45⁺ and/or Ter119⁺ cells. (**f**) Using a traditional singlet gate results in the exclusion of some cycling cells, that exhibit higher DNA levels. (**g**) The distribution of cells that are positive for markers of cell cycle in the 'doublet' and 'singlet' gates, compared to (**h**) the distribution of cells positive for generic markers. (**i**) Adjustment of the standard singlet gate allows for (**j**) retention of proliferating cells, whilst (**k**) keeping known doublets to a low level. These singlet / doublet gate placements have little impact on the frequency of (**l**) mature (non-cycling cells), but (**m**) are critical for the retention of progenitor cells, which have a higher frequency of cells in cycle

3. If proceeding to intracellular permeabilization on the same day: Incubate for 20 min at RT (subject to appropriate biosafety approvals), top up with 150 μL FACS, and proceed to Subheading 3.13 (intracellular permeabilization).

Table 1
Perm approaches summary

Intracellular target	Examples	Perm method	Harshness (1 = low, 4 = high)
Cytoplasmic factors and cytokines	IFN-γ, TNF	Cytoplasmic permeabilization	1
Nuclear and transcription factors	IRF8, FoxP3, Ki67	Nuclear and transcription factor (incl. FoxP3) permeabilization	2
Incorporated BrdU	BrdU, IdU	BrdU permeabilization	3
Phosphorylated targets	pSTAT1, pRb, pHH3, Cyclin B1	Methanol permeabilization	4

4. If proceeding to intracellular permeabilization on a subsequent day: Incubate for 20 min at RT, top up with 150 μL FACS, centrifuge (800 × *g*, 5 min, RT), resuspend in 200 μL of FACS buffer, keep overnight, and proceed to Subheading 3.13 (intracellular permeabilization).

3.13 Intracellular Permeabilization

After the samples have been surface stained, there are a number of options for how to perform intracellular staining. Below we have described approaches for intracellular labeling of BrdU/IdU, as well as intracellular labeling of pRb, Ki67, and pHH3. We have provided details of permeabilization procedures for intracellular labeling of cytokines and transcription factors in **Note 14** and Table 1. These approaches can be used in combination for optimal labeling of different types of intracellular targets, by performing one type of intracellular permeabilization and staining procedure (e.g., permeabilization, DNase digestion and anti-BrdU/IdU labeling) followed by another (e.g., methanol permeabilization and anti-pRb labeling). Considerations for combinations of staining approaches, and their compatibilities, are detailed in **Note 14**.

3.13.1 Fixation, Permeabilization, and DNase Digestion for Anti-BrdU or Anti-IdU Staining

These steps are required for antibody labeling of IdU or BrdU, but are not required for the detection of IdU by mass in mass cytometry.

1. Centrifuge at 800 × *g* for 5 min at RT, and discard supernatant (all centrifuge steps after fixation at 800 × *g* for 5 min).
2. Permeabilization: Resuspend in 50–100 μL of BD Cytoperm buffer (or Cytofix/Cytoperm buffer, *see* **Note 13**).
3. Incubate for 20 min at RT.
4. Top up with 1× BD perm wash buffer.
5. Centrifuge at 800 × *g* for 5 min at RT, and discard supernatant.

6. Resuspend cells in 100 μL of BD Cytoperm Permeabilisation Buffer Plus.
7. Incubate for 10 min on ice or at 4 °C.
8. Top up with BD perm wash buffer.
9. Centrifuge at 800 × *g* for 5 min at RT, and discard supernatant.
10. Refix: Resuspend in 50–100 μL of BD Cytofix/Cytoperm buffer.
11. Incubate for 5 min at RT in the dark.
12. Top up with 1× BD perm wash buffer.
13. Centrifuge at 800 × *g* for 5 min at RT, and discard supernatant.
14. Treat cells with DNase: Resuspend cells in 100 μL of DNase solution (at a predetermined optimal concentration).
15. Incubate for 1 h at 37 °C.
16. Top up with BD perm wash buffer.
17. Centrifuge at 800 × *g* for 5 min at RT, and discard supernatant.
18. *Repeat* **steps 16** and **17**.

3.13.2 Fixation and Methanol Permeabilization for Intracellular Phosphorylated Targets

This procedure is recommended for staining phosphorylated targets, such as pRb and pHH3, and is also suitable (though not optimal) for labeling of Ki67. These steps can be performed after cell fixation (Subheading 3.12), or after DNase digestion (Subheading 3.13.1) and anti-BrdU/IdU labeling (Subheading 3.14) where a second round of staining would be performed after methanol permeabilization to label intracellular pRb, Ki67, and pHH3.

1. Centrifuge at 800 × *g* for 5 min at RT, and discard supernatant (all centrifuge steps after fixation at 800 × *g* for 5 min).
2. Resuspend the pellet in residual volume (using a small volume pipette, or agitate the plate). It is critical that the sample is resuspended in residual volume, else the addition of methanol may cause clumping.
3. Slowly add 150 μL ice-cold 100% methanol.
4. Incubate at 4 °C for 10 min.
5. Top up with 150 μL FACS buffer.
6. Centrifuge at 800 × *g* for 5 min at RT.
7. Resuspend in 250 μL FACS buffer.
8. Centrifuge at 800 × *g* for 5 min at RT, and discard supernatant.
9. *Repeat* **steps** 7 and **8**.

3.14 Intracellular Staining

1. Resuspend in 50 μL of intracellular antibody (Ab) mix (in perm buffer, *see* **Note 15**).
2. Incubate for 45 min at RT.
3. Top up with 150 μL 1× perm buffer.
4. Centrifuge at 800 × *g* for 5 min at RT.
5. Resuspend in 250 μL of 1× perm buffer.
6. Centrifuge at 800 × *g* for 5 min at RT.
7. *Repeat* **steps 5** and **6**.
8. If proceeding to acquisition on a subsequent day: Resuspend in 250 μL of FACS buffer, store at 4 °C overnight (for longer storage, resuspend samples in 4% PFA, and store at 4 °C), and proceed to "final cell processing" (3.16 for mass cytometry, 3.17 for flow cytometry).
9. If proceeding to acquisition on the same day: Resuspend in 250 μL of FACS buffer and proceed to "final cell processing" (3.16 for mass cytometry, 3.17 for flow cytometry).

3.15 Final Cell Processing for Mass Cytometry

If performing flow cytometry, *see* **Note 16**.

1. Centrifuge at 800 × *g* for 5 min at RT.
2. Resuspend in 100 μL of DNA intercalator mix. If samples are fixed but not permeabilized; use 2% PFA + 0.25 μM iridium intercalator (1/2000 of 500 μM stock) in PBS; if samples are fixed and permeabilized; use 2% PFA + 0.125 μM iridium intercalator (1/4000 of 500 μM stock) in PBS.
3. Incubate for at least 20 min at RT.
4. If running samples immediately, proceed with instructions below. If acquiring samples on a subsequent day, store samples at 4°C for up to 1 week before washing. If acquiring samples longer than 1 week after staining, wash samples (*see* below) and cryopreserve samples in FCS with 10% DMSO.
5. Top up with 150 μL of ultrapure water.
6. Centrifuge at 800 × *g* for 5 min at RT.
7. Resuspend in 250 μL of ultrapure water.
8. Centrifuge at 800 × *g* for 5 min at RT.
9. *Repeat* **steps** 7 and **8**.
10. Resuspend in 250 μL of ultrapure water.
11. Transfer each sample through 40 μm filter cap into a FACS tube or 1.5 mL Eppendorf Tube.
12. Count using a hemocytometer (no requirement for trypan blue) and add ultrapure water to sample to achieve desired concentration.

13. Total cells should be at a concentration of 1×10^6 per mL for a typical CyTOF/Helios instrument.
14. Some reports suggest that cell concentration can be up to 2×10^6 per mL on the Helios.
15. Example: Cells/sample = cells/squares x d.f. $\times 10^4 \times$ vol (mL); cells/sample = $400/1 \times 1 \times 10^4 \times 0.250$ mL; cells/sample = 1×10^6 cells in 250 μL. Add 750 μL MilliQ water to sample to achieve 1×10^6 cells in 1 mL.
16. Calculate 11% of the volume of the existing sample, and add this volume of EQ beads to the sample. For example: For 1 mL of sample, add (11 × 1 mL/100=) 0.11 mL of EQ beads. The total volume will now be (1 mL + 0.11 mL=) 1.11 mL, where 10% of the solution is EQ beads.

3.16 Acquisition on Mass Cytometer

Acquisition on a mass cytometer is described in other protocols in this collection, and therefore will not be addressed specifically in this protocol.

3.17 Data Cleanup and Preparation and Singlet Discrimination

A number of steps are necessary to prepare raw data for analysis. Initially data recorded over time should be normalized using either the CyTOF acquisition software or the Matlab package [15] (https://github.com/nolanlab/bead-normalization) to mitigate the impact of signal drift on the data. Subsequently, the EQ beads should be gated out, and DNA^+ cells selected (Fig. 7a, b). In cases where both leukocytes (DNA^+) and erythrocytes (DNA^-) are being investigated, DNA should be plotted against an erythrocyte marker, such as Ter119, so that $DNA^+Ter119^-$, $DNA^+Ter119^+$, and $DNA^-Ter119^+$ cells can be selected (Fig. 7b). The next key step is the elimination of "doublets." Most doublets in mass cytometry occur because of the fusion, or partial fusion, of two ion clouds (representing two distinct cells). Because of this, most doublets are expected to exhibit a long event length, and higher DNA content than singlets. As such, these can be removed by plotting "event length" vs. "DNA," to exclude cells that are $Length^{hi}DNA^{hi}$ (Fig. 7c). Subsequently, live ($cisplatin^-$) cells (Fig. 7d) and $CD45^+$ (leukocyte) and/or $Ter119^+$ (erythrocyte) cells should be gated (Fig. 7e). In Fig. 7e, the gating used excludes a $CD45^+$, $Ter119^+$ population that is likely to be doublets.

We have found that cycling cells will increase in DNA intercalator content during S/G2/M-phase, and will exhibit a slightly longer event length compared to non-cycling cells. As such, the typical "singlet" gate (Fig. 7f) will exclude a large number of cycling cells. This can be demonstrated by comparing the Event Length and DNA profile of cells that express markers of cell cycle (pRb^{hi} cycling cells, IdU^+ S-phase, $pHH3^+$ M-phase), compared to cells that are negative for those markers (Fig. 7g, h). Because of this, we tested iterative gating placement for discriminating singlets and doublets in mass cytometry data (Fig. 7i). In each iteration, we looked for the frequency of IdU^+ cells in various lineages and differentiation states (Fig. 7j). Simultaneously, we investigated the frequency of doublet occurrence by looking for known doublet phenotypes, such as cells expressing both Ly6G (neutrophil marker) and B220 (B cell marker), as these two markers are not normally co-expressed, and belong to the most numerous populations in the BM (Fig. 7k). And finally we investigated the overall frequency of various mature (Fig. 7l) or progenitor (Fig. 7m) subtypes. As the gating restriction was relaxed in regard to DNA content, we recovered expected frequencies of IdU^+ cells in different lineages when compared with the same sample analyzed by flow cytometry (FCM) (Fig. 7j). However, modified gating strategies did cause a small increase in the number of known doublet populations ($Ly6G^+B220^+$) (Fig. 7k). Because the Helios system includes improvements in event discrimination compared to older models, we also examined this effect in data generated on the CyTOF 2. Inclusion of IdU^+ events while minimizing the contribution of doublets was more accurate with data generated on the Helios system, compared to data generated on the CyTOF 2 system (data not shown). The modified singlet/doublet gates did not substantially change the frequencies of mature (non-cycling) populations (Fig. 7l), but did increase the number of progenitors, such as monoblasts (that normally have high levels of IdU incorporation), to expected levels (Fig. 7m). Overall, when we profiled the whole leukocyte system, with appropriately set singlet/doublet discrimination gates, the frequency of typical gated populations, as well as the distribution of IdU incorporation in various lineages, was comparable between flow and mass cytometry.

1. All samples run should be normalized using either the CyTOF acquisition software inbuilt normalizer, or in the Matlab normalizer available from the Nolan lab [15] (https://github.com/nolanlab/bead-normalization).
2. After normalization, sample FCS files should be imported into an analysis program, such as FlowJo.
3. Gate out the EQ beads by plotting 140Ce against DNA (Fig. 7a, b).
4. Gate singlets by plotting DNA vs. event length—Taking care to not exclude DNA^{hi} cycling cells (Fig. 7c).
5. Gate live cells by plotting DNA vs. $cisplatin^-$ (Fig. 7d).
6. Depending on the types of cells being analyzed, samples can be further gated on CD45 to isolate leukocytes. If Ter119 is included in the panel (labels erythrocytes), then gates for leukocytes and erythrocytes can be drawn by plotting CD45 vs. Ter119 (Fig. 7e).

3.18 System Analysis Using FlowSOM and tSNE Analysis

A number of computational tools are available to aid in investigating the BM hematopoietic system. Here we describe a workflow that is well suited to the analysis of large and complex datasets. It is beyond the scope of this protocol to comprehensively detail each step in the analytical process, and a number of other protocols exist, including those in this protocol collection, to guide users in such analysis. Here we provide a summary of the steps necessary to perform this analysis, and have provided more comprehensive protocols available at www.github.com/sydneycytometry/capx.

These steps can be performed using the "Cytometry Analysis Pipeline for large and compleX data" (CAPX) scripts using the R programming language, where we have provided scripts and instructions at www.github.com/sydneycytometry/capx. However, these steps can also be performed in user interface software packages such as cytofkit [16] (https://bioconductor.org/packages/release/bioc/html/cytofkit.html) or FlowJo v10.05 (www.flowjo.com). Specific instructions in the use of other computational tools, including cytofkit, are available in this collection of protocols.

1. Export a relevant population of interest (POI) from FlowJo or similar (e.g., singlet, live, $CD45^+$, and/or $Ter119^+$ cells) as CSV (channel value) files or FCS files. *See* **Note 17** for a discussion of data transformation and export options. If working with multiple samples, export the same population from each sample (Fig. 8a).
2. Read CSV or FCS files into R/cytofkit/FlowJo (Fig. 8b).
3. OPTIONAL: If the number of cells in each sample is widely variable due to technical reasons, consider downsampling each sample to a comparable number (*see* **Note 18**).
4. Set selection for data transformation. If using CSV channel values, no transformation is required. If using FCS files, use arcsinh ratio of 15 for transformation of mass cytometry data (*see* **Note 17**).
5. Run FlowSOM on large dataset, and save resultant FCS and/or CSV files (Fig. 8c).
6. Subsample the large dataset after FlowSOM analysis (reimport into program if using cytofkit or FlowJo) to a total of

Fig. 8 (continued) and performing dimensionality reduction using tSNE, (**e**) generate coloured tSNE images using the tSNEplots script. (**c, d**) create new FCS or CSV files of the resultant clustered data. (**f**) Distribution of cell types in the whole BM, with (**g**) the marker expression patterns on those cell types. (**h**) After the removal of erythrocytes (cluster 7) from the dataset, and re-analysis using tSNE, the array of leukocyte cell types are evident, as are (**i**) the marker expression patterns on those cell types. Markers that are indicative of cell cycle, such as (**j**) pRb, Ki67, and (**k**) IdU are highest on immature cells that are low for CD45 and high for CD117. tSNE plots in (**k**) are from a separate experiment

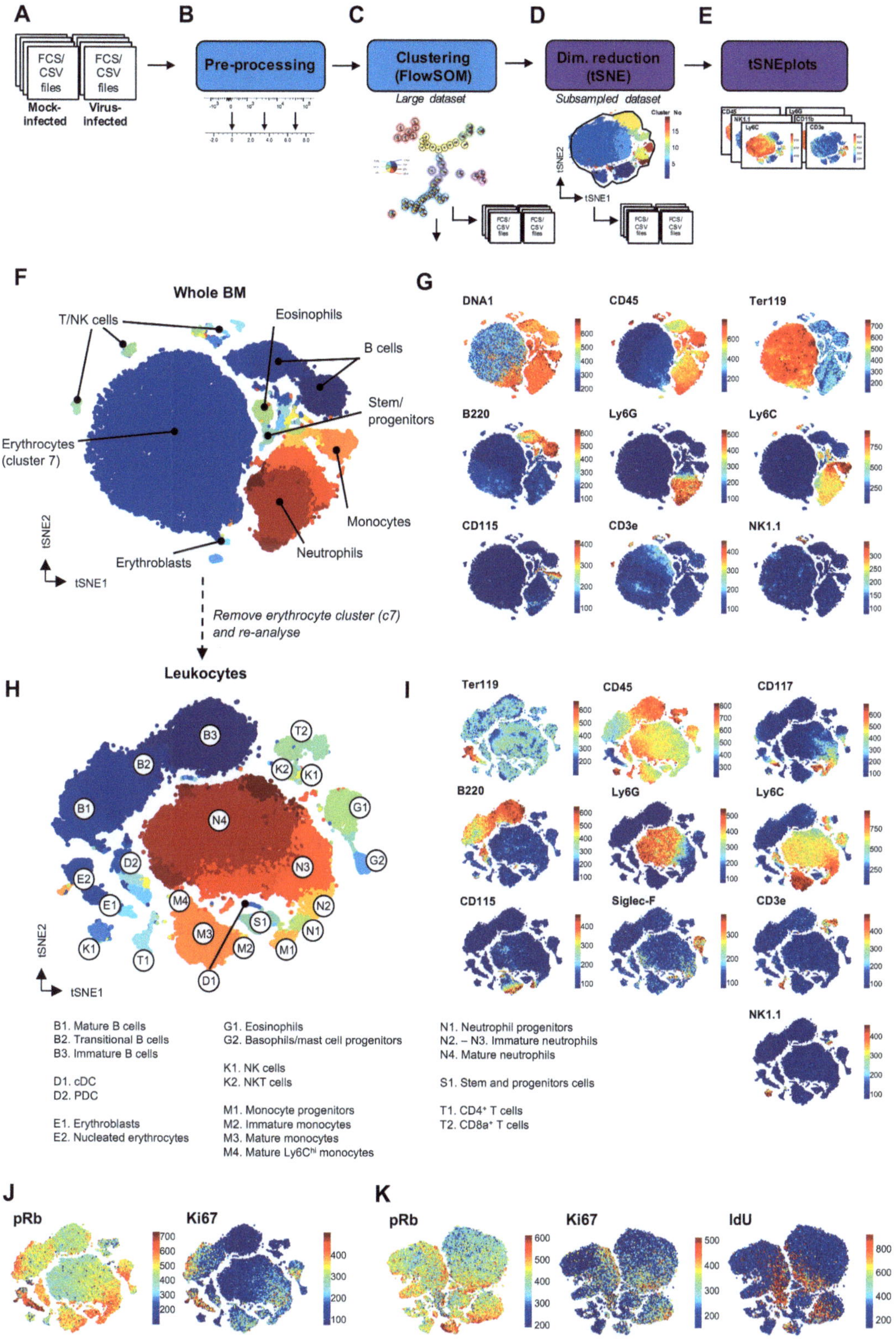

Fig. 8 Analysis of the haematopoietic system in murine bone marrow. An analytical approach to analysing BM data. (**a**) Export FCS or CSV files from FlowJo or similar software, (**b**) pre-process the data to prepare for clustering, (**c**) perform high-speed clustering using FlowSOM, (**d**) downsample the dataset to approx. 8×10^4 cells

7×10^4–8×10^4 cells. This can be done by randomly downsampling the entire dataset, downsampling each file by a specific value, or downsampling each sample based on weighted values. *See* **Note 18** for a discussion on downsampling targets for tSNE.

7. Run tSNE on the downsampled dataset, and save resultant FCS and/or CSV files (Fig. 8d).
8. Use either an in-built export function or the "SumTables" script (available at www.github.com/sydneycytometry/CAPX) to save expression level and cell frequency/number data of the large FlowSOM dataset and the smaller tSNE dataset.
9. Use the "tSNEplots" script (www.github.com/sydneycytometry/tSNEplots) to create tSNE plots colored by each marker for every sample (Fig. 8e).

3.19 Data Exploration

The result of the analysis pipeline described above is a large dataset where every cell is assigned a cluster number by the FlowSOM algorithm. Subsequently, this dataset is downsampled (to approx. 7×10^4 cells) and each cell given a "tSNE1" and "tSNE2" coordinate. Because the cluster numbers represent the same populations in both datasets, we can explore the tSNE plots (small dataset) to determine which clusters represent which biological populations, and then analyze those same clusters/populations in the large dataset with greater numerical and statistical power. Additionally, for any rare populations that are infrequent on the tSNE map, the cluster(s) representing these cell types can be extracted from the large dataset, downsampled, and analyzed in isolation in a new tSNE plot.

In the normal BM, the most numerous populations are erythrocytes (Ter119$^+$), neutrophils (Ly6G$^+$), and B cells (B220$^+$CD11c$^-$) followed by Ly6C^{hi} monocytes (Ly6C^{hi}CD115$^+$CD48$^+$CD11b$^+$) and eosinophils (Siglec-F$^+$) (Fig. 8f, g). Removing the erythrocyte cluster (cluster 7 in Fig. 8f) and replotting the data in a new tSNE analysis (Fig. 8h, i) allows for simpler visualization of other leukocyte subtypes, such as CD4$^+$ and CD8a$^+$ T cells, NK cells, NKT cells, cDC, Ly6C^{lo} monocytes, basophils, mast cell progenitors, and stem and progenitor cells. As cells progress from a stem cell phenotype, through intermediate progenitors and into mature cells, they typically follow a characteristic progression of CD117 downregulation and CD45 upregulation (Fig. 8h, i). Cells expressing CD117, but no mature lineage markers, are enriched for the stem and early progenitor cells, including Lineage-SCA-1$^+$CD117$^+$ (LSK) cells. Because monocyte and neutrophil progenitors share a similar phenotype, and CD115 expression is prone to downregulation, we often use CD48 to aid in separation of the CD48$^+$CD115$^+$ monocyte progenitors and

$CD48^{-}CD115^{-}$ neutrophil progenitors. Typically, the highest level of proliferation occurs in the intermediate progenitors of various lineages, and downregulates during maturation. This can be seen through high levels of markers of cell cycle such as pRb and Ki67 expression (Fig. 8j) as well as in IdU incorporation in cells that exhibit high levels of pRb and Ki67 (Fig. 8k). This analytical approach allows for exploration of complex hematopoietic datasets at any level of depth, by focusing into specific elements of the system using the cluster assignments.

4 Notes

1. A number of laboratories have reported a variety of metal contaminants in buffers after storage in glass bottles cleaned by laboratory dishwashing services. As such, a common recommendation is to only use buffers made up and stored in plastic containers. In our laboratory, we have not found such contaminants to be a problem; however, this should be tested in each location specifically.
2. The BrdU Flow Kit (BD Biosciences) contains reagents necessary for antibody labeling of incorporated BrdU or IdU by flow or mass cytometry. In particular, this kit contains BrdU solution (32.5 mM, 10 mg/mL solution, store at −80 °C) and 300 μL vials of 1 mg/mL DNase solution (store at −80 °C). Importantly, this kit provides fixative as a combination of PFA with saponin-based permeabilization reagent. Other vendors provide BrdU kits, but these have not been tested here.
3. Intracellular labeling of cytokines (e.g., TNF) requires PFA fixative and saponin-based permeabilization buffer. A number of suppliers provide Fixation Buffer (usually 2% or 4% PFA, BioLegend, BD, Thermo Fischer Scientific) and permeabilization buffers for intracellular staining, such as Intracellular Staining Permabilisation Wash Buffer (BioLegend), Perm/Wash Buffer (BD Biosciences), or Permeabilisation Buffer (Thermo Fischer Scientific). Alternatively, various suppliers provide "fix/perm" buffers that contain a mixture of fixative and permeabilization buffer (e.g., Cytofix/Cytoperm Buffer, BD Biosciences).

   Intracellular labeling of transcription factors (e.g., FoxP3, Ki67) is optimal following permeabilization using special buffers, such as the FoxP3/transcription factor staining kit (Thermo Fischer Scientific). This kit includes a Fixation/Permeabilization buffer which contains PFA and saponin, as well as additional compounds to aid in transcription factor labeling, which likely includes a form of detergent (such as Triton X-100) and DNase. Similar kits are available from other vendors, such as FOXP3 Fix/Perm Buffer Set (BioLegend) or Mouse FoxP3 Buffer Set (BD Biosciences).

4. In our example, we titrated fresh DNase to determine optimal concentration, but used this experimentally after storage at −80 °C. A superior approach is to freeze a concentrated solution of DNase, store at −80 °C overnight, and then titrate for efficacy to account for the effect of freezing.
5. It is critical to ensure that the timing of injections, or incubations, is consistent. For example, if 20 mice need to be injected in an experiment, the interval between injections will have to match the interval of anesthesia injections, so that each mouse is exposed to BrdU/IdU for 3 h. If 20 mice are harvested over 2 h (6 min each), then BrdU injections might start at 8 am, with a new mouse injected every 6 min. After the 2 h injection session, this leaves a 1 h break before starting a 2 h harvest experiment. Similarly, if performing ex vivo incubations, careful attention to timing is required to ensure each sample is incubated for 1 h, in among the rest of the tissue collection. Typically, we don't recommend exceeding 3 h in vivo or 1 h ex vivo incubations, to reduce the likelihood that cells that have incorporated BrdU or IdU proceed into S-phase in a second round of replication.
6. While BrdU is readily soluble at 10 mg/mL (32.56 mM) in PBS, and is typically injected at 5 mg/mL (16.28 mM), IdU is barely soluble in PBS at ~1.47 mg/mL (~4.15 mM). In order to match the amount of IdU given compared to BrdU, a ~4× larger injection volume would be required at a lower concentration. Alternatively, we found IdU to be readily dissolvable in DMSO at a maximum concentration of approximately 576.52 mg/mL (1.62 M). When diluted to a working concentration with PBS, we found that a 1/200 dilution of this stock (final IdU at 8.14 mM, 2.88 mg/mL; and DMSO at 1/200) maintained a stable solution, but the IdU came out of solution at higher concentrations than this. For example, a 8.14 mM solution of IdU would only require a 2× increase in volume to deliver an amount equivalent to BrdU (400 μL of 8.14 mM, 2.88 mg/mL IdU; compared to 200 μL of 16.28 mM, 5 mg/mL BrdU).
7. Although we have not tested it, the BrdU Flow Kit manual (BD Biosciences, http://www.bdbiosciences.com/ds/pm/others/23-12721.pdf) states that BrdU solutions, once thawed, are stable when stored at 2–8 °C for up to 4 months, and can be refrozen. This has not been validated for IdU solutions, where the presence of DMSO may impact longevity of solutions kept at 4 °C. In both cases, we would recommend making up solutions in volumes small enough to allow for a single use for each thawed aliquot.

8. In many experiments, we perform a perfusion of the circulatory system with PBS to remove any blood-borne leukocytes from the microvasculature in the central nervous system. In our experience, PBS perfusion has little effect on the numbers or phenotypes of cells seen in the BM.
9. It is well established that some cellular markers, such as surface expression of M-CSFR (CD115), are downregulated when cells are removed from their biological niche. Storing cells at 4 °C or on ice prevents this downregulation (*see* Fig. 5b).
10. The Sysmex XP-100 histogram analysis was configured for human blood cells, which are larger than the equivalent mouse cells. When we compared mouse leukocyte counts on the Sysmex XP-100 and hemocytometer, they were reasonably comparable, although the XP-100 was not able to distinguish lymphocytes, monocytes, and granulocytes from one another. However, erythrocyte and thrombocyte counts were inaccurate, and underreported the true numbers, as is typical of many automated cell counters designed for human blood counting. The default leukocyte concentrations are reported as 10^9 cells/L, which equates to 10^6 cells/mL.
11. Alternative procedure for viability and Fc blocking for flow cytometry.
    (a) Centrifuge at 300 × *g* for 3 min at 4 °C.
    (b) Resuspend in 50 μL of a fixable viability dye (e.g., 1/1000 dilution of UV-excitable LIVE/DEAD Blue stock in PBS) with purified anti-mouse CD16/32 (1/100 in PBS) (*see* **Note 12**).
    (c) Incubate for 30 min at 4 °C.
    (d) Top up with 200 μL PBS.
    (e) Centrifuge at 300 × *g* for 3 min at 4 °C.
12. The presence of the Fc receptors, CD16 and CD32, on myeloid cells can lead to nonspecific binding of antibodies to these cells. An antibody against CD16/32 can be used to block the Fc receptors, reducing the likelihood of nonspecific binding. This can be done using a purified anti-CD16/32 antibody, or a metal/fluorophore-conjugated antibody, if quantifying the levels of CD16/32 is also desired.
13. While many companies provide "fix/perm" buffers (containing a mixture of PFA and saponin-based permeabilization reagent) which simultaneously fix and permeabilize cells, we have found improved cell recovery and stability when fixing with PFA before permeabilizing. However, some reagents such as the fix/perm buffer used in the Thermo Fischer Scientific FoxP3 Fix/Perm kit contain additional reagents that aid in FoxP3 staining. In these situations, the fix/perm buffer

must be used for optimal staining, as the additional reagents are only present in the fix/perm mixture. To balance these two requirements, we usually recommend performing fixation (4% PFA for 20 min at RT) followed by permeabilization (for cytokines) or fix/perm (with the FoxP3 fix/perm buffer, for transcription factor labeling). Although typical fix/perm buffers provided for cytokine staining are a simple mix of PFA and saponin-based permeabilization reagent, to avoid complications, we often recommend users perform fixation followed by fix/perm in these situations as well.

14. Depending on which intracellular factors are being targeted in an assay, different permeabilization methods are appropriate (*see* Table 1). Most of these procedures are compatible with typical surface staining procedures, provided the cells have been adequately fixed before permeabilization. One exception is methanol permeabilization, which rapidly degrades any fluorophore that is not FITC or an Alexa-Fluor dye. CyTOF metals are resistant to methanol degradation.

    Cytoplasmic permeabilization

    Cytoplasmic factors require "gentle" permeabilization, which usually consists of a saponin-based buffer that creates reversible pores in the cellular membrane. Cells must be kept in this buffer for permeabilization, intracellular staining, and washing. Intracellular labeling of cytokines can be performed in the same step as antibody labeling of BrdU/IdU, as the BrdU/IdU permeabilization procedure described in this protocol is sufficient to permeabilize for cytokines. The procedure below describes permeabilization for generic cytoplasmic and/or cytokine staining using saponin-based permeabilization from the BD Biosciences kit.

    (a) Start after sample fixation.
    (b) Centrifuge at 800 × *g* for 5 min at RT, and discard supernatant (all centrifuge steps after fixation at 800 × *g*, for 5 min).
    (c) Resuspend in 50–100 μL of BD Cytofix/Cytoperm buffer.
    (d) Incubate for 20 min at RT.
    (e) Top up with 150–200 μL of 1× perm buffer.
    (f) Centrifuge at 800 × *g* for 5 min at RT.
    (g) Resuspend in 250 μL of 1× perm buffer.
    (h) Centrifuge at 800 × *g* for 5 min at RT.
    (i) *Repeat* **steps (g)** and **(h)**.
    (j) Proceed to intracellular staining.

    Nuclear and transcription factors (incl. FoxP3)

Nuclear factors require harsher permeabilization, usually a form of detergent, to permeabilize the nuclear membrane. Some transcription factors (such as FoxP3) are very difficult to stain for, and so specific kits are recommended. The cells are permeabilized with the specific buffer initially (usually a fix/perm buffer containing detergent, and possibly DNase to free up the binding sites of DNA-bound transcription factors), and then are stained and washed, typically in a saponin-based wash buffer. We recommend the FoxP3/Transcription Factor Staining Buffer Set from Thermo Fischer Scientific (00-5523-00). Other kits: BioLegend True-Nuclear Transcription Factor Buffer set (424401), BD Biosciences (human) Human FoxP3 buffer set (560098), BD Biosciences Mouse FoxP3 buffer set (560409). This procedure also works for cytoplasmic targets, however, may not be compatible with DNase digestion protocols for antibody labeling of BrdU/IdU (*see* above).

(a) Start after sample fixation.

(b) Centrifuge at 800 × *g* for 5 min at RT, and discard supernatant (all centrifuge steps after fixation at 800 × *g*, for 5 min).

(c) Resuspend in 50 μL of 1× FoxP3 Fixation/Permeabilization buffer.

(d) Incubate for 30 min at RT.

(e) Top up with 1× permeabilization buffer.

(f) Centrifuge at 800 × *g* for 5 min at RT and discard supernatant.

(g) Resuspend cells in 250 μL permeabilization buffer.

(h) Centrifuge at 800 × *g* for 5 min at RT and discard supernatant.

(i) Repeat **steps (g)** and **(h)**.

(j) Proceed to intracellular staining.

DNase digestion for BrdU/IdU labeling

Antibody labeling of BrdU or IdU that has been incorporated into cellular DNA requires nuclear permeabilization, combined with DNase digestion to expose antibody-binding sites. Typically this approach is compatible with most other surface and/or intracellular stains, despite some partial reduction in signal levels of surface-bound fluorophores.

Methanol permeabilization

Some phosphorylation targets are hidden within dimers, and while some detergent-based methods are sufficient to reveal these targets, methanol appears to be the most effective. As previously noted, methanol permeabilization rapidly degrades any fluorophore that is not FITC or an Alexa-Fluor, while CyTOF metals are resistant to methanol degradation.

Combining permeabilization methods

If multiple perm methods are required for a sample, then these can be performed sequentially from least to most harsh (*see* Table 1), with antibody labeling of each relevant intracellular target after the corresponding permeabilization approach, with fixation after each stain (e.g., fix, nuclear permeabilization, FoxP3 labeling, refix, methanol permeabilization, pRb labeling). Alternatively, multiple permeabilization approaches could be used sequentially, followed by staining of all intracellular factors at the end (e.g., fix, nuclear permeabilization, methanol permeabilization, anti-FoxP3 and anti-pRb labeling), although this is less desirable. These combinations should be optimized on a case-by-case basis. For more information on these considerations, *see* Chapter 10.

Critically, permeabilization approaches for anti-BrdU/IdU labeling and transcription factor labeling are not necessarily compatible in sequential assays, as both likely use DNase to expose the binding site of target epitopes. When we have used the FoxP3 fix/perm buffer (Thermo Fischer Scientific) followed by DNase treatment, we have found reduced labeling of intracellular factors compared to using the FoxP3 fix/perm kit alone. This suggests that the DNA is being over-digested. It is possible that the DNase digestion procedure described in this protocol for BrdU/IdU labeling would be sufficient to enable labeling of transcription factors, but we have not tested this.

15. Intracellular labeling should be performed in perm buffer, as the saponin-based perm buffers are reversible, and antibody may become trapped inside the cells if normal FACS buffer is used. Intracellular labeling following methanol fixation can be performed in FACS buffer, as methanol causes irreversible permeabilization.
16. Final cell processing for flow cytometry.
    (a) Resuspend in 100–200 μL FACS buffer.
    (b) Keep cells at 4 °C until acquisition on cytometer (for either same day or following 1–3 days).
    (c) Filter samples before running.
17. Data that is analyzed by algorithmic methods, such as FlowSOM and tSNE, is typically transformed in such a way that the relationships observed in log or logicle/bi-exponential space are captured in a linear space. Typically in mass cytometry, this conversion is performed using arcsinh transformation (examples in Fig. 5.8.12 of Ashhurst et al. [17]). This converts the data from a logarithmic distribution to a linear one, as opposed to simply changing the scale from log to linear. Additionally, the cofactor used dictates the extent of

low-end signal compression to near zero. While an arcsinh cofactor of 5 or 15 are typically used in mass cytometry, higher and more variable values are required for flow cytometry. A useful function available in FlowJo analysis software is the ability to export the data as a "CSV Channel Values." In this case, the distribution of cells on a biaxial plot (usually with a bi-exponential or similar scale) is captured and given a linear scale between the values of 0 and 1024, through a process of "binning." This allows the user to visually adjust the scale and low-end compression settings to a satisfactory point, rather than using a single cofactor for archsinh transformation.

18. It is possible to downsample each sample by a value that reflects the number of cells present in the original sample. For example, if two samples of BM contained 2×10^7 and 2×10^6 leukocytes per femur, respectively, and we downsample each to 4×10^4 for tSNE analysis, then we have artificially inflated the number of cells being displayed in the second sample, relative to the first. This might lead us to observe increases or decreases in populations that aren't reflective of the actual sample. If, however, each was downsampled in a fashion that matched the total cells/femur (e.g., downsample the first sample to 6×10^4 cells, and the second to 6×10^3 cells), then the relative number of cells is preserved.

Acknowledgments

We wish to thank Dr. Paula Niewold for critical reading of this manuscript, and assistance with many of the experiments that lead to the development of this protocol, and to Rebecca Burrell for assistance with experiments. We would like to thank Dr. Philip Hansbro, Dr. Nicole Hansbro, Dr. Markus Hofer, David Jung, Tamara Suprunenko, Dr. Iain Campbell, and Vickie Xie for provision of bone marrow samples from various disease models over the years. We would like thank Dr. Fabian Held for assistance in troubleshooting our tSNEplots script, as the use of such a tool expedited our analysis of bone marrow data. We would also like to express gratitude to Dr. Sebastian Stifter, Dr. Carl Feng, Dr. Ben Roediger, Dr. Sean Bendall, Dr. Greg Behbehani, Dr. Helen Goodridge, and Dr. Alberto Yáñez for discussions on the nature of bone marrow hematopoiesis over the years, as these discussions have been invaluable. We would like to thank the Sydney Cytometry Facility, Ramaciotti Facility for Human System Biology, and Caryn van Vreden for assistance with mass cytometry and provision of reagents.

Thomas Ashhurst was supported by an Australian Postgraduate Award (APA) during the early development of parts of this protocol, and is currently supported by the International Society for the Advancement of Cytometry (ISAC) Marylou Ingram Scholars program. Darren Cox was also supported by an APA. This work was funded by National Health and Medical Research Council (NHMRC) funding.

References

1. Behbehani GK, Bendall SC, Clutter MR et al (2012) Single-cell mass cytometry adapted to measurements of the cell cycle. Cytometry A 81(7):552–566. https://doi.org/10.1002/cyto.a.22075
2. Bendall SC, Simonds EF, Qiu P et al (2011) Single-cell mass cytometry of differential immune and drug responses across a human hematopoietic continuum. Science 332(6030):687–696. https://doi.org/10.1126/science.1198704
3. Dolbeare F, Gratzner H, Pallavicini MG et al (1983) Flow cytometric measurement of total DNA content and incorporated bromodeoxyuridine. Proc Natl Acad Sci U S A 80(18):5573–5577
4. Kass L, Varayoud J, Ortega H et al (2000) Detection of bromodeoxyuridine in formalin-fixed tissue. DNA denaturation following microwave or enzymatic digestion pretreatment is required. Eur J Histochem 44(2):185–191
5. Ye W, Mairet-Coello G, DiCicco-Bloom E (2007) DNAse I pre-treatment markedly enhances detection of nuclear cyclin-dependent kinase inhibitor p57Kip2 and BrdU double immunostaining in embryonic rat brain. Histochem Cell Biol 127(2):195–203. https://doi.org/10.1007/s00418-006-0238-6
6. Rothaeusler K, Baumgarth N (2006) Evaluation of intranuclear BrdU detection procedures for use in multicolor flow cytometry. Cytometry A 69(4):249–259. https://doi.org/10.1002/cyto.a.20252
7. van der Maaten L, Hinton G (2008) Visualizing data using t-SNE. J Mach Learn Res 9:2579–2605
8. van der Maaten L (2014) Accelerating t-SNE using tree-based algorithms. J Mach Learn Res 15:3221–3245
9. Amir el AD, Davis KL, Tadmor MD et al (2013) viSNE enables visualization of high dimensional single-cell data and reveals phenotypic heterogeneity of leukemia. Nat Biotechnol 31(6):545–552. https://doi.org/10.1038/nbt.2594
10. Van Gassen S, Callebaut B, Van Helden MJ et al (2015) FlowSOM: using self-organizing maps for visualization and interpretation of cytometry data. Cytometry A 87(7):636–645. https://doi.org/10.1002/cyto.a.22625
11. Weber LM, Robinson MD (2016) Comparison of clustering methods for high-dimensional single-cell flow and mass cytometry data. Cytometry A 89(12):1084–1096. https://doi.org/10.1002/cyto.a.23030
12. Challen GA, Boles N, Lin KK et al (2009) Mouse hematopoietic stem cell identification and analysis. Cytometry A 75(1):14–24. https://doi.org/10.1002/cyto.a.20674
13. Krutzik PO, Clutter MR, Nolan GP (2005) Coordinate analysis of murine immune cell surface markers and intracellular phosphoproteins by flow cytometry. J Immunol 175(4):2357–2365
14. Weiss J, Zimmermann F (1999) Tribromoethanol (Avertin) as an anaesthetic in mice. Lab Anim 33(2):192–193. https://doi.org/10.1258/002367799780578417
15. Finck R, Simonds EF, Jager A et al (2013) Normalization of mass cytometry data with bead standards. Cytometry A 83(5):483–494. https://doi.org/10.1002/cyto.a.22271
16. Chen H, Lau MC, Wong MT et al (2016) Cytofkit: a bioconductor package for an integrated mass cytometry data analysis pipeline. PLoS Comput Biol 12(9):e1005112. https://doi.org/10.1371/journal.pcbi.1005112
17. Ashhurst TM, Smith AL, King NJC (2017) High-dimensional fluorescence cytometry. Curr Protoc Immunol 119:5.8.1–5.8.38. https://doi.org/10.1002/cpim.37

Chapter 13

Mass Cytometric Cell Cycle Analysis

Gregory K. Behbehani

Abstract

The regulated proliferation of cells is a critical factor in tumor progression, antineoplastic therapies, immune system regulation, and the cellular developmental of multicellular organisms. While measurement of cell cycle state by fluorescent flow cytometry is well established, mass cytometry allows the cell cycle to be measured along with large numbers of other antigens enabling characterization of the complex interactions between the cell cycle and wide variety of cellular processes. This method describes the use of mass cytometry for the analysis of cell cycle state for cells from three different sources: in vitro cultured cell lines, ex vivo human blood or bone marrow, and in vivo labeling of murine tissues. The method utilizes incorporation of 5-Iodo-2′-deoxyuridine (IdU), combined with measurement of phosphorylated retinoblastoma protein (pRb), Cyclin B1, and phosphorylated Histone H3 (pHH3). These measurements can be integrated into a gating strategy that enables clear separation of all five phases of the cell cycle.

Key words Cell cycle, Mass cytometry, CyTOF, Iodo-deoxyuridine, Cyclin, Retinoblastoma protein, Phosphorylated Histone H3, Ki-67

1 Introduction

Alterations in the cell cycle are a critical aspect of normal cellular growth and differentiation and the regulation of almost every tissue in complex organisms. The cell cycle is also critical for understanding diseases of abnormal cell proliferation such as malignancies (and the therapies used to treat them), as well as the regulated proliferation of activated cells of an immune response. Fluorescent flow cytometry has long been used for the characterization of DNA and RNA content at the single cell level and these measurements were the first to enable the determination of DNA ploidy and cell cycle phase [1]. The combination of these DNA and RNA stains with the antibody-mediated measurement of incorporated halogenated nucleoside analogs (i.e., bromo-deoxyuridine, BrdU) enables a relatively precise characterization of cell cycle phase [2]. Such studies have been used for the characterization of malignant cell proliferation, immune cell activation, and the developmental regulation of cell proliferation. While very useful, these assays primarily

Helen M. McGuire and Thomas M. Ashhurst (eds.), *Mass Cytometry: Methods and Protocols*, Methods in Molecular Biology, vol. 1989,
https://doi.org/10.1007/978-1-4939-9454-0_13,

rely on bright fluorescent dyes at saturating concentrations that have spillover into adjacent measurement channels, which can significantly hamper measurement of other surface or intracellular markers. The measurement of BrdU incorporation can also further complicate the assessment of other antigens as this requires partial DNA degradation using acid or DNase, which can potentially damage antigens of interest.

Mass cytometry has both advantages and disadvantages with respect to cell cycle analysis. The major disadvantage is that there are no comparable DNA or RNA stains with the same level of resolution as those used in fluorescent flow cytometry (e.g., DAPI, Hoechst, Pyronin Y). The high resolution of these dyes stems in part from changes in their fluorescent properties that occur upon interaction with nucleotide bases, and this property cannot be replicated through the detection of metal atoms. Mass cytometry measurement of the cell cycle state thus requires an alternative measurement approach based on measurement of: 5-Iodo-2′-deoxyuridine (IdU) incorporation, Cyclin B1 levels, phosphorylation of the Retinoblastoma protein (pRb), and phosphorylation of histone H3 (pHH3) [3]. While slightly more complicated, this approach has two major advantages: first, cell cycle measurements can be combined with measurement of up to 35 or more additional antigens; second, IdU incorporation can be measured directly without the need for an antibody or for the degradation of DNA with acid or DNase.

Mass cytometric cell cycle analysis is well suited for the analysis of cell cycle state in highly complex cell mixtures, or for the correlation of cell cycle state with a large number of other functional variables within less complex cell populations. We have previously utilized this approach for the measurement of cell cycle state during normal hematopoiesis in human bone marrow [3], and transgenic murine models of telomerase deficiency [4]. We also utilized this method to demonstrate subtype-specific differences in leukemia stem cell S-phase fractions in patients with acute myeloid leukemia [5]. Such high parameter studies would be extremely difficult or impossible with current fluorescent cell cycle analysis methods. Other researchers have utilized this methodology to characterize differences in proliferation rates across different immune cell subsets [6], proliferation and chemotherapy response of tumors xenografted in to mice [7], and for monitoring of immunotherapy [8]. Many of the basic methods of this staining protocol are based on the standard phospho-flow cytometry methods originally developed by Krutzik and Nolan [9] and additional background for some of the methods can be found in their original protocol. Finally, it is worth noting that this basic analysis approach also works well for performing cell cycle analysis by fluorescent flow cytometry, allowing avoidance of bright DNA or RNA dyes [3].

This chapter expands on our previous methods chapter [10], with additional options for other staining approaches and background information.

2 Materials

All solutions for mass cytometry should be prepared with ultrapure water and all reagents should be tested for heavy metal contamination. Do not inject any concentrated solution that may contain heavy metal directly into the mass cytometer without first testing a 1/10th or 1/100th dilution of the solution. Care should be used in handling IdU, paraformaldehyde, and the Smart Tube buffer as all are toxic and potentially mutagenic; appropriate protection (e.g., gloves, eye and respiratory protection) should be employed when needed. Researchers should also be aware that standard laboratory dishwashers, autoclaves, and dishwashing detergents are frequently contaminated with heavy metals (particularly Barium), which can disrupt experiments or damage the mass cytometer. Use of either disposable plastic containers or hand-washed glassware is recommended.

2.1 IdU

1. Dry 5-Iodo-2′-deoxyuridine (IdU) powder can be purchased from various suppliers. The dry powder can be stored at 4 °C and used to periodically prepare concentrated stock solutions (*see* below).
2. IdU is not soluble at high concentrations in water; 50 mM stock solutions of IdU can be prepared in DMSO (5000× final concentration). 88.5 mg of IdU will make 5 mL of stock solution at 50 mM, which can be aliquoted into small volumes and frozen at −80 °C (though it is likely stable at −20 °C). This solution should be sterile filtered using a syringe filter prior to making aliquots. Stocks of up to 250 mM in DMSO can also be made for the purposes of creating IdU solutions for injection (*see* **Note 1**).
3. Before adding IdU stock solution to cells, it's advisable to dilute the stock solution to 100× final concentration in a prewarmed aqueous solution to ensure the IdU will readily mix with the cells without the requirement for extensive mixing that might disrupt the cells being studied.

2.2 Fixatives, Buffers, and Tubes

1. 16% Paraformaldehyde (PFA) solution: this must be methanol-free (i.e., not formalin). This is available in 10 mL ampules (Electron Microscopy Sciences). Upon opening an ampule, it can be transferred to a sealed, foil-wrapped tube, as exposure to air and light will cause it to loose activity in a few weeks. Discard any PFA that's been open for more than 1 month.

2. Alternative fixative: Smart Tube proteomic stabilizer (STPS; can be purchased from Smart Tube incorporated, San Carlos, CA). This comes as a working solution that can be added directly to cell samples at a sample:buffer ratio of 1:1.4.
3. Smart Tube lysis solution (Smart Tube incorporated, San Carlos, CA) for red blood cell lysis.
4. Pure methanol: this should be kept cold (−20 to 4 °C) in a sealed bottle. This should not have any drying agents added as these may contain heavy metals.
5. Culture media: typically, the standard culture media used for routine cell culture will work well, but some growth media may contain barium contamination, which should be verified before the start of the experiment.
6. Cell staining media (CSM): phosphate buffered saline (PBS), plus 0.5% bovine serum albumin (BSA), and 0.02% sodium azide, pH 7.4. This is made in 4 L batches by adding 20 g of BSA and 800 mg of sodium azide. Start with 3 L of sterile PBS and mix in the dry ingredients until dissolved. Then add additional PBS to a total volume of 4 L. Sterile filter this using a 0.2 μM bottle-top filter into rinsed, 500 mL glass bottles or sterile plastic containers.
7. Intercalator solution is made by addition of the Iridium-based intercalator solution (Fluidigm, 201192A or 201192B) to PBS at a final concentration of approximately 125 nM. To this add 1/10th volume of 16% PFA to achieve a final concentration of 1.5% PFA. This solution should be prepared fresh immediately before addition to the cells. The intercalator staining concentration and volume can be titrated for optimal staining of the particular cells of interest.
8. This protocol is written for use with standard 5 mL polystyrene FACS tubes, but a variety of other tubes can be used such as 1.5 mL Eppendorf-style microcentrifuge tubes, or 1.1 mL polypropylene "cluster" microcentrifuge tubes. For smaller volume tubes, additional washes may be required after staining steps.

2.3 Antibodies

1. A wide variety of antibodies can be utilized for mass cytometry analysis of cell cycle state; however, four key antibodies are routinely used: phosphorylated retinoblastoma protein (pRb) (S807/811), Cyclin B1, phospho-Histone H3 (pHH3) (S28), and Ki-67 that allow for determination of all five cell cycle phases when combined with measurement of IdU incorporation (*see* **Note 2**). These are detailed in Table 1 and **Notes 3–5**. Additionally, several other antibodies can be

Table 1
Common antibodies used for mass cytometry cell cycle assessment

Antibody	Clone	Manufacturer	Purpose	Notes
Essential antibodies				
p-Rb (S807/811)	J112906	BD Biosciences	G0/G1 Resolution	**Note 3**
Cyclin B1	GNS-1	BD Biosciences	G2 Resolution	**Note 4**
p-HH3 (S28)	HTA28	Biolegend	M-phase Resolution	**Note 5**
Optional antibodies				
Ki-67	SolA15	eBiosciences	Confirmation of G0/G1	**Note 6**
Cleaved-Caspase3 (D175)	C92-605 D3E9	BD Biosciences CST	Identification of apoptotic cells	**Note 7**
Cleaved-PARP (D214)	F21-852	BD Biosciences	Identification of apoptotic cells	**Note 8**
PCNA	PC10	BD Biosciences	Confirmation of G0/G1	**Note 9**
p-RP-S6 (S236/236)	N7-548	BD Biosciences	Confirmation of M phase Sample quality	**Note 10**
p-H2AX (S139)	JBW301	Millipore	Detection of DNA damage	**Note 11**
Cyclin A	BF683	BD Biosciences	Confirmation of G0/G1, G2 resolution	**Note 12**
Cyclin E	HE12	Invitrogen	Confirmation of G0/G1, G2 resolution	**Note 13**
pCDKl (Y15)	10A11	CST	Confirmation of M phase, G2 resolution	**Note 14**

used to subset cell cycle phases or provide additional confirmation of cell cycle state or checkpoint activation, the most useful are summarized in Table 1 and **Notes 6–14**. All antibodies should be titrated to determine the optimal staining concentration for the cells of interest, as the staining properties of these antibodies are dependent on the cell type being stained as well as the approximate cell cycle distribution of the cells (this is particularly true for anti-pHH3 which must be used at concentrations below antigen saturation).

2. All intracellular antibodies described in this protocol have previously been tested following PFA or STPS fixation and methanol cell permeabilization. We have not extensively tested staining under other fixation permeabilization conditions (e.g., saponin), though other permeabilization methods would likely be compatible with this method as these same antigens have been successfully analyzed with other protocols [7, 11].

3 Methods

3.1 Sample Collection and Processing

3.1.1 IdU Incubation and Processing of Cultured Cells In Vitro

1. Quickly and carefully place the desired number of cells for analysis into a separate culture container at least several hours before analysis. Each mass cytometry sample typically requires at least 500,000 cells (we typically collect a minimum of two million cells to allow for one million cells to be stained and analyzed twice if necessary; *see* **Note 15**).
2. At the desired time-point, add IdU to the cell culture media while cells are still growing under normal culture conditions. Try to minimize the time required to add IdU to the cells so that the cell cycle is not disrupted. We typically add IdU to a final concentration of 10 μM for 10–15 min incubations, but higher or lower concentrations can be used (particularly if longer IdU incubations will be performed). If starting with the 50 mM IdU stock solution, add 0.2 μL of IdU for each 1 mL of cell culture media (2 μL total for a 10 mL culture dish). Be sure to gently swirl or pipette the cells to ensure that the IdU is evenly distributed (the DMSO stock solution will sink to the bottom of an aqueous solution). Alternatively, it is preferable to pre-dilute the IdU (to 50–500×) in pre-warmed media so that a larger volume of aqueous solution can be added to the cells enabling much more rapid mixing of the IdU solution with the culture media.
3. Once IdU is added, the cells should be returned to the incubator for approximately 10 min. The duration of the incubation is not critical, but each sample of an experiment should be treated consistently (IdU incorporation works well with incubations in the 10–30 min range). Longer incubations will lead to higher intensity of IdU labeling in S-phase cells but reduced resolution of G2 and M cells, which may exhibit IdU that was incorporated at the end of S phase before these cells progressed to G2 or M when IdU incubations are prolonged (*see* **Note 16**).
4. **Optional**: If you plan to perform viability staining with cisplatin [12], it should be performed at the completion of the IdU incubation and before cell fixation (as this could lead to partial permeabilization of the cell membrane reducing the accuracy of the viability stain).
5. At the completion of the incubation, add 1/10th volume of 16% paraformaldehyde to achieve a final concentration of 1.5% PFA. The media should turn yellow. Leave the cells at room temperature for 10 min. During this time, cells can be transferred from the culture tube to a centrifuge tube.

6. At the end of the 10 min incubation, centrifuge the cells to pellet them (this is typically done at 300 × *g* for 5 min but this may be cell type specific; *see* **Note 17**).
7. Aspirate the media from the cells, removing as much as possible, preferably to a pellet of 50 μL or less. This will enable enough methanol to be added to achieve a final concentration of 90–95%.
8. **Optional**: If cells will require assessment of surface antigens that are disrupted by methanol exposure or if an alternative permeabilization method will be employed, cells can be washed twice with CSM and then snap frozen (using a dry ice methanol bath or liquid nitrogen) in CSM with DMSO added to a final concentration of 10% (*see* **Notes 18** and **19**).
9. Thoroughly vortex the cell pellet to resuspend the residual cells into a single-cell suspension.
10. While gently vortexing the cell pellet, rapidly add 1–2 mL of ice-cold methanol (this volume depends on the cell number and desired storage tube; final concentration of methanol should be 90–95%). It is very important to add the methanol to a completely resuspended cell pellet while vortexing. If the cells are not maintained in an even single-cell suspension, the cells will clump and will be incompatible with cytometry analysis. Once the methanol has been added, transfer the cells onto ice for at least 10 min.
11. Cells can then be transferred into one or more storage tubes and stored at −80 °C (in methanol) until analysis. We routinely store the cells at 5–10 million cells per mL in methanol, so up to 15 million cells will fit into a 1.5 mL Eppendorf microcentrifuge tube or a cryogenic freezing tube.

3.1.2 IdU Incorporation and Processing of Ex Vivo Human Cell Suspensions

For analysis of fresh primary human cell suspensions (typically, human peripheral blood or bone marrow aspirate), we will typically utilize Smart Tube proteomic stabilizer (STPS) solution. This solution utilizes a fixative cocktail, but is sufficiently gentle to allow subsequent lysis of red blood cells as well as detection of antigens known to be disrupted by standard PFA fixation. An alternative fixation procedure developed by Chow et al. [13] utilizing PFA fixation followed by Triton X-100 for red cell lysis can also work well for ex vivo sample fixation; however, this method tends to be slightly more disruptive to fixation-sensitive antigens.

1. Human blood or bone marrow is typically collected in green top (sodium heparin) blood tubes. Samples should be obtained as fresh as practically possible. For cell cycle analysis of human leukemia cells, we will typically collect bone marrow aspirates at the bedside and begin IdU incubation within 1–2 min of collection. Accurate resolution of all phases of the cell cycle

requires samples that have been processed within approximately 30 min of sample collection; however, measurement of more general markers of proliferation can be performed after an hour or more [14]. Samples should be kept at room temperature or 37 °C prior to IdU incubation, which should be performed at 37 °C.

2. As soon as possible after sample collection, cells are placed in a 37 °C incubator and a 100× (1 mM) solution of IdU in PBS is added to achieve a final concentration of 10 μM IdU. The 100× solution is made by dilution of the 50 mM DMSO stock into sterile PBS. This solution can be added to the empty green top tube prior to sample collection, or added after the sample has already been placed into the tube. Invert the sample several times after IdU addition to ensure adequate mixing (*see* **Note 20**).
3. Incubate the green top blood collection tube containing the sample with IdU added for 10–15 min at 37 °C. Longer incubations may be beneficial if samples have cooled to room temperature prior to incubation.
4. Following the 37 °C incubation, the STPS solution, or other fixative such as PFA [13], should be added at a ratio of 1 part sample to 1.4 parts STPS, or a final concentration of 1.5–4% PFA. Following addition of the STPS, invert the sample several times, then incubate at room temperature for 10 min.
5. Following incubation with STPS, the sample can be frozen at −80 °C and stored for up to 2 years (or more). For PFA fixation, multiple postfixation processing methods are available as detailed in Chow et al. [15].
6. At the desired time of sample analysis, the fixed cells should be thawed at 4 °C (this can be done rapidly in a circulating 4 °C water bath). Once thawed, red cells in the sample can be lysed with incubation with 10 volumes of Smart Tube lysis solution (per the manufacturer's protocol).

3.1.3 Sample Processing with In Vivo IdU Incorporation in Mice

1. Mice should be maintained under desired experimental condition up to and during IdU incorporation. Twenty to thirty minutes prior to sacrifice, each mouse should be given an i.p. injection of ~1 mL of IdU solution. This solution can be prepared by diluting 250 mM IdU (in DMSO) into sterile PBS to achieve a 1 mM IdU solution (with ~0.5% DMSO).
2. Mice can be sacrificed in accordance with any established protocol, but care should be taken to minimize physiologic stresses during sacrifice that might lead to disruption of the cell cycle state of the cells of interest.
3. The desired mouse tissues can then be harvested, using any desired application-specific protocol. Our group has harvested

the spleen, lymph nodes, and long bones (for preparation of bone marrow cells), although this protocol should work for most other tissues as well.

4. It is important to harvest the tissues as quickly as possible to preserve the intracellular signaling and cell cycle state. If tissues are harvested on ice, fixation can be performed with PFA while simultaneously warming the samples (in our experience, this prevents the cells from adversely responding to cooling process).
5. Fixation is performed by adding PFA to a final concentration of 1.5%. If the cell samples are at room temperature, fixation can be performed for 10 min. If the samples are on ice when PFA is added, the PFA should be added directly to the cold samples (to a final concentration of 1.5%) and the cold samples should be placed on a mixer at room temperature. As the fixation is temperature-dependent, fixation should be performed for 20 min if the starting temperature is near 0 °C and the samples warm to 20 °C.
6. Pellet the fixed cells by centrifugation at 500 × *g* (5 min at 4 °C) and aspirate the supernate. Wash the fixed cells two times in CSM (500 × *g* for 5 min at 4 °C). After these washes the cells are resuspended in CSM plus 10% DMSO, aliquoted (if desired) and then snap frozen (using liquid nitrogen or a dry ice alcohol bath). The cells can be stored at −80 °C for up to 2 years (or potentially longer; *see* **Note 19**).

3.1.4 Antibody Staining

All cell cycle antibody staining should be performed after cell permeabilization. If the planned permeabilization will disrupt surface antigens required for the experiment, a two-step stain can be utilized by first staining cells with antibodies directed against surface markers of interest, then permeabilizing the cells and staining for intracellular antibodies. This protocol is compatible with cellular barcoding either before or after cell permeabilization [16, 17]. With either type of barcoding, the barcoding should be performed after cell fixation and freezing (if done) but before antibody staining. If cell surface staining will be performed before permeabilization, we will typically perform a second fixation step followed by a methanol permeabilization after completion of the surface stain and two CSM washes (*see* **steps 5–10** of Subheading 3.1.1).

1. **Optional**: If performing cellular barcoding, this should be completed prior to the start of antibody staining [16, 17].
2. **Optional**: If performing a surface stain prior to intracellular permeabilization, wash fixed cells twice in CSM and aspirate to a residual volume. Add surface antibody cocktail at previously determined optimal antibody concentrations and staining volume. Incubate for 30–60 min with mixing to ensure

cells remain in suspension. After completion of staining, wash cells at least twice with CSM.

3. **Optional**: If performing a separate surface stain, we typically perform a second fixation (to fix surface antibodies in place) prior to permeabilization, though this is likely not required for the majority of surface antigen staining. If performed, we typically will fix by resuspending the cell pellet in 1 mL of CSM and then adding PFA to a final concentration of 1.5% for 10 min. After the 10 min incubation, completely fill the tube with CSM and centrifuge to pellet the cells and aspirate to a minimal volume.
4. If cells have not been methanol treated, resuspend cell pellet thoroughly and evenly in the residual CSM, then add 1 mL of ice-cold pure methanol while slowly vortexing to ensure an even single-cell suspension. Incubate on ice for 10 min. Alternatively, cells can be stored at −80 °C in methanol for up to several weeks. For large cell pellets (i.e., after barcoding), the volume of methanol should be increased to ensure cells are in a final concentration of >90% methanol.
5. Cells in methanol (generated in **step 4** above or generated in Subheading 3.1) should be transferred to a tube half-filled with PBS. Fill the remaining volume of the tube with CSM. Pellet cells by centrifugation for 5 min at 600×G, and aspirate supernate (*see* **Note 21**).
6. Wash cells once in CSM. If the volume of methanol added in **step 1** was more than 10% the volume of the tube being used, then a second CSM wash should also be performed to achieve a final methanol concentration of the residual buffer and cell pellet of less than 0.5%.
7. Add a staining cocktail containing the desired antibodies for cell cycle analysis (*see* Table 1) and any additional desired antibodies. We typically stain 1–2 million cells in 100 μL staining reaction for 40–50 min at room temperature, but other staining volumes, incubation times, or incubation temperatures can be used provided the antibodies are titrated for optimal staining under the same conditions.
8. After the completion of staining, wash cells at least twice with CSM.
9. Add at least 100–200 μL of intercalator solution for each million cells in the cell pellet. This can also be titrated for the specific cell types being stained. In our experience, the Ir intercalator staining is dependent on the number of cells being stained and the amount of residual protein. Mix or vortex cells gently to resuspend them evenly in the intercalator solution. Place the cells at 4 °C for at least 20 min. The cells will be stable in the intercalator solution at 4 °C for at least a

week. It is important to ensure the PFA used has not degraded, as the cells will not be stable in intercalator (or subsequently in water) if inactive PFA is used at this step.

10. Wash cells once in CSM and twice in pure water after intercalation. Resuspend the final cell pellet in pure water with calibration beads.
11. Analyze cells on CyTOF mass cytometer. Exact acquisition settings are dependent on which CyTOF version is being used and the cell types being studied.

3.2 Data Analysis

3.2.1 S-Phase Gating

1. The S-phase gate is typically the easiest to determine, but is also the most important, as almost all other cell cycle gates are made on the basis of a biaxial gate of IdU incorporation vs. a second cell cycle parameter. Several ways of making the standard S-phase gate are shown in Fig. 1a. This gate can be created in a biaxial plot of IdU (I127) vs. either: Ki-67, pRb, or Cyclin B1. For healthy cells under normal growth conditions, any of these plots should result in the same S-phase fraction (plotting IdU incorporation vs. Iridium intercalator signal will also work well in this setting); however, disruption of the cell cycle or S-phase can reduce the resolution of the IdU gate, and under these conditions, a plot of IdU vs. Ki-67 or pRb may be most useful.

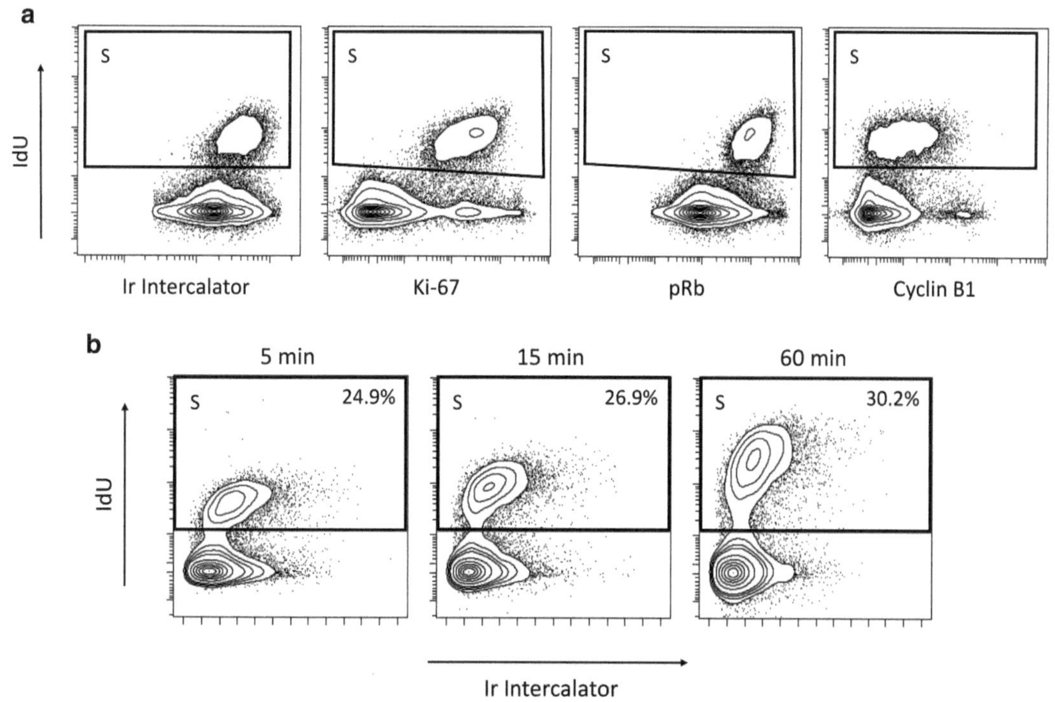

Fig. 1 S-phase gating based on IdU incorporation. The S-phase cell population can be gated using a variety of biaxial plots as shown in (**a**). The effect of increased IdU incubation time is shown in (**b**). Figure adapted from [10]

2. The incorporation of IdU into S-phase cells is dependent on the time of IdU incubation, the IdU concentration, and on the rate of nucleotide synthesis of the cells being studied. As shown in Fig. 1b, as the time of incubation increases, both the IdU signal and the fraction of IdU-positive events will increase. The increase in signal intensity is due to more IdU being incorporated as the constant rate of IdU incorporation continues over a longer time period. The increase in S-phase fraction occurs due to the entry of additional cells into S-phase during the course of the incubation. For most cell types, the fraction of G1 cells that enter S-phase or progress to G2 during the incubation is quite small during a typical 10–30 min IdU incubation; however, longer incubations >1–2 h can result in a significant fraction of the cells in G2 at the time of fixation having IdU incorporated into their DNA from the cell's previous S-phase.

3.2.2 G1 and G2-M Gating

1. While the discrimination of G0 cell from G1-M cells can be visualized by plotting IdU against pRb (Fig. 2a), the discrimination of G1 cells from G2-M cells is based on the level of Cyclin B1 (Fig. 2b). However, Cyclin B1 levels increase continuously from G1 to M phase, making it impossible to determine the exact phase boundaries between the G1, S, and

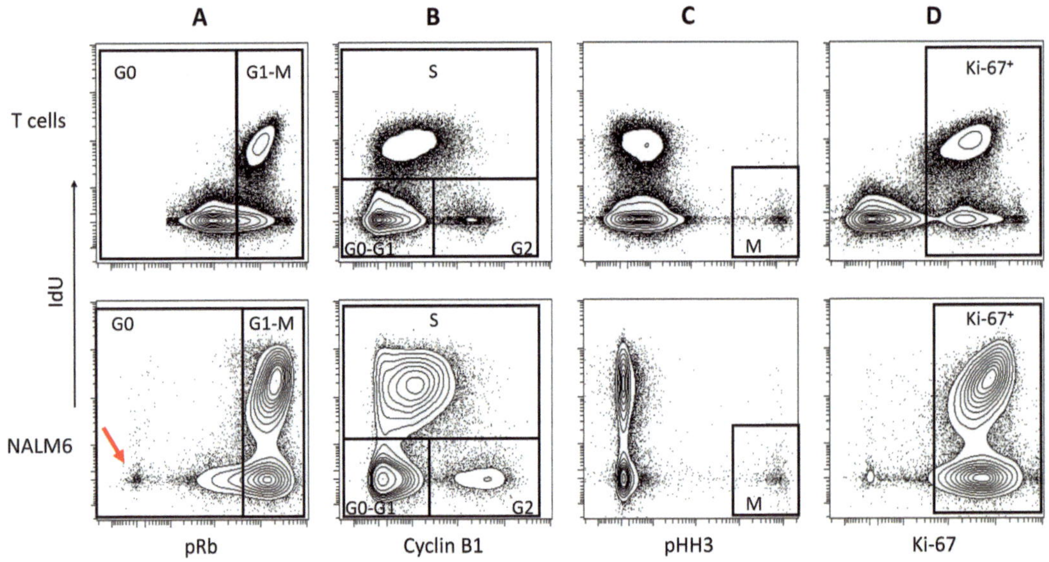

Fig. 2 Major cell cycle phase gates are shown for stimulated human T cells and the NALM6 leukemia cell line. (**a**) G0 cells are identified as having pRb levels of less than 95–99% of S-phase cells. A pRb-negative population is also commonly observed (red arrow); this appears to be composed primarily of apoptotic and necrotic cells, though senescent cells may also be present. (**b**) A biaxial plot of IdU incorporation vs. Cyclin B1 allows for gating of G0-G1 cells, S phase cells, and G2-M cells. (**c**) M-phase cells can be identified based on high levels of pHH3. (**d**) Ki-67⁺ cells are defined based on the level of Ki-67 in 95–99% of S-phase cells, and this cutoff varies across cell types. Figure adapted from [10]

G2 phases based on Cyclin B1 expression unless the S-phase cells can be clearly separated. Thus, this distinction is made based on a plot of IdU incorporation vs. Cyclin B1. This biaxial plot, shown in Fig. 2b, has three distinct populations: CyclinB1^{low} IdU^{-}, correlating to G0-G1 phase cells; Cyclin B1^{mid} IdU^{+}, correlating to S-phase cells; and CyclinB1^{high} IdU^{-}, correlating to G2-M-phase cells.

2. The G2 gate in Fig. 2b can be difficult to draw for two reasons. First, normal (i.e., healthy, untransformed) cells have a very low fraction of cells in G2 compared to commonly used cancer cell lines; thus, the small size of the G2-M phase cell fraction (~2–4%) can make identification of this distinct cell group difficult. Second, the amount of Cyclin B1 antigen in the G2 cells appears to be sensitive to inadequate fixation, with relatively low levels of Cyclin B1 observed in cells that have not been completely fixed. Thus, it is important to ensure good sample processing and that the Cyclin B1 antibody is well labeled with metal and that this antibody is conjugated to a relatively sensitive metal channel. We also routinely notice that normal human cells analyzed ex vivo have significantly lower levels of Cyclin B1 as compared to cultured cell lines.
3. If G2-M phase gating is difficult, back-gating with the M-phase population (Fig. 2c) may help to identify the boundaries of this population. Cyclin B1 levels are high early in M phase and drop during M phase progression; thus, if a significant number of M phase cells are present in the cell population of interest, displaying these cells in the plot of IdU vs. Cyclin B1 will show a large fraction of cells with G2 Cyclin B1 expression and a smaller fraction with very low (G0/G1) Cyclin B1 levels. Use of other cell cycle markers such as Ki-67 and pCDK1 can aid in defining the exact boundaries of the G2 cells in the Cyclin B1 vs. IdU plot.

3.2.3 G0/G1 Gating

1. The discrimination of G0 cells from those in G1 (Fig. 2a) is based on previous work demonstrating that Rb becomes phosphorylated at serine 807 and 811 by a CyclinC-CDK3 complex during the transition from G0 to G1 [18]. Importantly, in many cell types there is not a clear separation of the level of pRb between G0 and G1 cells, making it essential that this gate is determined from a plot of pRb vs. IdU incorporation (*see* Fig. 2a). We typically draw a pRb^{+} gate that includes 95–99% of S phase cells, though this rule may require modification in circumstances where cell cycle checkpoint activation is anticipated. For instance, cells with IdU staining and high levels of DNA damage (e.g., following genotoxin exposure) may exhibit reduced pRb staining, presumably due to activation of an S-phase cell cycle checkpoint.

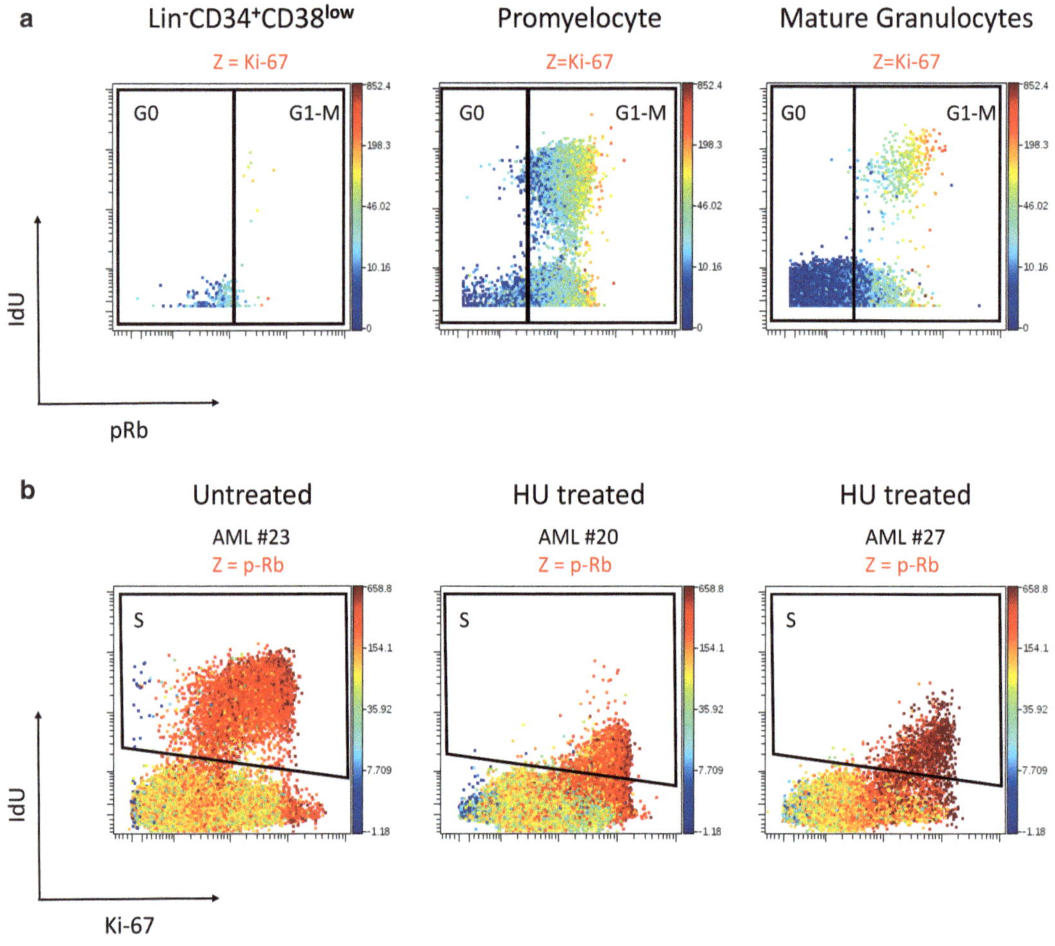

Fig. 3 Special situations in cell cycle gating. (**a**) When the pRb gate position is based on the pRb levels in S-phase cells, the exact gate position changes throughout cellular differentiation from stem and early progenitor cells (Lin$^-$CD34$^+$CD38^{low}) to mature granulocytes. Cells events are colored according to Ki-67 expression level from low expression in blue to high expression in red. (**b**) Samples of CD34$^+$ leukemia cells in patient with AML prior to treatment or during treatment with hydroxyurea (HU). HU treatment leads to a decrease in IdU incorporation and an increase in Ki-67 levels. Cell events are colored according to p-Rb expression level from low expression in blue to high expression in red. Figure adapted from [10]

2. We have also noted that the absolute level of pRb in S-phase cells varies in different cell types, and can vary with cellular differentiation. The G0/G1 gate should thus be examined for each distinct cell type being studied, and the biaxial gate of IdU vs. pRb should be adjusted as needed if the level of pRb in the S-phase cells is found to be different between cell types or stages of differentiation (*see* Figs. 2a, 3a, and 4a).
3. The IdU vs. pRb gate is typically drawn on the total cell population (Fig. 2a), and then applied to the population of G0-G1 cells gated from the IdU vs. cyclin B1 plot (Fig. 2b) as described

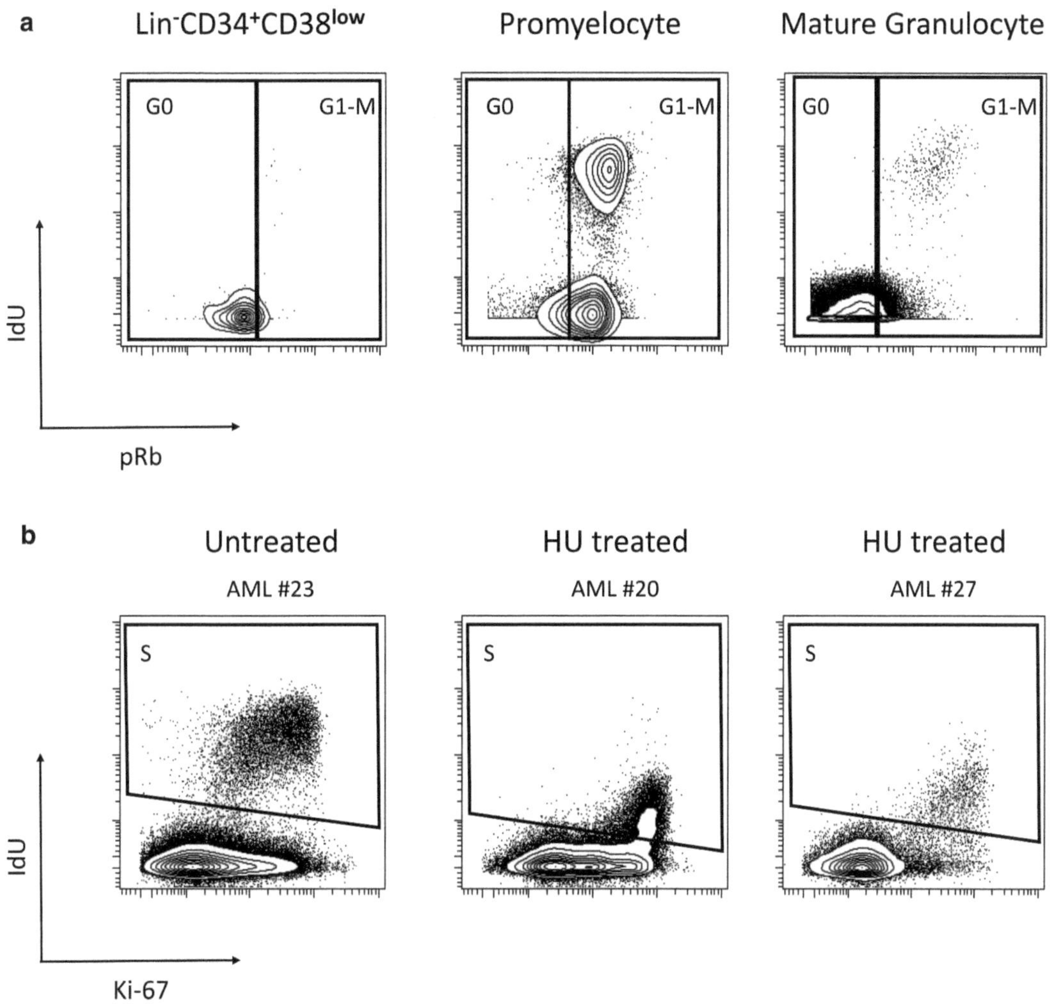

Fig. 4 Special situations in cell cycle gating. (**a**) When the pRb gate position is based on the pRb levels in S-phase cells, the exact gate position changes throughout cellular differentiation from stem and early progenitor cells ($Lin^{-}CD34^{+}CD38^{low}$) to mature granulocytes. (**b**) Samples of $CD34^{+}$ leukemia cells in patients with AML prior to treatment or during treatment with hydroxyurea (HU). HU treatment leads to a decrease in IdU incorporation and an increase in Ki-67 levels

above to arrive at the population of G0 cells (*see* also Table 2). This approach is similar to G0 estimation using Pyronin Y staining of total RNA and G0 fractions determined in this way correlate well with those estimated using Pyronin Y [3].

4. In most cell types, a pRb-negative population is also observed that is significantly dimmer than the majority of cells with levels of pRb below that of the S-phase cells (*see* Fig. 2a; red arrow). This cell population typically represents apoptotic or necrotic cells (though senescent cells may also be present in this population). The addition of antibodies against Ki-67, PCNA, cleaved-caspase3, and cleaved-PARP can all help further delineate the exact state of these cells (*see* **Notes 6–9**).

Table 2
Assignment of cell cycle phase based on gated cell populations

Cell cycle phase	IdU gate	Cyclin B1 gate	pRb	p-HH3
G0	Negative	Low	Low-Mid	Negative
G1	Negative	Low	High	Negative
S	Positive		High	Negative
G2	Negative	High	High	Negative
M	Negative			Positive

3.2.4 M-Phase Gating

1. Gating of M phase cells on the basis of histone H3 phosphorylation at serine 10 or 28 is well established [19, 20]. This assay utilizes antibodies against the S28 phosphorylation on the basis of the slightly higher specificity for M phase cells. M phase cells typically exhibit very high levels of pHH3, so this population is usually easily identifiable; however, these cells are normally quite rare (<1%). This gate is typically drawn around the IdU^- $pHH3^+$ cells in a plot of IdU incorporation vs. pHH3 (*see* Fig. 2c). In healthy untransformed cells, however, progression from S-phase to M-phase can occur quite rapidly. Thus, if IdU incubations are prolonged, cells in M phase at the time of cell fixation my exhibit IdU that was incorporated during the previous S-phase.
2. The pHH3 antibody requires careful titration and thoughtful selection of its position in the design of the antibody panel, since this antibody cannot be used at saturating concentrations (due to the extremely large amount of this antigen in M-phase cells; *see* **Note 5**).
3. M-phase cells also typically exhibit high levels of phosphorylated ribosomal protein S6 (S236/235), Ki-67, and phosphorylated MAMKAPK2 (T334). These markers can be used for confirmation of cell cycle phase assignment or as a control for appropriate cell processing and viability of the cells at the time of fixation.

3.2.5 Final Assignment of Cell Cycle States and Integration with Other Analyses

1. Final cell cycle phase assignment is typically done on the basis of the gate combinations shown in Table 2. The boundaries of positive and negative gates for each marker should be immediately adjacent to one another, such that all cells fall within either the positive or the negative gate for each marker. If this is done, the sum of the percentage of cells in each cell cycle phase will be very close to 100%.

2. We have previously validated this methodology in head-to-head comparisons with a fluorescent flow cytometer method utilizing IdU incorporation, Hoechst DNA staining, and Pyronin Y RNA staining [3]. However, it is likely that this methodology will yield different results from methods utilizing total DNA or RNA staining when cells are being perturbed or are arrested in cell cycle checkpoints. For instance, a cell that has completely arrested DNA synthesis after entering an S-phase checkpoint will not actively incorporate IdU (appearing to be in G0 or G1 by mass cytometry), but may have a total DNA content between 2n and 4n (indicative of S phase by fluorescent total DNA staining). Researchers should understand these differences when interpreting data generated by this assay. Use of other antibodies can enable assessment of cell cycle perturbations such as checkpoint response (e.g., p-H2AX, p-ATM) to enable interpretation of cell cycle data. Other analysis strategies can also be used to derive cell cycle phase information from the measurement of these same markers (*see* **Note 22**). Some examples of unique situations in cell cycle gating are shown in Figs. 3a, b and 4a, b.
3. The assay described in this protocol has been designed for measuring a "snapshot" of the cell cycle at the time of fixation. Assessment of the total cell cycle time or of precise cell cycle kinetics typically requires labeling with two different halogenated uridine analogs (e.g., CldU and IdU) [2]. This method is compatible with these protocols; however, assessment of uridine analogs that incorporate halogen atoms of less than 75 Daltons (FdU, CldU) cannot be measured directly by mass cytometry. While BrdU should theoretically be measureable, its low mass, higher ionization energy, isotopic distribution (51% ^{79}Br; 49% ^{81}Br), and the presence of a strong argon dimer signal at 80 Daltons, all complicate direct measurement of this reagent. It is, however, possible to utilize metal-conjugated antibodies directed against any of these uridine analogs to perform double-labeling experiments (provided the antibody chosen does not cross-react with IdU. The 3D4 clone of anti-BrdU antibody, for example, has been shown to label both BrdU and IdU (*see* Chapter 12).
4. When integrating cell cycle measurement into higher parameter analyses, it is typically best to first perform a SPADE [21] or viSNE [22] (or other clustering) analysis utilizing relevant markers of cell type first, followed later by analysis of cell cycle state within the clustered cell populations of interest. This is advantageous for several reasons. First, performing clustering using markers of cell type, and not markers of cell cycle, allows for a broad assessment of populations in the data, without grouping cells together based on cell cycle status. This approach

is necessary, as some of the cell cycle markers exhibit high expression levels that would create artifactual associations between immunophenotypically different cells within the same cell cycle state if used for clustering (e.g., S-phase monocytes and S-phase B cells co-existing within the same node or viSNE region). Second, the majority of cell cycle markers do not behave as continuously distributed variables (e.g., IdU expression is bimodal, p-Rb has three distinct populations), making the display of the bulk median expression level of these markers less helpful. Third, the exact marker expression levels that define the gate boundaries of each cell cycle phase can be specific for a given cell type or developmental stage within a cell lineage. This means that the same absolute expression level of a marker such as p-Rb could be indicative of different cell cycle states in different cell types, and clustering cell types into individual populations aids in this cell-type-specific approach.

4 Notes

1. We have noted the IdU solution will turn brown if left out at room temperature, so we typically make single-use aliquots.
2. Ki-67 is not essential for identification of cell cycle phase; however, it is quite useful given its broad use in cell biology and pathology. This marker also provides additional resolution of G0 and G1 cell cycle phases. While traditionally interpreted as positive or negative, we find that Ki-67 is best defined by drawing Ki-67 positive gates based on the level of Ki-67 in S-phase cells, similar to how the pRb^+ gate is drawn.
3. p-Rb (S807/811) increases during the transition from G0 to G1 [18] and is used for separation of G0 and G1 cells. This phosphorylation site is different from the classical Rb phosphorylation site (S780) associated with G1-S phase progression. pRb (S807/811) staining typically identifies 3 populations: negative cells that are apoptotic or senescent, pRb-mid cells which are G0 cells, and pRb bright which are cells in G1-M phase of the cell cycle. pRb requires a relatively bright metal channel that does not have significant isotopic contamination or high level of oxidation (as the signal can become quite bright). The best metal channels are Tb159, Ho165, Tm169, or La175.
4. Cyclin B1 is used to define G2/M cells (Cyclin B1 high, IdU negative). This antibody doesn't label well with metal, so it needs to be used on a relatively bright channel, but most channels above Eu151 should work well. Clear resolution of G2 cells with Cyclin B1 requires good cell fixation, and tends to be slightly better if cells are permeabilized with methanol

immediately after fixation. Note that normal cells (i.e., not transformed) tend to have much lower cyclin B1 staining and frequencies of G2 cells than cultured cell lines.

5. pHH3 staining is extremely bright on M-phase cells. This metal needs to be used on a channel without significant isotopic contamination and any antibody used on the mass channel 1 Dalton above may receive spillover from bright pHH3 signal on M-phase cells. Good metals to use for this antibody are La175 or Lu176. Because of the very large amount of antigen present on M-phase cells, it is difficult to saturate this antigen; as a result, staining reactions will be sensitive to the number of M-phase cells present (higher cell numbers will result in dimmer staining, lower cell numbers in brighter staining).
6. Ki-67 is used as a second measure of proliferation. It is generally quite consistent with pRb, but we have observed subtle differences between the two markers. Measuring both gives additional confidence to the assignment of G0 cells (i.e., cells negative for both Ki-67 and pRb). We typically use clone SolA15 which is actually raised against mouse Ki-67, but has good cross-reactivity with human Ki-67 allowing this reagent to be used for both murine and human cells (unlike many other human-specific anti-Ki-67 antibody clones).
7. Cleaved-Caspase 3 is used to identify apoptotic cells. In our experience, cleaved-Caspase 3 is slightly more specific for apoptotic cells than cleaved-PARP; however, cleaved-Caspase 3 typically gives slightly dimmer staining. For experiments where identification of all apoptotic cells is extremely important, we will use both cleaved-Caspase 3 and cleaved-PARP.
8. Cleaved-PARP is used to identify apoptotic cells. In our experience, cleaved-PARP gives slightly higher staining intensities than cleaved-Caspase 3; however, we have rarely observed small cell subsets with cleaved PARP staining that do not appear to be apoptotic by cleaved-Caspase 3 and other cell cycle measures. For experiments where identification of all apoptotic cells is extremely important, we will use both cleaved-Caspase 3 and cleaved-PARP.
9. Like Ki-67, PCNA is used as a second measure of proliferation to identify the G0-G1 transition. All three makers (pRb, Ki-67, and PCNA) can be used simultaneously if identification of G0 cells is essential; however, we have only rarely observed significant discrepancies between Ki-67 and PCNA.
10. p-RP-S6 signal is strongly positive on M-phase cells. We use this antibody to provide confirmation of M-phase assignment. This marker is also useful for quality control, as we have observed cells that were not fixed immediately after collection loose p-RP-S6 signal on M-phase cells, suggesting that other cell cycle or intracellular signaling measures may also be disrupted.

11. pH2AX staining helps to identify cells with DNA damage, which may indicate that these cells have entered a cell cycle checkpoint. Knowing which cells have DNA damage can help to explain alterations of expected correlations in cell cycle response (e.g., unexpected observation of low levels of pRb in IdU-positive cells), and it may be helpful to ignore p-H2AX-positive cells when determining some gate boundaries (e.g., G0/G1 gates).
12. Cyclin A staining can be used to help identify G0 cells (negative for Cyclin A) and to assist with G2 gating (G2 cells should be slightly lower for Cyclin A relative to Cyclin B1 [3, 11]. However, the resolution of this marker is not as good as Cyclin B1 or p-Rb, making it less useful for cell cycle assignment than these other markers.
13. Cyclin E staining can also help with resolution of G0 vs. G1 (G0 cells should be negative) and is useful for resolution of G2 phase cells (which should be negative for Cyclin E) [3, 11]. Like Cyclin A, however, the lower staining intensity of this antigen by mass cytometry limits its utility.
14. p-CDK1 staining should peak in G2 phase and be lost upon transition to M phase; it can thus be helpful as a confirmatory marker of these cell cycle phases. The resolution of this marker is also not very good, making it less useful than the other markers of these cell cycle phases.
15. The minimum number of cells required for mass cytometry analysis varies based on the exact number of centrifugation steps, the type of centrifuge tubes employed, and the number of washes performed. For the protocol described here (utilizing standard 5 mL polystyrene FACS tubes), 500,000 cells is a good minimum number as the mass cytometer generates data from 30–50% of the cells that are injected, necessitating a proportional increase in the number of starting cells (as compared to fluorescent flow cytometry). Utilizing smaller, polypropylene tubes, and reducing the number of staining and wash steps can allow for the analysis to be performed with lower starting cell numbers (as low as 100,000). Cells fixed according to this protocol will be stable for at least several months, and we have successfully analyzed cells 2–3 years after fixation.
16. While we have not observed any evidence of DNA damage or cell death with short IdU incubations, we have observed poor cell proliferation and toxicity with prolonged IdU incubations. We suspect that toxicity may occur during the replication of DNA strands containing IdU that had been incorporated during the previous S-phase.

17. All centrifugation steps prior to methanol permeabilization can be performed at 300–600 × *g* (or whatever centrifugal force is known to be optimal for the cells of interest) for 5 min. After methanol permeabilization, centrifugation should be performed at least 600 × *g*. We recommend use of a refrigerated swinging bucket centrifuge (e.g., Sorvall Legend XTR [Thermo/Fisher], or similar) set to 4 °C.

18. Throughout this protocol, a "wash" refers to resuspending the cell pellet in 10–20 volumes (typically about 5 mL) of CSM followed by centrifugation to pellet the cells (typically 300–600 × *g* for 5 min at 4 °C) and aspiration of the supernate. If the volume of the tube being used is not large enough to allow for 10 volumes of CSM to be added, perform additional washes until the total dilution is greater than 200-fold after all washes. Note that washing is much more important in mass cytometry than fluorescent flow cytometry, since residual antibody in solution can create significant background, measurement error, and may increase wear on the mass cytometer's detector.

19. It is not required to freeze samples after fixation; however, given the time required to perform mass cytometry experiments, and the frequent need to collect multiple time-points for cell cycle studies, freezing is typically required for practical purposes. If freezing is not desired, samples can be taken straight into the staining steps after fixation and two CSM washes.

20. The Smart Tube proteomic stabilizer can also be purchased as part of Smart Tubes that are designed to work with an automated processing machine, the Smart Tube base station. These tubes have the fixative solution contained in a glass vial that is broken open by the base station at a user-specified time to start fixation. The user can also program the base station to perform incubations at any desired temperature or duration allowing these tubes to be used to perform the incubations for IdU incorporation (*see* Subheading 3.1.2, **steps 3** and **4**) prior to fixation. Following the 10 min fixation, the Smart Tube base station will cool the sample to 4 °C slowing the fixation reaction and allow the user several hours to move the sample to −80 °C.

21. The BSA in the CSM will precipitate if added to a solution with a high concentration of methanol, so it's important to reduce the methanol concentration before adding the CSM to the cell suspension.

22. Alternate gating approaches, in addition to the biaxial plots described above, can be used. Measurement of these same markers enables cell cycle state assignment through the use of clustering methods such as SPADE [21] or viSNE [22]. When

the core markers cell cycle markers (shown in Table 1) are used as the basis for clustering in these analysis algorithms, clear populations that correlate with each cell cycle state are readily identifiable.

Acknowledgments

This chapter has been adapted from a previous chapter [10].

References

1. Darzynkiewicz Z, Juan G, Srour EF (2004) Differential staining of DNA and RNA. Curr Protoc Cytom Chapter 7:Unit 7 3. https://doi.org/10.1002/0471142956.cy0703s30
2. Pozarowski P, Darzynkiewicz Z (2004) Analysis of cell cycle by flow cytometry. Methods Mol Biol 281:301–311. https://doi.org/10.1385/1-59259-811-0:301
3. Behbehani GK, Bendall SC, Clutter MR et al (2012) Single-cell mass cytometry adapted to measurements of the cell cycle. Cytometry A 81(7):552–566. https://doi.org/10.1002/cyto.a.22075
4. Raval A, Behbehani GK, Nguyen le XT et al (2015) Reversibility of defective hematopoiesis caused by telomere shortening in telomerase knockout mice. PLoS One 10(7):e0131722. https://doi.org/10.1371/journal.pone.0131722
5. Behbehani GK, Samusik N, Bjornson ZB et al (2015) Mass cytometric functional profiling of acute myeloid leukemia defines cell-cycle and immunophenotypic properties that correlate with known responses to therapy. Cancer Discov 5(9):988–1003. https://doi.org/10.1158/2159-8290.cd-15-0298
6. Strauss-Albee DM, Fukuyama J, Liang EC et al (2015) Human NK cell repertoire diversity reflects immune experience and correlates with viral susceptibility. Sci Transl Med 7(297):297ra115. https://doi.org/10.1126/scitranslmed.aac5722
7. Chang Q, Ornatsky OI, Koch CJ et al (2015) Single-cell measurement of the uptake, intratumoral distribution and cell cycle effects of cisplatin using mass cytometry. Int J Cancer 136(5):1202–1209. https://doi.org/10.1002/ijc.29074
8. Wei SC, Levine JH, Cogdill AP et al (2017) Distinct cellular mechanisms underlie anti-CTLA-4 and anti-PD-1 checkpoint blockade. Cell 170(6):1120–1133.e1117. https://doi.org/10.1016/j.cell.2017.07.024
9. Krutzik PO, Nolan GP (2003) Intracellular phospho-protein staining techniques for flow cytometry: monitoring single cell signaling events. Cytometry A 55(2):61–70. https://doi.org/10.1002/cyto.a.10072
10. Behbehani GK (2018) Cell cycle analysis by mass cytometry. In: Lacorazza HD (ed) Cellular quiescence: methods and protocols. Springer, New York, pp 105–124. https://doi.org/10.1007/978-1-4939-7371-2_8
11. Darzynkiewicz Z, Gong J, Juan G et al (1996) Cytometry of cyclin proteins. Cytometry 25(1):1–13. https://doi.org/10.1002/(sici)1097-0320(19960901)25:1<1::aid-cyto1>3.0.co;2-n
12. Fienberg HG, Simonds EF, Fantl WJ et al (2012) A platinum-based covalent viability reagent for single-cell mass cytometry. Cytometry A 81(6):467–475. https://doi.org/10.1002/cyto.a.22067
13. Chow S, Hedley D, Grom P et al (2005) Whole blood fixation and permeabilization protocol with red blood cell lysis for flow cytometry of intracellular phosphorylated epitopes in leukocyte subpopulations. Cytometry A 67(1):4–17. https://doi.org/10.1002/cyto.a.20167
14. Devine RD, Sekhri P, Behbehani GK (2018) Effect of storage time and temperature on cell cycle analysis by mass cytometry. Cytometry 93:1141–1149. https://doi.org/10.1002/cyto.a.23630
15. Will B, Zhou L, Vogler TO et al (2012) Stem and progenitor cells in myelodysplastic syndromes show aberrant stage-specific expansion and harbor genetic and epigenetic alterations. Blood 120(10):2076
16. Behbehani GK, Thom C, Zunder ER et al (2014) Transient partial permeabilization with saponin enables cellular barcoding prior to surface marker staining. Cytometry A 85(12):1011–1019. https://doi.org/10.1002/cyto.a.22573

17. Zunder ER, Finck R, Behbehani GK et al (2015) Palladium-based mass tag cell barcoding with a doublet-filtering scheme and single-cell deconvolution algorithm. Nat Protoc 10(2):316–333. https://doi.org/10.1038/nprot.2015.020. http://www.nature.com/nprot/journal/v10/n2/abs/nprot.2015.020.html#supplementary-information
18. Ren S, Rollins BJ (2004) Cyclin C/Cdk3 promotes Rb-dependent G0 exit. Cell 117(2):239–251. https://doi.org/10.1016/s0092-8674(04)00300-9
19. Goto H, Yasui Y, Nigg EA et al (2002) Aurora-B phosphorylates histone H3 at serine28 with regard to the mitotic chromosome condensation. Genes Cells 7(1):11–17. https://doi.org/10.1046/j.1356-9597.2001.00498.x
20. Hirata A, Inada K, Tsukamoto T et al (2004) Characterization of a monoclonal antibody, HTA28, recognizing a histone H3 phosphorylation site as a useful marker of M-phase cells. J Histochem Cytochem 52(11):1503–1509. https://doi.org/10.1369/jhc.4A6285.2004
21. Qiu P, Simonds EF, Bendall SC, et al (2011) Extracting a cellular hierarchy from high-dimensional cytometry data with SPADE. Nat Biotechnol 29(10):886–891. http://www.nature.com/nbt/journal/v29/n10/abs/nbt.1991.html#supplementary-information
22. Amir el AD, Davis KL, Tadmor MD et al (2013) viSNE enables visualization of high dimensional single-cell data and reveals phenotypic heterogeneity of leukemia. Nat Biotechnol 31(6):545–552. https://doi.org/10.1038/nbt.2594

Chapter 14

Picturing Polarized Myeloid Phagocytes and Regulatory Cells by Mass Cytometry

Mikael Roussel, Todd Bartkowiak, and Jonathan M. Irish

Abstract

The immune monocyte/phagocyte system (MPS) includes numerous cell subsets of the myeloid lineage including monocyte, macrophage, and dendritic cell (DC) populations that are heterogeneous both phenotypically and functionally. Previously, we characterized these diverse MPS phenotypes with multi-parametric mass cytometry (CyTOF). In order to expansively characterize monocytes, macrophages, and dendritic cells, a CyTOF panel was designed to measure 35 identity-, activation-, and polarization-markers. Here we provide a protocol to define a reference map for the myeloid compartment, including sample preparation, to produce reference cell subsets from the monocyte/phagocyte system. In particular, we focused on monocyte-derived macrophages that were further polarized in vitro with cytokine stimulation (i.e., M-CSF, GM-CSF, IL-4, IL-10, IFNγ, and LPS), as well as monocyte-derived DCs, and myeloid-derived suppressor cells (MDSCs), generated in vitro from human bone marrow and/or peripheral blood.

Key words Mass cytometry, Macrophage polarization, Dendritic cells, Myeloid-derived suppressor cells, Myeloid regulatory cells

1 Introduction

The monocyte/phagocyte system (MPS) is a complex cellular compartment that includes phenotypically and functionally heterogeneous populations of cells, including monocytes, macrophages (MΦ), and dendritic cells (DC) [1]. A distinct phenotypic definition of myeloid cells remains contentious due to a number of factors. For instance, a lack of consistency between cellular expression of markers was first identified in mice and correlates on human myeloid cells (e.g., while murine macrophages are defined as F4/80^{high}, the human F4/80 homolog is expressed on eosinophils). Further, many of the markers of interest expressed on human myeloid cells (e.g., CD14, CD11b, CD33, HLA-DR, CD64) are shared between various myeloid cell subsets and none are lineage specific. Moreover, myeloid cells are highly plastic with respect to phenotype and function and depend upon various environmental signals for differentiation

Helen M. McGuire and Thomas M. Ashhurst (eds.), *Mass Cytometry: Methods and Protocols*, Methods in Molecular Biology, vol. 1989,
https://doi.org/10.1007/978-1-4939-9454-0_14, © Springer Science+Business Media, LLC, part of Springer Nature 2019

and/or polarization. This complexity of phenotypic definition is highlighted by the growing literature on monocyte, DC, and macrophage nomenclature [1–5].

At the protein level, characterization of these heterogeneous cell types has been largely accomplished with "low resolution" approaches (e.g., morphological evaluation and immunohistochemistry), wherein only one or a few proteins were used to identify populations. As an example, CD68 and CD163 are frequently proposed to characterize macrophage subtypes. High-resolution approaches such as mass cytometry (also known as cytometry by time-of-flight, or CyTOF) are invaluable tools that will advance our understanding of cellular diversity and function, and identify potential targets for novel therapies [6–8]. CyTOF combined with high-dimensional analysis, in particular visualization of t-distributed stochastic neighbor embedding (viSNE), spanning-tree progression analysis of density-normalized events (SPADE), and marker enrichment modeling (MEM), are robust methods to identify numerous and novel subsets from heterogeneous populations [9–15].

During the last months, myeloid cell subsets were defined in humans within various healthy and tumor tissues. Beside definition of DC- [16, 17] or monocyte-subsets [18, 19], large numbers of tumor-associated macrophage (TAM) phenotypes were revealed in renal cell carcinoma (17 TAM phenotypes were distinguished) and lung adenocarcinoma [20, 21]. Notably, mass cytometry definition of myeloid cells subsets was also linked to patient outcomes, in particular, in stage IV melanoma monocytes ($CD14^{pos}CD16^{neg}HLA\text{-}DR^{high}$) were predictive in response to anti-PD-1 immunotherapy [22]. Ex vivo preparation of myeloid counterparts might help to guide the high-dimensional analysis and to define a reference cartography (Fig. 1) of myeloid heterogeneity [23].

Here we provide a protocol to define a reference map for the myeloid compartment including procedures and sample preparation amenable to the creation of reference cell subsets from the monocyte/phagocyte system, in particular, monocyte-derived macrophages further polarized using in vitro cytokine stimulation (i.e., M-CSF, GM-CSF, IL-4, IL10, IFNγ, or LPS), as well as monocyte-derived DCs, and myeloid-derived suppressor cells (MDSCs), generated in vitro from human bone marrow and/or peripheral blood [23].

2 Materials

2.1 In Vitro Polarization of Monocyte-Derived DC, MΦ, and MDSC

1. A source of human monocytes (*see* **Note 1**).
2. Six-well tissue-culture-treated plates.
3. Tissue culture media (RPMI-1640) with 10% FBS.
4. Percoll or Ficoll leukocyte isolation media.

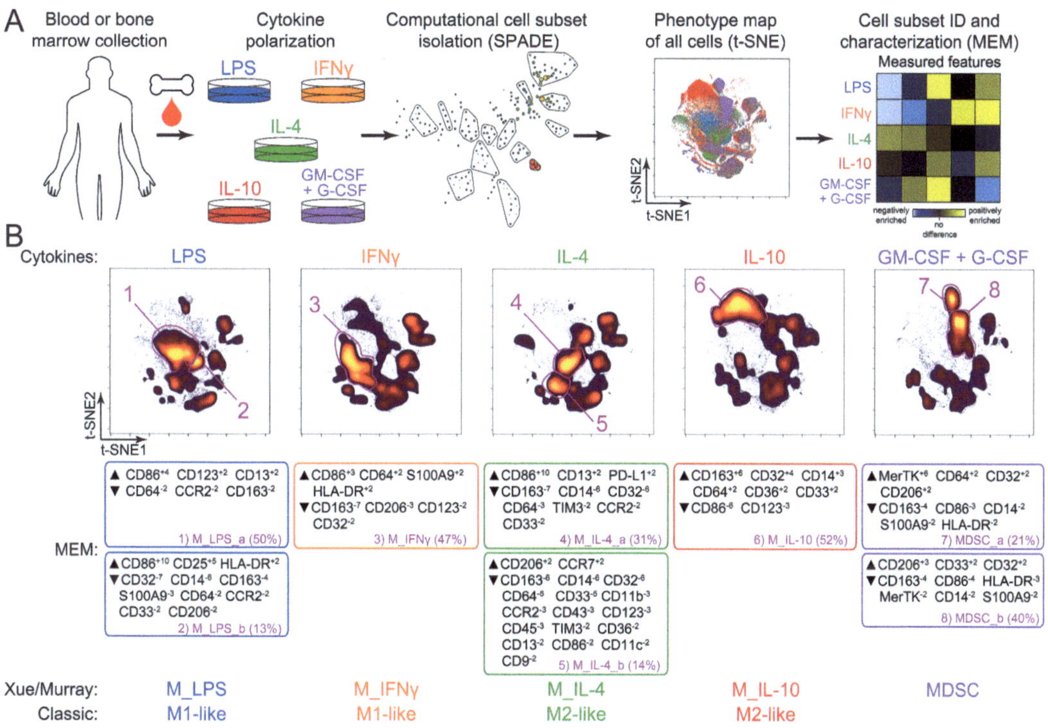

Fig. 1 Single-cell analysis platform for interrogation of human myeloid cell polarization phenotypes. (**a**) Patient-derived tissue (human blood or bone marrow) can be single-cell dissociated and cultured in a variety of cytokine stimulation conditions (LPS, IFNγ, IL-4, IL-10, GM-CSF/G-CSF) for various lengths of time in order to polarize myeloid populations into distinct phenotypes. Polarized cell populations may then be computationally isolated using minimum spanning tree algorithms (SPADE) and phenotypically identified using t-SNE and Marker Enrichment Modeling analysis. (**b**) viSNE maps (top) of polarized myeloid populations reveal distinct subsets of myeloid populations after cytokine stimulation, correlating with enrichment of phenotypic markers (MEM labels; middle) correlating to classically defined populations of macrophages or myeloid-derived suppressor cells (bottom)

5. Cytokines, growth factors, and stimulants: M-CSF, GM-CSF, IL-4, IL-10, LPS, IFNγ, TNFα, PGE2, Pam3CSK4 (a Toll-like receptor 2 agonist).

2.2 Antibody Staining and Mass Cytometry

1. Metal isotope-conjugated antibodies (*see* **Note 2**).
2. Staining buffer (PBS + BSA 0.5%).
3. Phosphate buffered saline (PBS).
4. Viability marker: e.g., cisplatin Cell-ID™ (Fluidigm).
5. Paraformaldehyde (PFA): 16%.
6. Ice-cold methanol at −20 °C.
7. Double-distilled water (ddH_2O).
8. Cell intercalator dye (e.g., Iridium; Cell-ID™ Intercalator-Ir, Fluidigm).

9. CyTOF Calibration Beads (Fluidigm).
10. Mass cytometer (Fluidigm).

2.3 Data Analysis

1. Software for gating.
2. A pipeline for high-dimensional analysis (e.g., Cytobank, Cytofkit, or "R" algorithms).

3 Methods

The purpose of the protocol is to derive DC, MDSC, and MΦ populations from peripheral human monocytes and then to polarize these baseline MΦ. These polarization schema were inspired by Xue et al. [24].

3.1 Prepare Blood-Derived Monocytes

1. Prepare a Buffy coat by pooling at least 40 mL of human peripheral blood (*see* **Note 3**).
2. Centrifuge for 15 min at 600 *g* with no brake.
3. Decant plasma (*see* **Note 4**) and resuspend the buffy coat by 4× with RPMI + 10% FBS.
4. Add 1× Ficoll and overlay with diluted buffy coat.
5. Proceed with Ficoll gradient separation by centrifuging at 400 *g* for 30 min.

3.2 Monocyte Separation by Plastic Adherence (Optional, *See* Note 1)

1. Use 6-well, tissue-culture-treated plates (surface 9.5 cm^2).
2. After Ficoll separation, resuspend PBMC at 1×10^6 cells/mL in serum-free RPMI.
3. Add 2 mL of the suspension into 6-well plates; prepare wells for various conditions including Monocyte, DC, Macrophage baseline, Macrophage Unstimulated, and as many wells as the chosen stimulation conditions (in replicate if possible due to potential cell mortality).
4. Incubate at 37 °C, 5% CO_2 for a minimum of 3 h.
5. Gently shake and discard non-adherent cells by washing twice with pre-warmed culture medium.
6. Collect only the cellular fraction of the wells labeled "Monocyte" after 5 min of Accutase treatment. For the other wells, start the differentiation as described in Subheading 3.3.
7. Discard the media (and freeze the supernatant as control).
8. Wash twice with 1 mL of PBS and decant.
9. Add 250 μL of Accutase pre-warmed at 37 °C.
10. Incubate and watch every 30 s (max 2 min).

11. Add 2 mL of media.
12. Pipette and transfer into a FACS tube, then begin Subheading 3.4 for the staining.

3.3 *In Vitro* Monocyte-Derived DC, MDSC, and MΦ Polarization

3.3.1 Polarization of Dendritic Cells

1. Add RPMI-1640 supplemented with 10% FBS and 1% PenStrep solution (1.5–2 × 10^6 cells/mL, 3 mL/well) to the "DC" well(s).
2. Add 800 IU/mL rhGM-CSF (40 ng/mL) and 500 IU/mL rhIL4 (40 ng/mL).
3. Incubate at 37 °C, 5% CO_2 for 3 days.
4. At day 3: Change the medium (centrifuge any floating cells).
5. At day 6: For terminal moDC maturation, add rhTNF (800 IU/mL or 10 ng/mL) for 48 h.
6. At day 8: Collect the cellular fraction in the wells after Accutase treatment (after maturation of some adherent cells).
7. Discard the media (freeze supernatant if desired).
8. Wash with 1 mL of PBS and decant.
9. Add 250 μL of Accutase pre-warmed at 37 °C.
10. Place in 37 °C in incubator and watch every 30 s (max 2 min).
11. Add 2 mL of media.
12. Wash in PBS, and proceed to staining (Subheading 3.4).

3.3.2 Generation of MDSC (*See* Notes 5 and 6)

1. Add RPMI-1640 supplemented with 10% FBS and 1% PenStrep solution (1.5–2 × 10^6 cells/mL, 3 mL/well) to the "MDSC" well(s).
2. Add rhGM-CSF (40 ng/mL) and rhG-CSF (40 ng/mL) and/or rhGM-CSF (40 ng/mL) and rhIL-6 (40 ng/mL).
3. Place in an incubator at 37 °C, 5% CO_2 for 4 days.
4. At day 4: Collect the cellular fraction in the wells after Accutase treatment (after maturation of some adherent cells).
5. Discard the media and freeze the supernatant if desired.
6. Wash with 1 mL of PBS 1 mL and decant.
7. Add 250 μL of Accutase, pre-warmed at 37 °C.
8. Incubate at 37 °C and watch every 30 s (max 2 min).
9. Add 2 mL of RPMI culture media.
10. Wash in PBS, and proceed to staining (Subheading 3.4).

3.3.3 Generation of Macrophages

1. Add RPMI-1640 supplemented with 10% FBS and 1% PenStrep solution (1.5–2 × 10^6/mL, 2 mL/well) to the "macrophage" well(s).
2. Add 50 IU/mL rhM-CSF.

3. Incubate at 37 °C, 5% CO_2 for 3 days.
4. At day 3: Follow the stimulation step (Subheading 3.3.3) for macrophages other than the "Macrophage baseline" wells.
5. For the Macrophage baseline well: Collect the cellular fraction in the well after Accutase treatment.
6. Discard the media and freeze the supernatant if so desired.
7. Wash once with 1 mL PBS.
8. Discard PBS.
9. Add 250 μL Accutase pre-warmed at 37 °C.
10. Incubate at 37 °C, 5% CO_2, and watch every 30 s (max 2 min).
11. Add 2 mL of RMPI tissue culture media and collect the cells.
12. Wash wells one time in PBS and proceed to staining (Subheading 3.4).

3.3.4 Macrophage Polarization

1. Three days after initiating cytokine stimulation (Subheading 3.3.3), remove and freeze the supernatants.
2. During the following 3 days, stimulate cells with RPMI-1640 supplemented with 10% FBS and 1% PenStrep solution and Mock/Carrier, rhIFNγ (200 IU/mL), rhIL-4 (1000 IU/mL), TNF (800 IU/mL), Pam3CSK4 (P3C, 1 μg/mL), or prostaglandin E2 (PGE2, 1 μg/mL) (*see* **Note 7**).
3. At day 6 post initiation, collect the "Macrophage baseline" well after Accutase treatment.
4. Discard the media and freeze the supernatant if so desired.
5. Wash one time 1 mL PBS.
6. Discard PBS.
7. Add 250 μL Accutase pre-warmed to 37 °C.
8. Incubate at 37 °C, 5% CO_2 and watch every 30 s (max 2 min).
9. Add 2 mL of media.
10. Wash in PBS, and proceed to staining (Subheading 3.4).

3.4 Antibody Staining and Mass Cytometry Analysis

3.4.1 Prepare Staining Cocktail (*See* Note 8)

1. Make enough cocktail for the number of samples in PBS + BSA, plus add an extra 10% to account for pipette error. Samples can be stained in a final volume of 100 μL.

3.4.2 Live Cell Surface Stain

1. Transfer 30 μL of cells to a new FACS tube.
2. Add 80 μL of premade staining cocktail.
3. Vortex to mix.
4. Stain at room temperature for 30 min.

5. Wash 2× with 2 mL PBS + BSA.
6. Centrifuge at 200 *g*, for 5 min.
7. Repeat the staining if needed for secondary antibodies.

3.4.3 Cell Fixation

1. Decant the supernatant from Subheading 3.4.2.
2. Resuspend the cells in 200 μL PBS.
3. Fix by adding 25 μL of 16% PFA (for a final concentration of 1.8% PFA).
4. Vortex to mix.
5. Incubate for 10 min at room temperature.
6. Wash 2× with 2 mL PBS.
7. Centrifuge at 900 *g* for 5 min.

3.4.4 Cell Permeabilization

1. Decant supernatants from the last step in Subheading 3.4.3.
2. Resuspend the cells in the residual volume left after decanting by vortexing vigorously.
3. Add 1 mL of ice-cold methanol (−20 °C).
4. Vigorously vortex immediately.
5. Pipet as needed to break up clumps.
6. Incubate cells at −20 °C for at least 10 min. Cover the tube to avoid evaporation. Cells can be left overnight at −20 °C or for weeks at −80 °C.
7. Wash the cells 2× with 2 mL PBS.
8. Vortex to mix.
9. Centrifuge at 900 *g* for 5 min.

3.4.5 Intracellular Staining (If Required)

1. Resuspend the cells in staining media to a total volume of 40 μL.
2. Add 80 μL of premade staining cocktail.
3. Vortex to mix.
4. Stain for 30 min at room temperature.
5. Wash 2× with 2 mL PBS + BSA.
6. Centrifuge at 900 *g* for 5 min.
7. Repeat staining if needed for secondary antibodies.

3.4.6 Nucleic Acid Staining

1. Wash the samples from the previous step (Subheading 3.4.5) with 2 mL PBS + BSA.
2. Centrifuge at 900 *g* for 5 min. Decant.
3. Resuspend the cells in 200 μL PBS.
4. Add 4 μL 50× Iridium nucleic acid intercalator.

5. Vortex to mix.
6. Incubate for at least 15 min at room temperature. Cells can be left at 4 °C for several hours.

*3.4.7 Running Samples on CyTOF (**See Note 9**)*

1. Wash the samples with 1 mL double deionized water.
2. Centrifuge at 900 *g* for 5 min.
3. Dilute the sample in 1× CyTOF calibration beads (400 μL to 1 mL according to the number of cells).
4. Run samples on a CyTOF cytometer according to the manufacturer's protocol.

4 Notes

1. Monocytes can be obtained from buffy coats followed by cell sorting, plastic adherence, or elutriation.
2. Antibodies can be (1) bought from Fluidigm pre-conjugated to metal isotopes, (2) bought from another vendor and self-conjugated using the Fluidigm Maxpar conjugation kit, or (3) used in indirect staining with an anti-FITC, anti-PE, anti-APC, or anti-biotin metal-tagged antibodies.
3. The volume of blood or number of monocytes requested for the whole experiment depends on the number of experimental conditions and should be calculated before starting. Also take into account that a substantial number of monocytes and macrophages will adhere to the plastic dish and be lost in processing.
4. If molecule analyses are planned at different time points, spin the plasma at 1500 *g* for 10 min before aliquoting in 500 μL at −20 °C. These aliquots will constitute the reference point.
5. Peripheral blood or bone marrow may be used, but give rise to different suppressive myeloid cells both matching an MDSC phenotype [23, 25, 26].
6. Suppressive function of the cells should be assessed [4, 23].
7. Supernatant from a cell line or primary cells culture can also be used.
8. Make separate staining cocktails for surface and intracellular markers. Up to 4 staining cocktails might be necessary if secondary antibodies are employed in the panel. Transfer cells to new tubes for each staining cocktail so that volumes are precise (important for comparing between samples in particular for phosphoproteins).
9. Example data files available online: http://flowrepository.org/id/FR-FCM-Z2Z8

References

1. Guilliams M, Ginhoux F, Jakubzick C et al (2014) Dendritic cells, monocytes and macrophages: a unified nomenclature based on ontogeny. Nat Rev Immunol 14:571–578. https://doi.org/10.1038/nri3712
2. Ancuta P (2015) A slan-based nomenclature for monocytes? Blood 126:2536–2538. https://doi.org/10.1182/blood-2015-10-675470
3. Ziegler-Heitbrock L, Ancuta P, Crowe S et al (2010) Nomenclature of monocytes and dendritic cells in blood. Blood 116:e74–e80. https://doi.org/10.1182/blood-2010-02-258558
4. Bronte V, Brandau S, Chen SH et al (2016) Recommendations for myeloid-derived suppressor cell nomenclature and characterization standards. Nat Commun 7:12150. https://doi.org/10.1038/ncomms12150
5. Murray PJ, Allen JE, Biswas SK et al (2014) Macrophage activation and polarization: nomenclature and experimental guidelines. Immunity 41:14–20. https://doi.org/10.1016/j.immuni.2014.06.008
6. Engblom C, Pfirschke C, Pittet MJ (2016) The role of myeloid cells in cancer therapies. Nat Rev Cancer 16:447–462. https://doi.org/10.1038/nrc.2016.54
7. Ginhoux F, Schultze JL, Murray PJ et al (2016) New insights into the multidimensional concept of macrophage ontogeny, activation and function. Nat Immunol 17:34–40. https://doi.org/10.1038/ni.3324
8. Greenplate AR, Johnson DB, Roussel M et al (2016) Myelodysplastic syndrome revealed by systems immunology in a melanoma patient undergoing anti-PD-1 therapy. Cancer Immunol Res 4:474–480. https://doi.org/10.1158/2326-6066.CIR-15-0213
9. Bendall SC, Simonds EF, Qiu P et al (2011) Single-cell mass Cytometry of differential immune and drug responses across a human hematopoietic continuum. Science 332:687–696. https://doi.org/10.1126/science.1198704
10. Spitzer MH, Nolan GP (2016) Mass Cytometry: single cells, many features. Cell 165:780–791. https://doi.org/10.1016/j.cell.2016.04.019
11. Saeys Y, Van Gassen S, Lambrecht BN (2016) Computational flow cytometry: helping to make sense of high-dimensional immunology data. Nat Rev Immunol 16:449–462. https://doi.org/10.1038/nri.2016.56
12. Diggins KE, Greenplate AR, Leelatian N et al (2017) Characterizing cell subsets using marker enrichment modeling. Nat Methods 14:275–278. https://doi.org/10.1038/nmeth.4149
13. Amir EAD, Davis KL, Tadmor MD et al (2013) viSNE enables visualization of high dimensional single-cell data and reveals phenotypic heterogeneity of leukemia. Nat Biotechnol 31:545–552. https://doi.org/10.1038/nbt.2594
14. Qiu P, Simonds EF, Bendall SC et al (2011) Extracting a cellular hierarchy from high-dimensional cytometry data with SPADE. Nat Biotechnol 29:886–891. https://doi.org/10.1038/nbt.1991
15. Diggins KE, Gandelman JS, Roe CE et al (2018) Generating quantitative cell identity labels with marker enrichment modeling (MEM). Curr Protoc Cytom 83(2018):10.21.1–10.21.28. https://doi.org/10.1002/cpcy.34
16. Alcántara-Hernández M, Leylek R, Wagar LE et al (2017) High-dimensional phenotypic mapping of human dendritic cells reveals inter-individual variation and tissue specialization. Immunity 47(6):1037–1050.e6. https://doi.org/10.1016/j.immuni.2017.11.001
17. See P, Dutertre CA, Chen J et al (2017) Mapping the human DC lineage through the integration of high-dimensional techniques. Science 356:eaag3009. https://doi.org/10.1126/science.aag3009
18. Schulz D, Severin Y, Zanotelli VRT et al. (2019) In-Depth Characterization of Monocyte-Derived Macrophages using a Mass Cytometry-Based Phagocytosis Assay. Scientific reports 9:1925 https://www.ncbi.nlm.nih.gov/pmc/articles/PMC6374473/ PMID: 30760760
19. Sander J, Schmidt SV, Cirovic B et al (2017) Cellular differentiation of human monocytes is regulated by time-dependent interleukin-4 signaling and the transcriptional regulator NCOR2. Immunity 47:1051–1066.e12. https://doi.org/10.1016/j.immuni.2017.11.024
20. Chevrier S, Levine JH, Zanotelli VRT et al (2017) An immune atlas of clear cell renal cell carcinoma. Cell 169:736–738.e18. https://doi.org/10.1016/j.cell.2017.04.016
21. Lavin Y, Kobayashi S, Leader A et al (2017) Innate immune landscape in early lung adenocarcinoma by paired single-cell analyses. Cell 169:750–757.e15. https://doi.org/10.1016/j.cell.2017.04.014
22. Krieg C, Nowicka M, Guglietta S et al (2018) High-dimensional single-cell analysis predicts

response to anti-PD-1 immunotherapy. Nat Med 9:2579–2514. https://doi.org/10.1038/nm.4466

23. Roussel M, Ferrell PB, Greenplate AR et al (2017) Mass cytometry deep phenotyping of human mononuclear phagocytes and myeloid-derived suppressor cells from human blood and bone marrow. J Leukoc Biol 102:437–447. https://doi.org/10.1189/jlb.5MA1116-457R
24. Xue J, Schmidt SV, Sander J et al (2014) Transcriptome-based network analysis reveals a spectrum model of human macrophage activation. Immunity 40:274–288. https://doi.org/10.1016/j.immuni.2014.01.006
25. Marigo I, Bosio E, Solito S et al (2010) Tumor-induced tolerance and immune suppression depend on the C/EBPbeta transcription factor. Immunity 32:790–802. https://doi.org/10.1016/j.immuni.2010.05.010
26. Lechner MG, Liebertz DJ, Epstein AL (2010) Characterization of cytokine-induced myeloid-derived suppressor cells from normal human peripheral blood mononuclear cells. J Immunol 185:2273–2284. https://doi.org/10.4049/jimmunol.1000901

Chapter 15

Quantitative Measurement of Cell-Nanoparticle Interactions Using Mass Cytometry

Andrew J. Mitchell, Angela Ivask, and Yi Ju

Abstract

Mass cytometry is a technique that uses inductively coupled plasma mass spectrometry (ICP-MS) to quantify the isotopic composition of cells in suspension. Traditionally it has been used in conjunction with antibodies labeled with stable lanthanide isotopes to investigate cellular heterogeneity. Here we describe its use to quantify uptake of metal nanoparticles by cells in suspension.

Key words Mass cytometry, CyTOF, Nanoparticle, Endocytosis, Phagocytosis, Inductively coupled plasma mass spectrometry

1 Introduction

1.1 Measuring Nanoparticles

The burgeoning use of nanomaterials in industrial and biomedical applications has led to a corresponding need to understand the processes occurring when nanosize (<100 nm) particles interact with biological systems. Because of their size nanoparticles (NP) can have a range of biological effects that are distinct from larger particles composed of similar materials. Depending on nanoparticle properties such as size, shape, surface charge, and composition, these effects may lead to toxicity, or alternatively may be harnessed for use in nanomedicine. A particular focus in the field has been the characterization of NP interactions with cells and the correlation between such interactions and subsequent cellular effects [1]. Thus, characterization of NP–cell interactions is important not only to understand the toxicology of NPs [2] but also for the rational development of engineered nanomaterial theranostics [3, 4].

A number of technical approaches have been used to characterize cell–NP interactions (reviewed in [5]). Two central methods are flow cytometry and inductively coupled plasma mass spectrometry

The original version of this chapter was revised. The correction to this chapter is available at https://doi.org/10.1007/978-1-4939-9454-0_21

Helen M. McGuire and Thomas M. Ashhurst (eds.), *Mass Cytometry: Methods and Protocols*, Methods in Molecular Biology, vol. 1989, https://doi.org/10.1007/978-1-4939-9454-0_15, © Springer Science+Business Media, LLC, part of Springer Nature 2019

(ICP-MS). With flow cytometry, NPs are typically labeled with a fluorochrome and co-incubated with cells. The extent of interactions between cells and NPs is then determined by quantifying cell-associated NP fluorescence. Although flow cytometry provides high-throughput single cell data it suffers from drawbacks that include interference from autofluorescence, limited quantitation, and lack of ability to differentiate cell surface attached nanoparticles from those internalized by cells [5–7]. In contrast, ICP-MS based approaches enable the detection of stable isotopes, which may be intrinsic to the NP itself as with solid silica (^{28}Si), silver (^{107}Ag, ^{109}Ag), or gold (^{197}Au) particles, or may be exogenous labels. Traditional bulk ICP-MS methods have the capability of absolute quantitation of elemental isotopes but do not provide information on the amounts of those isotopes at the single cell level, and this limitation has spurred the development of single-particle ICP-MS. With this approach, particles such as cells are introduced sequentially into an ICP-MS system and time-resolved analysis of the resulting isotope signals allows individual cell-associated signals to be identified [8, 9]. When this technique is been applied to analysis of cell-associated isotopes it has been termed mass cytometry. Of the detection technologies that are in widespread use for mass cytometry, time-of-flight mass spectrometry (ICP-TOF-MS), which is used by CyTOF and Helios instruments, offers distinct advantages for the analysis of interactions between cells and NPs, and has been used to determine uptake of a range of metal-containing NPs [10–12]. In particular, the ability to quantify multiple isotopes simultaneously allows independent identification of nanoparticles, cells, and physiological cellular parameters [5]. In this chapter we illustrate the use of ICP-TOF-MS, specifically a Helios model mass cytometer, to quantify association of metal-containing NP with suspension cell lines in culture (*see* **Note 1**)

1.2 Quantifying Cell–Nanoparticle Interactions

Interactions between nanoparticles and cells are studied using a range of in vitro systems, and specific experimental considerations of these studies are beyond the scope of this chapter. Typically, however, cultured primary or immortalized cells are grown in multi-well plates and cocultured with NPs. Cells are washed to remove noninteracting NPs, and then analyzed to determine cell-associated NPs. Ideally, the absolute number of NPs per individual cell, as well as the localization of NPs (surface vs. internalized), can be determined. Depending on the experimental system, mass cytometry detection can fulfill these requirements.

Mass cytometry analysis involves passing suspensions of nanoparticle-exposed cells into the ICP-TOF system, where isotope signals associated with NPs and cells are detected simultaneously. These isotopes may be intrinsic to the NPs themselves (e.g., Ag or Au) [10, 11], be high molecular weight elemental payloads (e.g., Pt-containing pro-drug) [12] or exogenous probes that have been conjugated to NPs. Since CyTOF and Helios mass cytometers are tuned to exclude analysis of biological isotopes, endogenous elements cannot be used to determine the presence cells. Therefore, to ensure that particulate "events" represent single cells, an iridium-containing DNA intercalator is used to specifically mark cells [13]. Cells may additionally be stained for viability using cisplatin [14]. Since mass spectrometry detection is quantitative, the absolute amount of isotope per cell may be calculated through the use of exogenous standards [10, 11]. Furthermore, the number of nanoparticles per cell can be calculated if the isotope signal from an individual NP is known. Finally, for some NP types it is possible to determine surface-association vs. internalization by "chemically etching" of surface particles [5, 11].

2 Materials

1. Test nanoparticles (*see* **Note 2**).
2. Ultrapure water (*see* **Note 3**).
3. Cell culture medium specific for the suspension cell line used. For example, RPMI 1640 medium containing 4.5 g glucose/L supplemented with Na-pyruvate (1 mM), L-glutamine (2 mM), penicillin-streptomycin (100 U/mL and 100 μg/mL, respectively), and 10% FBS is suggested as growth and test medium for the T-lymphocyte cell line Jurkat.
4. Dulbecco's modified phosphate buffered saline (DPBS).
5. Non-adherent cell line. For example: T-lymphocyte cell line Jurkat (clone E6-1, ATCC® TIB-152™), maintained at 37 °C with 5% CO_2 and 95% relative humidity in supplemented RPMI medium (*see* **step 3**).
6. 4% formaldehyde in DPBS: Add 10 mL of 16% electron microscopy-grade formaldehyde and 4 mL of 10× DPBS to 26 mL ultrapure water. Freeze aliquots at −20 °C. Store thawed aliquots at 4 °C and use for up to one week (*see* **Note 4**).
7. 0.025 μM Iridium (Ir) intercalator/fixative solution in 4% formaldehyde: Add Ir intercalator (Fluidigm—#201192B) to 4% formaldehyde at a dilution of 1/4000. Prepare freshly before use.
8. Standard solution of the same metal that is to be used as a NP marker, prepared in 2% HNO_3 (*see* **Note 5**).
9. Helios mass cytometer (Fluidigm, CA, USA—*see* **Note 6**).

3 Methods

3.1 Cell/Nanoparticle Coculture

1. Seed cells into a 6-well cell culture plate in 2 mL supplemented RPMI 1640 medium at 6×10^5 cells/mL (*see* **Note 7**).
2. Incubate cells for 24 h (37 °C, 5% CO_2) to allow adaptation and division of cells.

3. Prepare a 100-fold concentrated NP suspension compared to the final NP concentration in the test (e.g., prepare 100 μg mL^{-1} NP suspension to achieve 1 μg mL^{-1} in the final test) in water. Add 1 part of nanoparticle suspension to 99 parts of cells on 6-well microplates (1/100 dilution) (*see* **Note 8**).
4. Incubate at 37 °C for between 1 and 24 h (37 °C, 5% CO_2) (*see* **Note 9**).
5. Wash unattached NPs from cells by repeating centrifugation-resuspension in DPBS: Cells are transferred to microcentrifuge tubes, collected by centrifugation at 150 × *g* for 3 min and resuspended in DPBS. The centrifugation-resuspension cycles should be repeated four times (for adherent cells *see* **Note 10**). If etching is to be performed *see* **Note 11** as well as discussion in ref. 5.
6. Cells are finally resuspended in 500 μL DPBS (for adherent cells *see* **Note 12**).

3.2 Preparation of Cells for Mass Cytometry

1. Optional: If viability staining is required, it can be performed at this stage (*see* **Note 13**)
2. Wash cells twice with DPBS by centrifugation (350 × *g*, 5 min RT) and aspirate supernatant.
3. Add 0.5 mL of iridium-intercalation fixation solution to cell pellet and resuspend by carefully pipetting 2–3 times.
4. Fix cells at 4 °C for a minimum of 4 h (*see* **Note 14**).
5. Add 1 mL of water to cell suspension and mix 2–3 times with a pipette. Centrifuge at 1000 × *g* for 10 min at room temperature. Following centrifugation, aspirate supernatant. Repeat centrifugation (1000 × *g*, 10 min at room temperature) and aspirate supernatant (*see* **Note 15**).
6. Resuspend cells in water at a concentration of ~5×10^5 cells/mL and filter through 70 μm nylon mesh (*see* **Note 16**).

3.3 Acquisition of Cell Data

1. Prior to running samples, tune the mass cytometer by optimizing plasma torch X/Y position, gases, detector voltage, and current using the manufacturer's standard protocols.
2. Define an event mode protocol and ensure mass channels are active for the appropriate NP-associated isotope(s), as well as ^{191}Ir and ^{193}Ir. Include ^{195}Pt if Cisplatin viability staining has been performed.
3. Set thresholding to the following values: Lower convolution threshold = 200, min event duration = 10, max event duration = 150 (*see* **Note 17**).
4. Clean sample lines by running either wash solution (0.01% HF) or 2% HNO_3 followed by water (*see* **Note 18**). In masses-per-reading (MPR) mode with an integration of 1 second, analyte isotope dual signal should be <2000 counts (*see* **Note 19**).

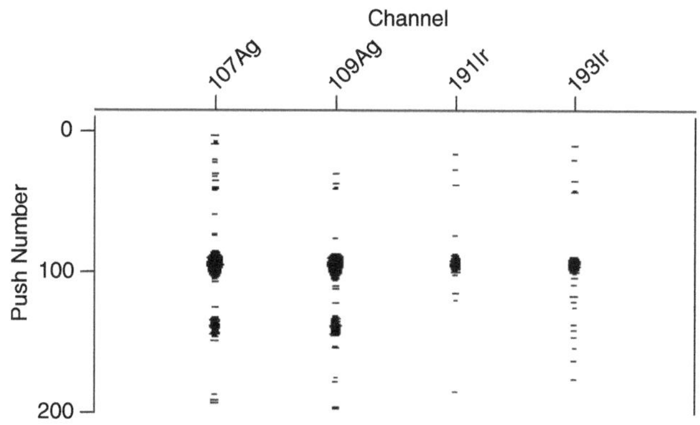

Fig. 1 Instrument screenshot from Helios during data acquisition. Data shows typical rain plot with X-axis as channel/isotope and the Y-axis push number/time. "Speckling" in channels represents individual soluble ions striking the detector while darker "clouds" result from particulates (either cells or nanoparticles). The data here show a cell with associated Ag-NP signal (upper series) and a non-cell-associated Ag-NP (lower series)

5. Run samples, ensuring that event rate <300/s (*see* **Note 20**) and collect a minimum of 5000 events (*see* Fig. 1). Following acquisition of sample, the Helios software automatically generates flow Cytometry standard 3.1 files (*see* **Note 21**).
6. Wash between samples with water (*see* **Note 22**).
7. Following acquisition of data from cell samples, clean sample lines by running wash solution or 2% HNO_3 followed by water. In MPR mode with an integration of 1 second, isotope dual signal should be <2000 counts.

3.4 Determination of Isotope Transmission Efficiency

1. The transmission efficiency (TE) [15] of the NP-associated isotope is determined by running a solution of the natural abundance metal of known concentration (*see* **Notes 5** and **23**).
2. Assess background signal in isotope channel by running water and previewing in MPR mode, with an integration of 1 s per reading. Isotope dual signal should be <2000 counts per reading. Collect MPR file of background signal for the NP-associated isotope by running water with integration of 1 s per reading for 100 readings.
3. Collect MPR file for NP-associated natural abundance metal, running standard solution with integration of 1 s per reading for 100 readings (*see* **Note 24**).
4. Clean sample lines by running 2% HNO_3 followed by water (*see* **Note 25**). Signal should be <2000 while running water.

5. Using spreadsheet software, open the MPR file collected in **steps 2** and **3** above and calculate the mean signal per second for NP-associated isotope (Obs $counts_A$) (*see* **Note 26**).
6. Calculate the expected analyte isotope (A) counts (Exp $counts_A$) per second using the following equation:

$$\text{Exp counts}_A = \left(\frac{C_A}{m_A}\right) \times \frac{\%NA_A}{100} \times Na \times I$$

where C_A and m_A are the concentration (g/L) and atomic mass, respectively, of the natural abundance element (i.e., containing all naturally occurring isotopes), $\%NA_A$ is the percent natural abundance of the test isotope, Na is Avogadro's number (6.022×10^{23}), and I is injection speed of sample into the instrument (3×10^{-5} L/min).

7. Using the observed isotope counts (from **step 5**) and the expected counts (from **step 6**), calculate the transmission efficiency (TE):

$$TE = \frac{\text{Obs counts}_A}{\text{Exp counts}_A}$$

3.5 Determination of Absolute Amount of Metal in Cell Population

1. Using flow cytometry analysis software (e.g., FlowJo (Flowjo, USA), FCS express (De novo software, USA) or others), identify nucleated single cells by plotting ^{193}Ir signal vs. event length and selecting, or "gating" the main population to a histogram showing NP-associated isotope signal (Fig. 2, also *see* **Note 27**).
2. Determine the median isotope signal of the gated cell population (Obs $cell_A$) (*see* **Note 28**).
3. In the gated cell population calculate the expected cell-associated isotope signal (Exp $cell_A$) using observed signal (Obs $cell_A$) and the TE (*see* **Note 29**):

$$\text{Exp cell}_A = \frac{\text{Obs cell}_A}{TE}$$

4. Calculate the total mass of natural abundance element in the cell population using the following equation:

$$\text{Cell M}_A = \text{Exp cell}_A \times \left(\frac{100}{\%NA_A}\right) \times \left(\frac{1}{Na}\right) \times m_{NAA}$$

where Cell M_A is the total mass (g) of all isotopes of the NP-associated element in each gated cell event, $\%NA_A$ is the percent natural abundance of specific reporter isotope, Na is Avogadro's number, and m_{NAA} is atomic mass of the natural abundance element.

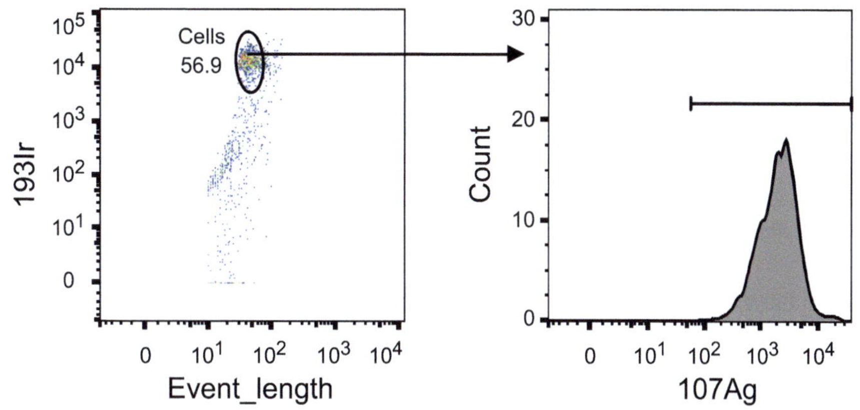

Fig. 2 Gating and analysis of cell data. Using flow cytometry analysis software (in this example FlowJo 10.4), ^{193}Ir vs. event length is plotted and the major population gated to a univariate histogram showing the reporter isotope (in this case ^{107}Ag). The median ^{107}Ag of this population is determined and used for calculation of absolute Ag content per cell

3.6 Acquisition of Signals from Single Nanoparticles and Determination of NP Number per Cell

This step is performed only when the equivalent number of nanoparticles in each cell is to be calculated. The goal is to quantify the isotope signal from a single nanoparticle and compare this information with isotope signal from each cell event.

1. Define an event mode protocol with the appropriate NP-associated isotope(s) and set thresholding to the following values: Lower convolution threshold = 200, min event duration = 3, max event duration = 150 (*see* **Note 30**).
2. Prepare NP suspension at <5 × 10^5/mL in water (*see* **Note 31**).
3. Acquire a minimum of 5000 events. Following acquisition of a sample, the Helios software automatically generates flow Cytometry standard 3.1 files (*see* **Note 32**).
8. Clean sample lines by running wash solution or 2% HNO_3 followed by water. In MPR mode with an integration of 1 s, isotope dual signal should be <2000 counts.
9. Analyze the obtained data using flow cytometry analysis software by plotting isotope vs. event length. Define a region on the main population and gate to a histogram showing NP-associated test isotope signal in a similar manner to cell data. Nanoparticles should be distinguishable from noise on a bivariate plot of test isotope vs. event length (Fig. 3).
10. Determine the median isotope signal of the gated particle population (Obs NP_A).
11. Calculate the median NP/cell ratio (NCR) by dividing the median measured cell-associated isotope signal (Obs $cell_A$) by the measured NP-associated isotope signal: $NCR = \frac{\text{Obs cell}_A}{\text{Obs NP}_A}$ (*see* **Note 33**).

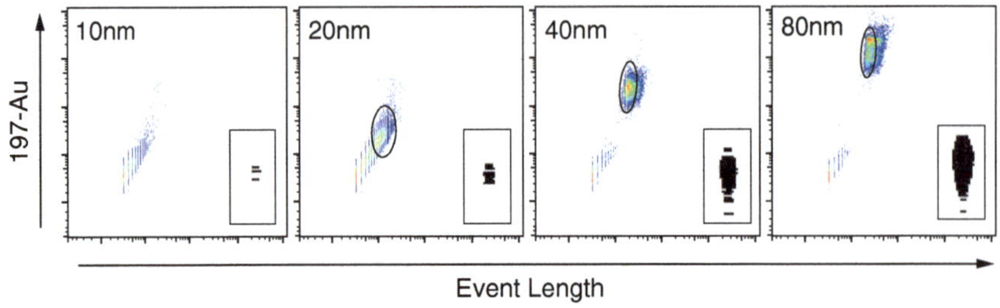

Fig. 3 Representative mass cytometry data of solid Au nanoparticles. Solid Au nanoparticles of the indicated diameter were run on a Helios mass cytometer and data was analyzed using FlowJo software. Event length vs. ^{197}Au signal is shown. Single nanoparticles of >40 nm diameter are readily distinguishable (elliptical regions) while particles of 10 nm diameter were not distinguishable from noise. 20 nm diameter particles overlap with noise. Boxed insets show screenshots of the appearance of typical particles with the corresponding diameters in a rain plot

4 Notes

1. Although we describe a procedure for the quantification of interactions between a certain cell line and specific nanoparticles, many different combinations of NPs and cell lines will work effectively and some details are given in subsequent notes. Among cell lines, suspension cell lines (e.g., B- and T-lymphocytes), primary cells (e.g., peripheral blood mononuclear cells PBMCs), as well as adherent cell lines (macrophages, fibroblasts, keratinocytes, lung cells, hepatocytes) can be used. The critical requirement for NP is that they contain an isotope of an element that can be detected by mass cytometer (*see* **Note 2**).
2. A range of NP types are suitable for use in this method. The main requirement for a NP is that it contains an isotope of an element that is detectable by the mass cytometer (for example, >75 AMU for a Fluidigm Helios platform). It is preferable that NP contain either one naturally occurring (e.g., Au, Tb, Ho, Tm) or two (e.g., Ag, Eu, Lu) isotopes. Use of elements with multiple isotopes potentially leads to decreased signal/noise. Nanoparticles of Zr, Ag, and Au have been used as these can be purchased commercially, and have high signal which ensures that the number of atoms in individual NPs can be calculated independently. NPs should be of diameter <100 nm since larger particles may not fully decompose in the plasma torch, potentially leading to nonlinearity as well as detector damage from high signal. In addition to NPs entirely consisting of the specific isotope, isotopic labeling can also be used for other types of NPs. For example, lanthanide isotopes can be used to label maleimide-DOTA nanoparticles.

3. Use only 18 MΩ ultrapure water and not other sources as these may contain trace metals. All buffers, including water, should be tested for contamination since they may contain isotopes that will interfere with analysis, or may decrease the life of the detector. To test buffers, prepare a 1/10,000 dilution in ultrapure water and run solution/TOF mode, comparing to a water blank, to identify the presence of contaminants.
4. Formaldehyde is toxic and should be handled with care using appropriate personal protective equipment and waste disposed of according to local regulations.
5. In the method presented here a single isotope of a NP-associated metal is ultimately used to determine uptake of nanoparticles by cells. However, soluble natural abundance metal (containing all naturally occurring isotopes of the element) can be used to experimentally determine the key instrument parameter transmission efficiency (TE—outlined in Subheading 3.4) for the isotope of interest. Natural abundance metal can be used as a standard because the relative percentages of specific isotopes in natural abundance elements are known (a useful reference site for this is: http://periodictable.com/Properties/A/IsotopeAbundances.html). The TE can then be used to calculate both the absolute cell-associated isotope and NP-associated isotope signals, as well as total cell- and NP-associated natural abundance metal (Subheadings 3.5 and 3.6). Soluble metal standard is typically prepared in 2–5% HNO_3. This is because many metals can attach to the sample delivery tubing of the instrument, which can lead to an artifactually low signal reaching the instrument detector. Dilutions of metal isotopes should be prepared in 2% HNO_3 to a level that is assumed to be beyond the limit of detection of the instrument (~1 ppt).
6. The description in this section is based on a Helios model mass cytometer (Fluidigm) running software version 6.5.358. Older model instruments (e.g., CyTOF1 and CyTOF2) are suitable but have different mass responsiveness, sample injection rates and may have different software steps.
7. Different cell lines require specific individual seeding densities, which should be determined in preliminary experiments. For suspension cell lines, a 24 h adaptation period is recommended before addition of NPs. Should adherent cells be used, a 24 h period is essential for their adaptation and attachment to microplates. For adherent cells, 60–80% confluency is recommended after 24 h incubation time.
8. For suspension cells, directly add NP suspension to well at appropriate dilution without removing supernatant. However,

if adherent cells are used, growth medium can be removed and NPs added to the cells in fresh growth medium.

9. The appropriate incubation conditions depend on the purpose of the experiment and on a number of variables including the cell line as well as NP size/shape/surface chemistry, and should ideally be determined empirically in preliminary experiments. For example, highly phagocytic cells such as macrophage cell lines will generally internalize large numbers of NPs within 1 hour while non-phagocytic cell lines with "stealth" particles may require longer times.
10. For adherent cells, wash the cells by removing the growth medium and adding equal amount of DPBS. Washing with DPBS should be repeated for four times.
11. Etching of surface-associated NPs can be done in order to distinguish cell surface-associated and intracellular NPs. Etching protocols may differ for different NPs (reviewed in [5]). As an example, etching of Ag NPs is performed using Tripotassium hexacyanoferrate (III) (a redox agent to oxidize Ag0 to Ag+) and Sodium thiosulfate (immediately coordinates with Ag+). Etchant cannot penetrate through cell membrane and will thus affect only NPs externally bound to cells. 2× etchant agent: weigh Tripotassium hexacyanoferrate (III) and hexacyanoferrate (III) and dissolve in DPBS to give 20 mM solution. For suspension cells, mix previously NP-exposed cells 1:1 with the etchant agent, incubate for 1 min, and centrifuge at 200 × *g* for 3 min. Remove the supernatant, add equal volume of DPBS and repeat the centrifugation-resuspension cycle. For adherent cells, add 1:1 DPBS and 2× etchant agent at 22 °C for 1 min, followed by two washes with DPBS.
12. Before collecting NP-exposed adherent cells, release of the cells form microplates is necessary: detach cells from plates by adding 700 μL of 1× trypsin with 0.5 mM EDTA and incubate at 37 °C for 5 min. Add 700 μL culture medium and transfer cell suspension into a 1.5 mL centrifuge tube. Centrifuge the cells at 200 × *g* for 5 min and resuspend in DPBS.
13. Cell viability in mass cytometry can be determined by staining with cisplatin. Cisplatin is largely excluded from viable cells but readily crosses the membranes of dead cells and binds to intracellular protein and nucleic acids. For cisplatin staining: carefully aspirate supernatant from pelleted cells and add 400 μL of a 1/4000 (1.25 μM) solution of stock cisplatin (Fluidigm #01064) in DPBS at room temperature for 5 min, and wash cells twice by centrifugation (350 × *g*, 5 min) with 500 μL DPBS containing 1% bovine serum albumin. Note that cisplatin is cytotoxic—appropriate safety precautions should be followed,

personal protection equipment should be worn, and cisplatin waste should be treated as cytotoxic waste.

14. Adequate fixation of cells is critical since they are ultimately resuspended in water prior to running on the instrument, and if fixation is inadequate they will disintegrate rapidly leading to cell loss. A minimum of 4% formaldehyde for a period of 4 h is required. It is often convenient to fix overnight at 4 °C and cells are generally stable in fixative for up to a week.
15. Fixed cells in water do not sediment efficiently, so postfixation washes are performed at a higher rcf and for a longer time than non-fixed cells.
16. A cell concentration of ~5 × 10^5/mL will lead to an event rate of <300/s during acquisition. Filtering of samples immediately prior to running is important to avoid clogging of instrument sample lines. Either commercial nylon cell-filtration units may be used, while a low-cost solution is screen printing mesh of the correct pore size cut into small squares.
17. These thresholding settings are typical for analysis of cells that have been stained with Iridium intercalator. The key parameter is the minimum event length of ~10 "pushes" (~130 μs)—i.e., 10 consecutive pushes with detectible signal in any channel. For correctly stained cells the duration of a single cell event is generally ~15–25 pushes. Events below this are likely to be noise, debris, or non-cell-associated NP, while events greater than this are likely to represent multiple cells entering the plasma in rapid succession, causing fusion of ion clouds.
18. Lanthanides are generally easily removed from sample tubing using instrument wash solution but other metals, such as Ag and Pt, require HNO_3.
19. CyTOF and Helios mass cytometers sample at a rate of 76.8 kHz. A background of <2000 counts per second is recommended since the ion cloud generated by a typical cell has an event length of ~15–25 pushes (~200–350 μs) and for a background of 2000 counts per second the average NP-associated isotope count for individual cells is <1.
20. At event rates >300/s there may be insufficient time between cells passing through the plasma torch for the ion clouds from individual cells to remain separate. Fused ion clouds will be categorized as a single "event" by the thresholding algorithm leading to erroneous data.
21. Data can be collected in two formats: flow cytometry standard (FCS) 3.1 or the Fluidigm proprietary IMD format. Files in the FCS format have data in which triggering algorithms have been applied to identify individual cell/NP "events." This format is compatible with many commercial and open source analysis packages including FlowJo, FCS Express, Cytobank,

FlowingSoftware, etc. In contrast, IMD files store a raw form of mass cytometry data that is only readable by proprietary Fluidigm Helios software. If IMD files are stored, it is possible to reanalyze data with modified event triggering parameters and generate new FCS files.

22. Sample carryover can be assessed in real time by "previewing" acquisition. Water should be run until the event rate <1/s, which typically takes 2–3 min. If streaking or contamination persists, then wash solution or 2% HNO_3 can be used to clean lines.

23. Instrument sensitivity can vary based on tuning and the TE varies across the mass detection range. Therefore, TE should be empirically determined for target elements at the time of the run. If a solution of the target element is not available, an element of a similar mass may be used as a surrogate (for example: Iridium results from tuning may be used to determine Pt transmission since they have similar atomic masses and therefore can be assumed to have similar TE).

24. Concentrated solutions of metals have a strong tendency to attach to sample tubing leading to high background and between-sample contamination and can cause damage to the detector. If in any doubt as to the concentration a solution, prepare serial tenfold falling dilutions in water to a concentration that is expected to be below the limit of detection of the instrument (~1 ppt). These are then run from highest dilution (least concentrated) to lower dilutions until an appropriate signal (~10,000–200,000 counts per second) is achieved. Data from a complete standard curve of soluble metal does not need to be collected as only a single value will be used for the determination of transmission efficiency.

25. Cleaning between samples, or following runs is performed by running 2–5% HNO_3. With heavily contaminated sample lines this may take several hours. Lines may be cleaned with the plasma off by placing the nebulizer/sample capillary into the nebulizer rest and running solution into a 50 mL tube.

26. When determining mean values of data in MPR mode, assess the values from readings over the full time of collection and only include those readings where the isotope signal has stabilized. To determine the mean signal per second for the NP-associated isotope, subtract the mean background signal from that of the standard.

27. Iridium intercalator binds to nucleic acids; therefore events with high iridium signal are defined as cells. Either of the naturally occurring isotopes of Iridium (^{191}Ir or ^{193}Ir) can be used but ^{193}Ir is more abundant and gives a higher signal. If association with nonnucleated cells (e.g., red blood cells) is desired, then another marker specific for these cells must be used in

place of Ir intercalator, such as a metal-conjugated antibody against human CD235a. Events with low Iridium signal may include non-cell-associated NP or cellular debris. Events with high DNA content and event length greater than the main population generally represent the fusion of two or more ion clouds resulting from multiple cells passing into the plasma in rapid succession and should be excluded from analysis by gating.

28. Many biological processes at the cellular level, including NP uptake, approximate log-normal distribution. The median is a robust measure of central tendency for log-normal distributions, while the arithmetic mean is generally inappropriate.

29. An assumption of this calculation is that the transmission efficiency of an isotope in particulate form is similar to that for a solution. The major influence on this assumption is the particle transport efficiency (PTE—i.e., the percentage of particles that are introduced into the sample stream that are ultimately registered by the detector), which ideally should be 100%, and is typically ~60–80% for Helios model mass cytometers, and between 15–30% for the CyTOF or CyTOF2. In contrast, traditional single-particle ICP-MS systems have PTE of ~1–10%. For further discussion on PTE *see* [16].

30. If the NP-associated element has multiple naturally occurring isotopes, include all major isotopes in the protocol since event mode triggers event detection based on signal in all channels. Because NP may contain much lower levels of metals than iridium-stained cells the minimum event length is decreased to the lowest value allowed by the Helios software. Note that the dominant contributor to event length is signal intensity and not cell/particle size as might be assumed.

31. Suspensions of metal nanoparticles have a strong tendency to attach to sample tubing leading to high background and between-sample contamination, and can cause damage to the detector. If in any doubt as to the concentration of a solution/suspension, prepare serial tenfold falling dilutions in water to a concentration that is expected to contain no NP. These are then run from highest dilution (least concentrated) to lower dilutions until an appropriate event rate (10–300/s) is achieved.

32. Carefully monitor the rain plot during acquisition of NP signals. Discrete particles should be identifiable and there should be minimal background “speckling” resulting from free ions or small nanoparticles that are just below the threshold detection. High background signal hitting the detector in a stochastic manner can lead to multiple consecutive positive pushes, which can be erroneously classified as events. If in doubt it can be worth acquiring a dilution series of NP and verifying that the

event rate changes correspondingly. It is also useful to take representative screenshots of rain plot data for reference.

33. If it is possible to experimentally quantify NP isotope signal, and determining the number of nanoparticles associating with cells is all that is required, then it is not necessary to determine the absolute amount of metal per cell (as described in Subheading 3.5). If nanoparticle-associated isotope signal is not able to be quantified on a particle-by-particle basis, then the absolute amount of isotope per cell can be used in combination with alternative approaches to quantifying signal from NP. These include: (a) a combination of particle counting and traditional ICP-MS on bulk NP preparations to determine average NP-associated isotope content, or (b) for solid spherical NP, the isotope mass and number of atoms may be calculated using the particle diameter and density. In either case the NP:cell ratio can be calculated by dividing the amount of isotope per cell by the amount of isotope per NP.

Acknowledgments

The authors acknowledge Hannah Kelly for providing gold nanoparticles. Funding from Future Industries Institute of the University of South Australia (Foundation Fellowship of Ivask) and Estonian Research Council grant PUT748 is acknowledged. This work was performed in part at the Materials Characterisation and Fabrication Platform (MCFP) at the University of Melbourne and the Victorian Node of the Australian National Fabrication Facility (ANFF). This work received funding from South Australian Government PRIF program Project "International Cluster on Nanosafety" of Nicolas H. Voelcker and Enzo Lombi.

References

1. Albanese A, Tang PS, Chan WCW (2012) The effect of nanoparticle size, shape, and surface chemistry on biological systems. Annu Rev Biomed Eng 14(1):1–16. https://doi.org/10.1146/annurev-bioeng-071811-150124
2. Love SA, Maurer-Jones MA, Thompson JW et al (2012) Assessing nanoparticle toxicity. Annu Rev Anal Chem 5(1):181–205. https://doi.org/10.1146/annurev-anchem-062011-143134
3. Gustafson HH, Holt-Casper D, Grainger DW et al (2015) Nanoparticle uptake: the phagocyte problem. Nano Today 10(4):487–510
4. Such GK, Yan Y, Johnston AP et al (2015) Interfacing materials science and biology for drug carrier design. Adv Mater 27(14):2278–2297. https://doi.org/10.1002/adma.201405084
5. Ivask A, Mitchell AJ, Malysheva A et al (2017) Methodologies and approaches for the analysis of cell–nanoparticle interactions. Wiley Interdiscip Rev Nanomed Nanobiotechnol 10:e1486. https://doi.org/10.1002/wnan.1486
6. Mitchell AJ, Pradel LC, Chasson L et al (2010) Technical advance: autofluorescence as a tool for myeloid cell analysis. J Leukoc Biol 88(3):597–603. https://doi.org/10.1189/jlb.0310184

7. Vanhecke D, Rodriguez-Lorenzo L, Clift MJ et al (2014) Quantification of nanoparticles at the single-cell level: an overview about state-of-the-art techniques and their limitations. Nanomedicine 9(12):1885–1900
8. Comi TJ, Do TD, Rubakhin SS et al (2017) Categorizing cells on the basis of their chemical profiles: Progress in single-cell mass spectrometry. J Am Chem Soc 139(11):3920–3929. https://doi.org/10.1021/jacs.6b12822
9. Mueller L, Traub H, Jakubowski N et al (2014) Trends in single-cell analysis by use of ICP-MS. Anal Bioanal Chem 406(27):6963–6977
10. Yang Y-SS, Atukorale PU, Moynihan KD et al (2017) High-throughput quantitation of inorganic nanoparticle biodistribution at the single-cell level using mass cytometry. Nat Commun 8:14069
11. Ivask A, Mitchell AJ, Hope CM et al (2017) Single cell level quantification of nanoparticle-cell interactions using mass cytometry. Anal Chem 89:8228. https://doi.org/10.1021/acs.analchem.7b01006
12. Dai Y, Guo J, Wang TY et al (2017) Self-assembled nanoparticles from phenolic derivatives for cancer therapy. Adv Healthc Mater. https://doi.org/10.1002/adhm.201700467
13. Ornatsky O, Bandura D, Baranov V et al (2010) Highly multiparametric analysis by mass cytometry. J Immunol Methods 361 (1–2):1–20. https://doi.org/10.1016/j.jim.2010.07.002
14. Fienberg HG, Simonds EF, Fantl WJ et al (2012) A platinum-based covalent viability reagent for single-cell mass cytometry. Cytometry A 81A(6):467–475. https://doi.org/10.1002/cyto.a.22067
15. Tricot S, Meyrand M, Sammicheli C et al (2015) Evaluating the efficiency of isotope transmission for improved panel design and a comparison of the detection sensitivities of mass cytometer instruments. Cytometry A 87(4):357–368. https://doi.org/10.1002/cyto.a.22648
16. Pace HE, Rogers NJ, Jarolimek C et al (2011) Determining transport efficiency for the purpose of counting and sizing nanoparticles via single particle inductively coupled plasma mass spectrometry. Anal Chem 83(24):9361–9369. https://doi.org/10.1021/ac201952t

Part V

Data Analysis

Chapter 16

Data-Driven Flow Cytometry Analysis

Sherrie Wang and Ryan R. Brinkman

Abstract

The emergence of flow and mass cytometry technologies capable of generating 40-dimensional data has spurred research into automated methodologies that address bottlenecks across the entire analysis process from quality checking, data transformation, and cell population identification, to biomarker identification and visualizations. We review these approaches in the context of the stepwise progression through the different steps, including normalization, automated gating, outlier detection, and graphical presentation of results.

Key words Flow cytometry, Data analysis, Bioinformatics

1 R/Bioconductor

More than 50 approaches to automate flow cytometry (FCM) data analysis are available (Table 1). The overwhelming majority have been developed and released as freely available, open-source tools using the R programming language [1]. These tools have been developed for high-throughput workflows, and are not generally amenable to graphical user interface manual interaction with individual files during the analysis process. However, these tools can be integrated into commercial tools familiar to users, facilitating adoption. For example, the flowWorkspace package can export automated gating results in a format readable by FlowJo (FlowJo Inc., Ashland OR). Many of the approaches have been released through the Bioconductor repository which enforces strict requirements on cross-platform compatibility and functional documentation. Algorithms for data analysis are provided as packages that generally address a single step in the analysis pipelines, with interoperability enforced through Bioconductor. This allows users to substitute new approaches to the same challenge as the field advances, an advantage over monolithic tools that attempt to solve a single or even multiple problems in isolation.

Helen M. McGuire and Thomas M. Ashhurst (eds.), *Mass Cytometry: Methods and Protocols*, Methods in Molecular Biology, vol. 1989, https://doi.org/10.1007/978-1-4939-9454-0_16,

Table 1
Bioinformatic tools for high-throughput data analysis

Package name	Use	References	Technical notes
Preprocessing			
FCS Trans	Manipulate file formats	PMC3932304	R package for FCS to .txt conversion
fdaNorm	Adjust data to account for batch effects	PMC2648208	R/Bioconductor software to adjust data to account for batch effects like laser drift
guassNorm	Adjust data to account for batch effects	PMC3648208	R/Bioconductor software to adjust data to account for batch effects like laser drift
flow	Variance stabilization	http://www.bioconductor.org/	R/Bioconductor package that removes mean variance correlations from cell populations
flowCore	Read/Write, process (transform, compensate) of flow data. The basic flow infrastructure	PMC2684747	R/Bioconductor core infrastructure for representing cell populations and parent/child relationships among them
flowBeads	Automated analysis of bead data	http://www.bioconductor.org/	R/Bioconductor package that provides gating and normalization specific to bead data
flowBin	Combining multitube flow cytometry data by binning	http://www.bioconductor.org/	R/Bioconductor package that combines flow cytometry data multiplexed into tubes by common markers
CATALYST	Pipeline for preprocessing of mass cytometry data	http://www.bioconductor.org/	R/Bioconductor package that includes normalization, single-cell deconvolution and compensation for Mass cytometry data
MetaCyto	Pipeline for analyzing cytometry data	http://www.bioconductor.org/	R/Bioconductor package that provides preprocessing, automated gating, and meta-analysis of cytometry data
flowQB	Quality control of cytometer sensitivity	http://www.bioconductor.org/	Automatically calculates detector efficiency (Q), optical background (B), and instrinsic CV of the beads
flowStats	Advanced statistical methods and functions, specialized and general gating algorithms	http://www.bioconductor.org/	R/Biconductor software that collects several algorithms together for normalization and gating
flowUtils	Import gates, transformation and compensation	http://www.bioconductor.org/	R/Bioconductor package to support Gating-ML specification to exchange gate coordinates between software

flowTrans	Estimate parameters for data transformation	PMC3243046	R/Bioconductor infrastructure to optimize parameter choice for different transformations
flowWorkspace	Import manually gated data from FlowJo workspaces, represent manual and automated gating hierarchies efficiently	PMC3992339	R/Bioconductor core infrastructure that makes manually gated data accessible to BioConductors computational flow tools by importing pre-processed and gated data from FlowJo
ncdfFlow	Advanced method for large dataset processing	PMC3992339	R/Bioconductor package that overcomes memory limitations when working with large datasets by storing FCS data in netCDF files on disk
plateCore	Analyze multiple plates	PMC2777006	R/Bioconductor package that enable automated negative control-based gating and plate-based analysis
flowAI	Identify Outlier Events	http://www.bioconductor.org	R package that removes spurious events based on time vs. fluorescence
flowClean	Identify outlier events	http://www.bioconductor.org	R package that removes spurious events based on time vs. fluorescence
flowQ	Identify outlier samples (e.g., wells drying out, reagent issues)	PMC2768034	R/Bioconductor package that provides infrastructure to generate interactive HTML quality report
QUALIFIER	Identify outlier samples	PMC3499158	R/Bioconductor software that uses manual gates to perform an extensive series of statistical quality assessment checks on gated cell subpopulations
Automated gating			
Unsupervised			
ACCENSE	Unsupervised cell population identification	PMC3890841	R/Matlab software for dimensionality reduction with density-based partitioning

(continued)

Table 1
(continued)

Package name	Use	References	Technical notes
FLOCK	Unsupervised cell population identification	PMC3084630	Stand alone software for clustering using an adaptive multi-dimensional mesh to estimate local density followed by hierarchical merging of adjacent regions based on density differentials
FlowSOM	Unsupervised cell population identification	PMID: 25573116	R/Bioconductor software that uses Self-Organizing maps
flowClust	Unsupervised cell population identification	PMC2701419	R/Bioconductor software for clustering using t-mixture model with Box-Cox transformation with support for Bayesian priors
flowFP	Unsupervised cell population identification	PMC2777013	R/Bioconductor software for fingerprint generation via multivariate probability distribution
flowMeans	Unsupervised cell population identification	PMC21182178	R/Bioconductor software for k-means clustering and merging using the F-R statistics
flowMerge	Unsupervised cell population identification	PMC2798116	R/Bioconductor software that combines flowClust and entropy-based or Mahalanobis distance-based cluster merging
flowPeaks	Unsupervised cell population identification	PMC3400953	R software for unsupervised clustering using k-means and mixture model
flowType	Unsupervised cell population identification	PMC3998128	R/Bioconductor software for combinatorial gating of high-dimensional populations and correlative analysis against clinical outcomes
SPADE	Unsupervised cell population identification	PMC3196363	Matlab/standalone/R/Bioconductor tool for density-based sampling, k-means clustering and minimum spanning trees
SWIFT	Unsupervised gating for rare cell population	PMID: 24677621	Iterative weighted sampling procedure with splitting and merging to retain discrimination of extremely small subpopulations
X-shift	Unsupervised cell population identification	PMC4896314	Using fast KNN estimation of cell event density and automatically arranges populations by marker-based classification systems

flowMatch	Cell population matching	PMC3471348	R/Bioconductor software to match clusters across samples for producing robust meta-clusters
flowMap-FR	Cell population matching	PMC5014134	R/Bioconductor software to match cell population clusters across samples using the F-R statistics
NetFCM	Semiautomated web-based method for flow cytometry data analysis	PMID: 25044796	Semiautomatic gating strategy that uses clustering and principal component analysis(PCA) together with other statistical methods to mimic manual gating approaches
Supervised			
flowDensity	Supervised cell population identification	PMID: 25378466	R/Bioconductor software for supervised gating to match manual analysis for clinical trials and diagnosis
OpenCyto	General framework to construct reproducible automated gating pipelines and simplify data processing	http://opencyto.org/	R/Bioconductor infrastructure for hierarchical automated gating that maintains relationships among cell populations
X-cyt	Supervised cell population identification	PMC3839720	R script that partitions each sample with initialization by user template, then optimizes on the parameters via estimation-maximization
SamSpectral	Supervised cell population identification	PMC2923634	R/Bioconductor software for efficient spectral clustering using density-based down-sampling
Data analysis			
Citrus	Identify most important cell populations correlated with outcome of interest	PMID:2497804	Regularized supervised learning algorithms to identify stratifying clusters that are best predictors of a known experimental endpoint of interest
COMPASS	Identify combinatorial subsets of polyfunctional T cells	http://rglab.github.io/COMPASS/	R software that is multivariate extension of MIMOSA that jointly models all combinatorial polyfunctional cell subsets

(continued)

Table 1
(continued)

Package name	Use	References	Technical notes
MIMOSA	Identify responders and nonresponders to stimulation in intracellular cytokine staining assay data	PMC3862207	R/Bioconductor software to detect antigen-specific changes in marginal or specific cell subsets
Biomarker discovery			
flowType	Unsupervised cell population identification	PMC3998128	R/Bioconductor software for combinatorial gating of high-dimensional populations and correlative analysis against clinical outcomes
RchyOptimyx	Identify most important cell populations correlated with outcome of interest	PMC3988128	R/Bioconductor software that optimize cellular hierarchies to preserve correlation with external variables and summarize large data sets in simple plots
Visualization and post-processing			
flowViz	Visualization (e.g., histograms, dot plots, density plots, gating hierarchies and layouts)	PMC2768483	R/Bioconductor software that employs trellis graphics and can be adapted to provide useful visualizations
flowPlots	Graphical displays with statistical tests for gated ICS flow cytometry data	http://www.bioconductor.org	R/Bioconductor software that provides analysis plots and data class for gated flow cytometry data

Core infrastructure widely used by other packages is provided by the flowCore R/Bioconductor package [2] that implements a computationally efficient data structure for reading and saving flow cytometry (FCM) data, and provides a systematic FCS file parsing. This in turn encourages new algorithms development and the use of combinations of tools in complex work flows [2]. It also includes a range of methods for data processing, including compensation, transformation, and gating.

When hundreds of FCS data files are generated by high-throughput instruments, processing can become a challenge as the memory limit can be reached when reading in all the data. ncdfFlow is a Bioconductor package that is designed to overcome memory limitations when working with large datasets by accessing FCS data on disk in a way that circumvents having to read the entire file into memory [3].

2 Data Acquisition and Quality Assessment

Technical issues resulting from instrumental or procedural variations during acquisition can bias the statistics of the obtained cell subpopulation and can impact the quality of the cytometry data and the subsequent analysis results. Clogs can results in abrupt changes in the fluorescence in the time domain analysis. Other issues such as unstable data acquisition can result in a shift in means of the populations analyzed, which can pose challenges for gating. These data should be identified or potentially removed by the user, either automatically or manually, before being passed to the downstream analysis. Data quality assessment aims to detect and as appropriate flag for review or remove such abnormalities in measurements which likely do not possess any underlying biological causes, and thus are likely caused by technical errors.

QUAliFiER is a tool that can be used for quality assessment on manually gated data [4]. The typical workflow for QUAliFiER is: importing data, extracting cell population statistics, defining quality assessment tasks, performing outlier calling, and generating a quality assessment report. QUAliFiER uses grouping and conditioning variables defined in the associated study metadata to apply filters to the population statistics and carry out outlier detection.

The flowQ package is designed for quality assessment for pregated data. It uses a generic framework that accounts for different quality assessment criteria of different experimental setup [5]. flowAI and flowClean are the two currently available algorithms that remove these outliers. flowAI detects and removes outlier events from the analysis of flow rate, signal stability, and dynamic range of FCM instruments. flowClean analyses frequency changes within a sample during acquisition for outlier detection. The result of running flowAI and flowClean on data from Flow Repository

(repository ID: FR-FCM-ZZGS) shows flowAI removes the spurious events with substantially shorter time but at the cost of removing a large amount of cell events, whereas flowClean removes fewer outliers events but with a relatively longer runtime (Fig. 1). The runtime for flowClean is 5 min for a 56 Mb file, while flowAI requires 15 s. Users often need to run thousands of files at a time for cleaning. Computational cost of each individual file should be kept at a minimum to ensure efficient utilization of resources and time overall.

flowQB [6] is another R/Bioconductor package that provides quality control, however at the instrumental level. It automatically calculates flow cytometer's detection efficiency (Q) and background

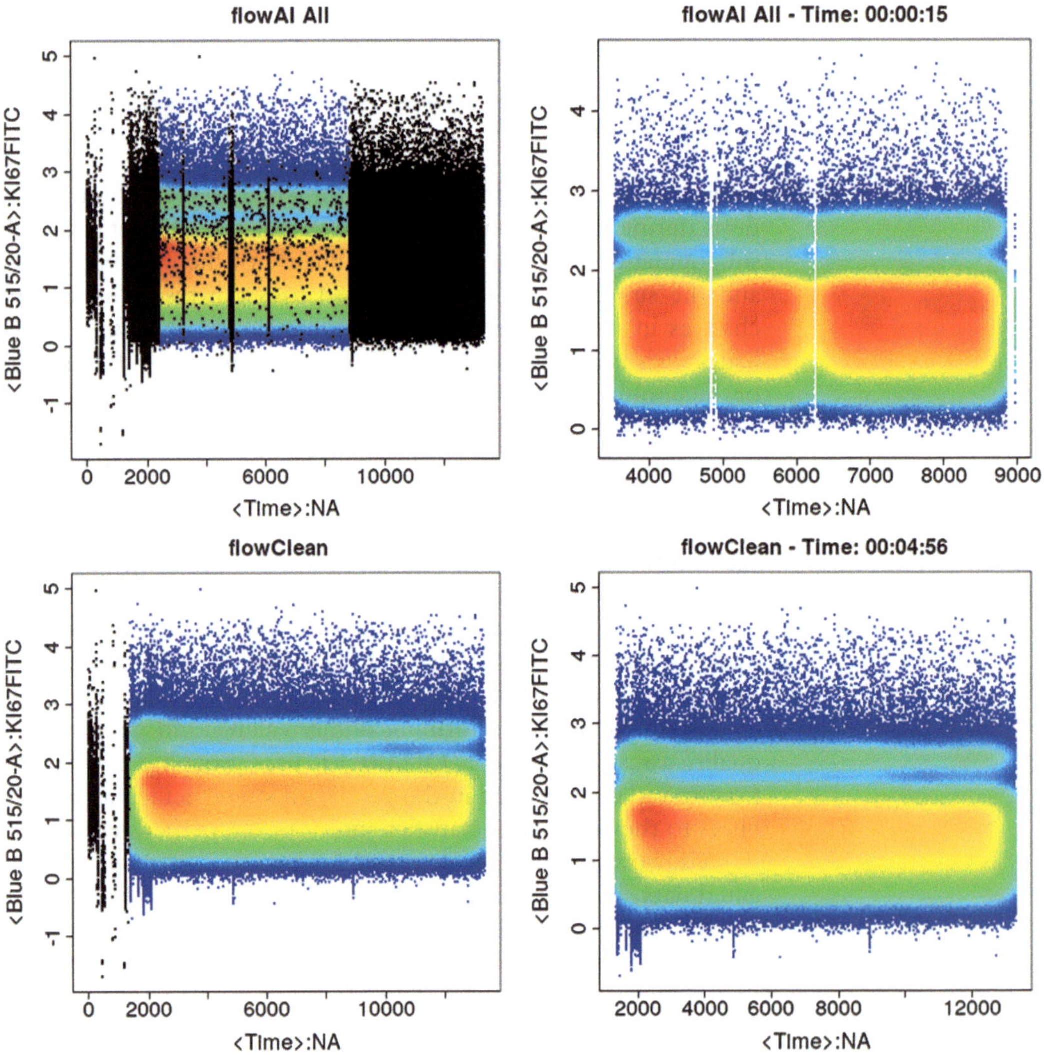

Fig. 1 flowAI and flowClean running on default parameters. The black dots indicate the cells being removed. The right column shows data after outlier removal

illumination (B) to estimate the precision of measurements at different signal levels. flowQB provides a mean for comparing different instruments and different channels on an instrument.

3 Data Transformation and Normalization

Data needs to be properly compensated, transformed, and normalized to ensure accuracy of any subsequent gating analysis. Compensation is necessary to correctly account for the contribution of each fluorochrome to each channel in conditions of spectral overlap. flowCore and flowUtils package have compensation and transformation functions for data preprocessing. The often-used transformation methods that handle negative values and display normally distributed cell types are logicle, hyperlog, and arcsine [7]. On occasion users require the normalization of data to eliminate technical variances that make matching populations across samples difficult. gaussNorm and fadNorm are two methods developed for this purpose and are integrated in the flowStats Bioconductor package [8]. Both methods aim to normalize single channel fluorescence data by finding and matching a common feature across samples. The common feature is assumed to be well aligned in ideal conditions and can thus be used as a reference to determine and remove technical variations by minimizing differences between the features across samples. A new fdaNorm algorithm [9] focuses on local normalization on specific cell subsets exhibiting variability and improves the peaks alignment and performance.

DeepCyTOF [10], an algorithm for semiautomated gating of cell samples uses deep learning technique with domain adaptation. It can also be used to overcome strong batch effects by calibrating target samples to reference sample.

4 Automated Gating

The most time consuming and subjective component of FCM data analysis is cell population identification and most of the efforts of the computational biology community have been focused on developing algorithms for automated gating [11, 12]. FCM data has unique requirements for computational efficiency, robustness with respect to different antigen/marker expression patterns, ability to determine true population number, and the ability to detect and handle outliers [13]. To date over 20 tools have been developed for automated gating (Table 1), and these can generally be divided into unsupervised and supervised approaches.

4.1 Unsupervised Gating

Clustering methods detect cells population that display similar biomarker expression in high-dimensional space and group them together. These algorithms generally do not require user input, but often allow the specification of some parameters such as the expected number of cell populations, to tune the results to a desired outcome. Unsupervised methods allow the discovery of unknown cell populations in high-dimensional data from large datasets that is not possible with manual analysis due to the exponentially increasing number of possible combinations of markers that must be investigated. Many approaches have extended advancements from other data types including k-means, random forests, self-organizing maps, and spectral clustering. Some tools have been developed for specific uses cases. For example, SWIFT is designed to gate on a large number of small clusters, and thereby can effectively separate rare populations [14], but has poorer performance on larger cell populations [15].

4.2 Supervised Gating

Supervised approaches are beneficial when users have some prior experimental expectations, for example, if a user wants to replicate an existing manual process to robustly target cell populations of interest in a specific way. This is useful for validating novel biomarkers discovered through high-dimensional cell discovery approaches. flowDensity [16] and openCyto [17] facilitate high reproducibility by automating the manual gating process. While flowDensity can process data in an unsupervised manner, the approach is designed to use customized one-dimensional density thresholds for each cell populations to mimic experts hierarchical gating order. However, unlike manual gating where the placement of gate boundaries is inherently subjective, thresholds are adjusted in a data-dependent manner for each sample (Fig. 2) [16].

OpenCyto is a framework that uses hierarchical gating on cell populations. Users define a template specifying the hierarchical relationships of the cell subpopulations and markers used. Most importantly, OpenCyto provides a larger framework for automated gating pipeline that fulfills all its analysis components, including preprocessing, cell population identification, population matching, and correlation with outcome variables [17]. It uses a plug-in framework that allows users incorporate gating algorithms into the pipeline.

4.3 Performance

Due to the lack of an unbiased gold standard, automated methods are often evaluated on their ability to best mimic a reliable manually gated cell populations as the current standard of practice. The performance of each method can be evaluated through measures of accuracy, precision, sensitivity, runtime, quality, and stability. High precision implies a low presence of false positives, and high sensitivity implies low false negatives. Sensitivity and precision can be summarized

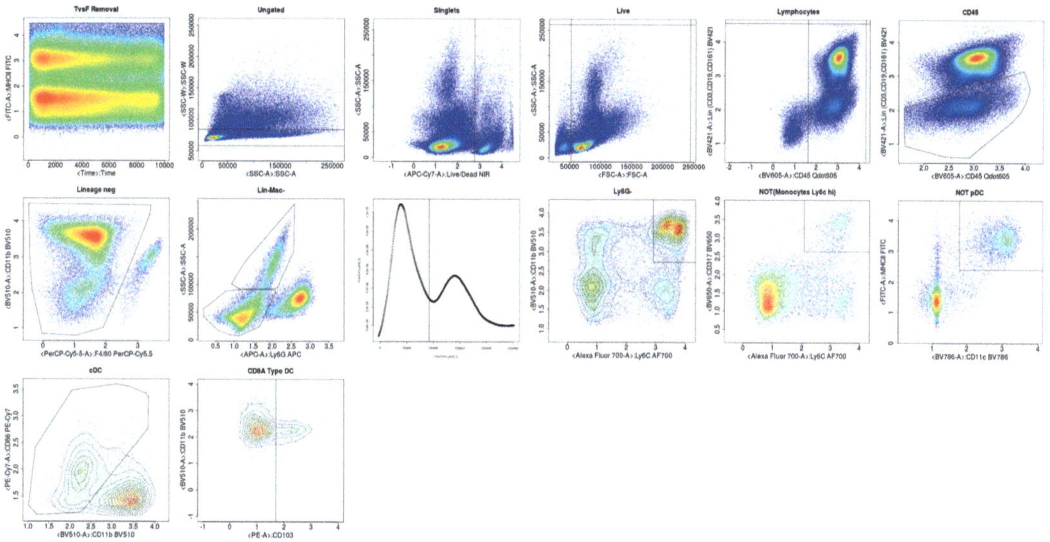

Fig. 2 flowDensity applied to International Mouse Phenotyping Consortium (IMPC) data. Colors moving from blue to red indicate increase density

in F-measure which is the harmonic mean of precision and recall. Fast runtime is important in performing interactive, exploratory analysis of large data sets. Runtime depends on algorithms subsampling, number of processor cores, and hardware specifications. Runtimes within a couple of minutes per FCS file scale to studies with hundreds of examples running on single processing core. Alternatively, users can configure processing to run on multicore settings where each CPU core takes up the task of running one FCS file through automated gating, as generally each file is considered independently. As such, several FCS files can run in parallel, which greatly reduces computing time of the overall program. Several software packages for multicore processing are available through CRAN (e.g., doMC, doParrallel).

Comprehensive community-based evaluations through the FlowCAP series of challenges [15, 18, 19] and other studies [20] have shown that automated approaches meet or exceed the performance of manual analysis for even simple datasets. To provide guidance in applying supervised methods of automated analysis to cytometry data for researchers and bioinformaticians, FlowCAP (Flow Cytometry: Critical Assessment of Population Identification Methods) has organized a series of challenges to facilitate a community-based evaluation of various FCM tools. FlowCAP I and II studied algorithms' ability for identifying cell population and sample classification, using manual gates as the evaluation baseline. The challenges identified flowMeans as the best-performing method taking into account of both speed and accuracy. One significant finding from the studies was that combining results from different cell population identification method produces a higher F-measure than any individual method for any dataset. As such, aggregate ensemble approach was found to be superior to any method used alone. However, a knowledge gap has been identified in the ability of algorithms to robustly identify rare cell population and deal with technical variables.

FlowCAP-3 was designed to explore the performance of algorithms addressing this area. It focused on the reproducibility of FCM tools, aiming to identify methods that could reproduce centralized manual gates with minimum bias and low variability and identification of rare populations [15]. The FlowCAP-3 study identified flowDensity and openCyto as the two co-best-performing supervised algorithms. It also recognized the difficulties in matching results from different centers even when given standardized reagents and analysis.

Weber et al. [20] reported a performance-comparison study of several clustering algorithms on their ability to detect all major immune cell populations and detect a single rare cell population, with a focus on CyTOF datasets. The study used F1 measure for the evaluation. F1 measure (harmonic mean of precision and recall) is the modified version of F-measure for the evaluation of unsupervised algorithms. Similar to F score, F1 score also ranges between 0 to 1, with 1 indicating a perfect imitation of the reference gated populations. However, unlike F score which used average weighted by size and assigns more importance to large cell populations, F1 score is calculated from unweighted averages in order to account for large and small population equally. Use of F or F1 score is a superior method in the evaluation of algorithms' performance than simply comparing population percentages, as population percentages do not give information on accuracy. Two populations can have exactly same population percentages yet not overlap with each other (i.e., low accuracy). For data sets containing multiple cell populations of interest, FlowSOM and flowMeans demonstrated good performance on high-dimensional immunological data sets. FlowSOM had the fastest runtime, usually several orders of magnitude faster than its counterparts. For single rare cell population detection, X-shift has the best F1 measure, followed by flowMeans, and FlowSOM among the top five. Overall, FlowSOM was shown to have the best performance taking into account of speed and accuracy and was identified as the best unsupervised clustering method for automated detection of cell populations in high-dimensional cytometry [20].

5 Probability Binning

Alternative to gating, probability binning is a method that analyzes FCM data in bins that contains nearly equal number of events. This method is extended into multiple dimensions to generate multivariate probability functions of FCM data, as incorporated in the R/Bioconductor package flowFP [21]. These specific multivariate probability functions are referred to as "fingerprints." Statistical tests can then be performed directly on the elements of a fingerprint to separate statistically informative subregions (or bins) that correlates most significantly with an experimental question. These bins can then subsequently be used as gates similarly to automated method.

6 Biomarker Identification

After identifying cell population in an individual sample, users might be interested in studying the difference in cell populations across groups of samples with biological variation. For application in discovery, the goal is to identify and describe cell populations that are correlated with an external variable. The current best-performing approach to biomarker discovery incorporates flowType, used in combination with flowDensity, the only two pipelines that showed significant performance in FlowCAP-IV [15, 18, 19]. The goal in FlowCAP-IV was to predict outcome associated with FCS data files, in this case, time of progression to AIDS among HIV+ cohorts. The results can be generalized to any outcome across datasets. flowType identifies all cell types in a sample by combining all possible partitions of cells, identified either manually or using automated gating [22]. It has been particularly successful in a type of exploratory analysis where correlation of different cell populations to external outcome is investigated.

FloReMi, available through Github (https://github.com/SofieVG/FloReMi), builds upon flowDensity and flowType to extract all features for each sample and then select features that most correlates with the survival time with minimal redundancy. As such, it is able to identify subpopulations of cells that can be used to predict phenotype of a patient. The components of the FloReMi pipeline include preprocessing, identifying, and selecting informative cell subsets and predicting survival time, all of which work independent from each other. It is thus possible to evaluate each component individually and substitute them with outside algorithms for similar tasks [23].

7 Visualization

Visualization is an essential part of data analysis in that it allows for communicating, exploring, and discovering possible significant outliers, cell populations. Visualization helps researchers study high-dimensional data in a way that provides biological insights. It is often desirable to reduce dimensionality in the complex data sets

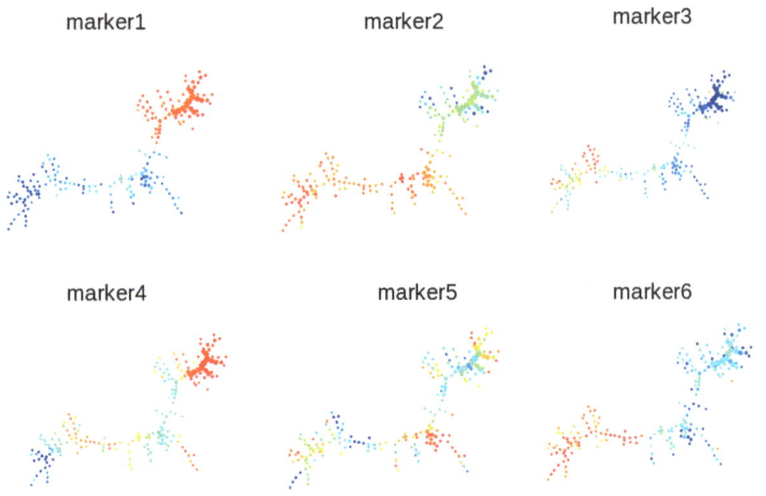

Fig. 3 SPADE tree color-coded based on the median expression intensities of cell markers within each node, where each node represents a cluster; each of the six trees represents the distribution of the different marker expression across cell clusters; red represents a high marker expression while blue a low expression; dataset is from the flowDensity package

in order for users to compare cell populations identified through high-dimensional space on a two-dimensional computer screen.

SPADE colors and connects similar immunophenotypes together in the form of a spanning tree, enabling visualization of single-cell data and inference of relationships between different cell types [24]. Biologists can then annotate the cell clusters by drawing inferences from cell marker expression. An example of a SPADE output is shown in Fig. 3.

t-SNE reduces dimensionality by arranging cell populations in ways that can preserve the spatial relationships of the cell populations in high-dimensional space [25], that is, cells that are close together in the high-dimensional space are in proximity in the 2D or 3D scatter map, thus enabling the visualization of different subpopulations (Fig. 4). Users can color code the t-SNE map based on gated results from either manual, supervised, or unsupervised methods. For detailed tutorials on t-SNE, users can go to https://github.com/Irrationone/graspods_scrnaseq_clustering_tutorial.

Both t-SNE and SPADE create a 2D (t-SNE can also produce 3D) visualization of higher order relationships and are both suited for cell population identification [24, 25]. In dealing with massive data, some algorithms implement subsampling to reduce computational time, however, often with the trade-off of cell population identification accuracy, especially for rare populations [20]. t-SNE performs a random down-sampling and SPADE density-dependent down-sampling. t-SNE has no clustering, and manual interpretation of distinct clusters for cell population identification is required.

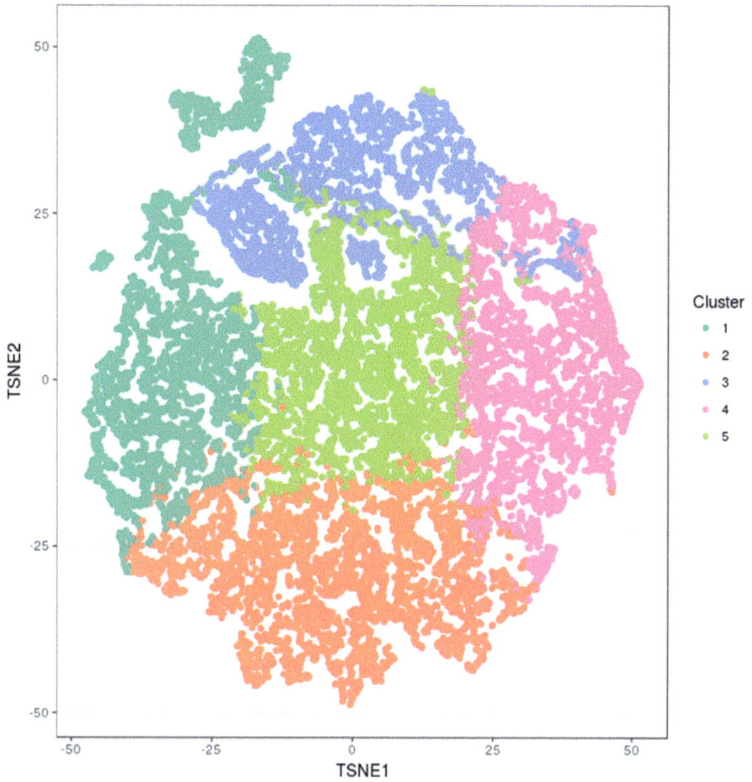

Fig. 4 Running t-SNE on two-dimensional space, where each data point represents one cell. Data is color-coded based on clustering. Data used is from the flowDensity package

Alternatively SPADE has a built-in clustering algorithm and project the clusters on a spanning tree and allows for additional inference of cellular hierarchy. However, manual interpretation is still required.

Citrus is a visualization tool that can aid in understanding high-dimensional data, and can help in the study of the correlation of cellular responses to external experimental endpoints. The algorithm splits data into many subdivisions and uses statistical tests to find differences among sample groups (control group and patient group) [26]. Citrus uses both predictive and correlative models to detect associations between clusters and experimental endpoints. The predictive model studies the properties of each cluster that are most predictive of the experimental endpoint. Citrus creates a series of predictive models with varying number of properties and then cross validates each model for their predictive accuracy (Fig. 5). Users can then select the model that generates the most predictive accuracy based on the error model rate and further exam those cluster properties identified by Citrus and their phenotypes.

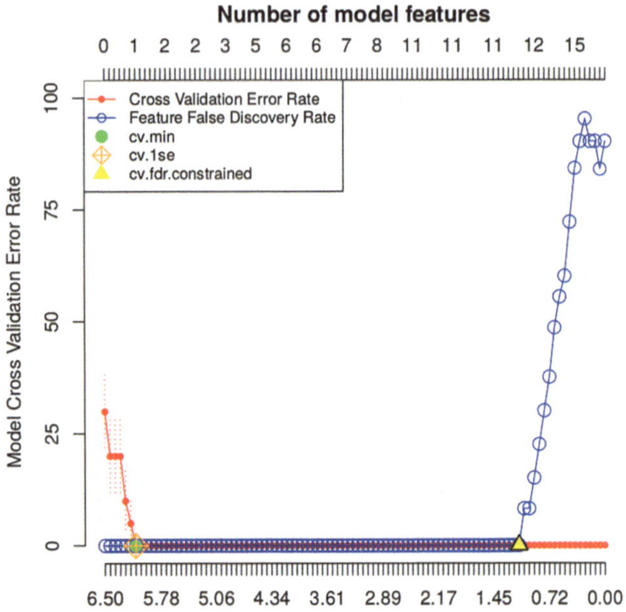

Fig. 5 Model error rate generated after cross validation of each predictive model. *cv.min* is the minimum error rate model. *cv.1se* is the simplest error rate model having 1 standard error of the minimum. *cv.fdr.constrained* is the model that contains the largest number of features while maintaining minimum false discovery rate. Figure reproduced from the Citrus simulated data, which contains 10 healthy patients sample groups and 10 diseased patients sample groups

RchyOptimyx is another visualization method that assigns importance to each cell type in biomarker identification studies by associating cell abundance to clinical outcomes, such as disease status or patient survival, and simplify the identified phenotypes [27]. RchyOptimyx provides dimensionality reduction for the description of a set of cell populations based on marker importance (Fig. 6).

Users should not confuse the hierarchies generated by RchyOptimyx to those generated from SPADE. SPADE and RchyOptimyx are different from each other in terms of algorithms and applications [28]. RchyOptimyx can group cell populations that have common parents which also exhibit functional similarities. In this way, it can automatically annotate cell populations identified by other methods (including SPADE) and list a single-cell hierarchy based on marker importance. Its functional strength lies in optimization of gating strategies. In contrast, SPADE uses the distance between mean/median fluorescence intensities to connect clusters together and then requires manual annotation of results based on experts knowledge. It is mostly used to overview different immune populations and surface marker or intracellular signaling molecule expressions [29, 30].

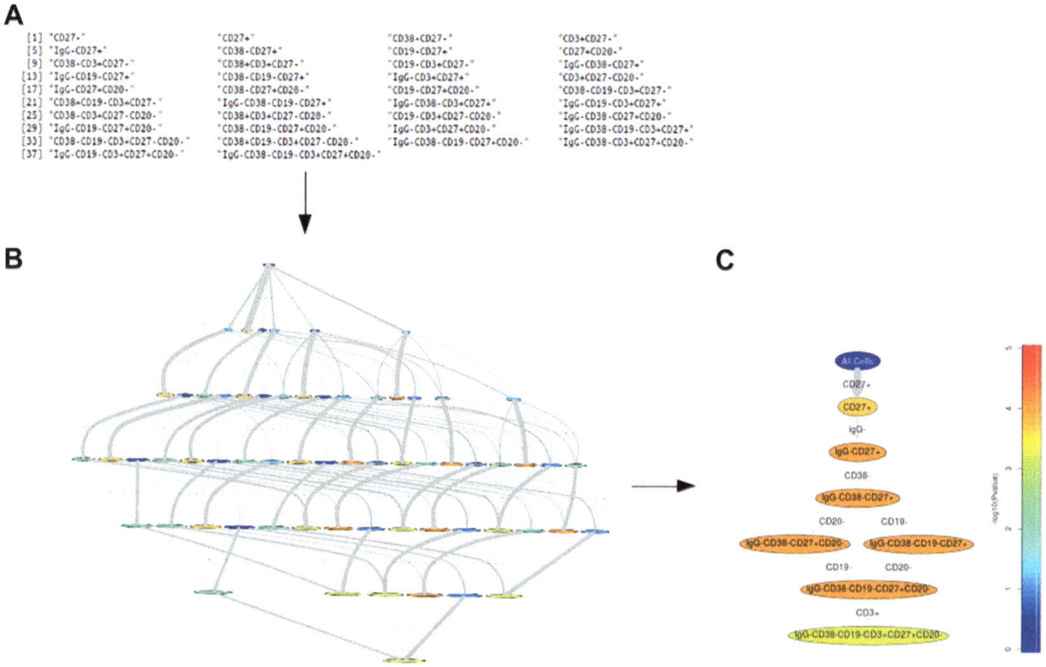

Fig. 6 A typical RchyOptimyx pipeline involves using flowType to extract all cell proportions (**a**) and then return a list of significantly different phenotypes and their scores. The phenotypes and their scores can be generated by any manual or automated algorithms. (**b**) shows a complete hierarchy describing all possible gating strategy for gating that cell population. (**c**) shows the relationship of number of markers in each phenotype with the significance of correlation with the clinical outcome. Data in (**b**) and (**c**) are reproduced from RchyOptimyx package data HIVData tube 2

8 Metadata Annotation

The ability to properly understand the results of analysis of even simple studies requires many components of the data generation and analysis be adequately described. The importance of metadata annotation only increases with complexity of studies, promoting the need for standardized reporting of experimental variables. The MIFlowCyt standard was developed with the aim of promoting effective curation, integration, analysis, interpretation, and sharing for cytometry data, and has been adopted by publishers [31]. One important component of data annotation is the semantic labeling of cell populations. flowCL semantically labels cell populations based on their surface markers profile. By doing so, flowCL provided means for standardized annotation for identified cell types. After identifying immunophenotypes through either manual or automated method, users might be interested to find out the cell type based on each cell marker's relative abundances. FlowCL does this by query against the cell ontology (CL) source for describing cell markers for hematopoietic cell lines [32]. CL is the cell ontology source that represents biological cell types with a structured description of each cell type. This is important in data integration when trying to match gating results from different sources. However, the manual exploitation of surface marker information in CL and the use of it requiring drawing inferences in multiple axes can pose great challenges for application on large dataset. flowCL addresses these challenges in ways that provide unambiguous labeling and optimize CL for application in automated cell labeling.

9 Learning

The use of computational algorithms described in this review requires users are familiar with R language. Most of the mentioned packages are available at Bioconductor website (https://www.bioconductor.org), a repository of open-source bioinformatic tools.

9.1 Install R

R is an open-source software that is freely available for all operating systems. Users can download R at its official website https://www.r-project.org or download through Bioconductor website at https://bioconductor.org/install. It is highly recommended users install RStudio (https://www.rstudio.com), the most widely used open-source integrated development environment (IDE) for R. IDE facilitates software development environments for computer programmers. The RStudio platform has a simple user interface and well suited for complex data handling and script writing.

9.2 First Step in Learning R

Starting off, users should learn the basic complexities of R, i.e., its data structures and functions. There are many free resources online for learning R, including courses [33, 34], tutorials, and forums. Users can get support from online forums such as Stack Overflow by asking or searching for specific problem-based questions. There are also R user groups available at most local universities and they often hold workshops.

9.3 Learn R for Flow Cytometry

Some recommended materials for learning how to conduct high-throughput data computational analysis using R are listed in Table 2. For package-specific usage, users should refer to the package-specific vignette, which is only available through Bioconductor. Vignettes contain thorough step-by-step instructions and code on how to use the software. It is often the first step in learning a new algorithm by reproducing the examples in the vignette. The package of each algorithm comes with example data sets, so that users can reproduce the results in the example by simply copying the code into RStudio or R.

FCM bioinformatics is a rapid evolving field. To keep up with the latest tool development and news, users are recommended to follow the literature including within the journals Cytometry A,

Table 2
Online resources for learning R for bioinformatics

Material	References
FCM Data Analysis Using R Workshop	https://bioinformatics.ca/workshops/2013/flow-cytometry-data-analysis-using-r-2013
Basic FCM Workshop	https://bioconductor.org/help/course-materials/2011/BioC2011/LabStuff/BasicFlowWorkshop.pdf
Statistics and R for High-Throughput experiments in life science	www.edx.org/course/statistics-r-harvardx-ph525-1x-0
Introduction to R and Bioconductor	https://github.com/Bioconductor/BiocIntro/tree/Bioc2017
Bioconductor workflow for Single-cell RNA-seq Data Analysis: Dimensionality Reduction, Clustering, and Pseudotime Ordering	https://github.com/fperraudeau/bioc2017singlecell
Conference BioC2017: Where Software and Biology Connect	http://bioconductor.org/help/course-materials/2017/BioC2017/

Bioinformatics where many tools to date have been published and BioConductor, which separately tracks packages in this domain by publishing and maintaining a broad range of analytical and graphical methods. Each published Bioconductor package has a maintainer, who is the first responder for helping solve any package-related issues such as those related to installation, compatibility, and software updates.

Acknowledgments

This work was supported by GenomeCanada (252FLO Brinkman), NSERC, GenomeBC, and NIH (1 R01 GM118417-01A1).

References

1. Kvistborg P, Gouttefangeas C, Aghaeepour N et al (2015) Thinking outside the gate: single-cell assessments in multiple dimensions. Immunity 42(4):591–592. https://doi.org/10.1016/j.immuni.2015.04.006
2. Hahne F, LeMeur N, Brinkman RR et al (2009) flowCore: a Bioconductor package for high throughput flow cytometry. BMC Bioinformatics 10:106. https://doi.org/10.1186/1471-2105-10-106
3. Robinson JP, Rajwa B, Patsekin V et al (2012) Computational analysis of high-throughput flow cytometry data. Expert Opin Drug Discov 7(8):679–693. https://doi.org/10.1517/17460441.2012.693475
4. Finak G, Jiang W, Pardo J et al (2012) QUAliFiER: an automated pipeline for quality assessment of gated flow cytometry data. BMC Bioinformatics 13:252. https://doi.org/10.1186/1471-2105-13-252

5. Gentleman R, Hahne F, Kettman J et al. (2017) flowQ: quality control for flow cytometry. R package version 1.38
6. Spidlen J, El Khettabi F, Moore W et al. (2017) flowQB: automated quadratic characterization of flow cytometry instrument sensitivity: Q, B and CV instrinsic calculations. R package version 2.6.0. https://www.bioconductor.org/packages/release/bioc/html/flowQB.html
7. O'Neill K, Aghaeepour N, Spidlen J et al (2013) Flow cytometry bioinformatics. PLoS Comput Biol 9(12):e1003365. https://doi.org/10.1371/journal.pcbi.1003365
8. Hahne F, Khodabakhshi AH, Bashashati A et al (2010) Per-channel basis normalization methods for flow cytometry data. Cytometry A 77(2):121–131. https://doi.org/10.1002/cyto.a.20823
9. Finak G, Jiang W, Krouse K et al (2014) High-throughput flow cytometry data normalization for clinical trials. Cytometry A 85(3):277–286. https://doi.org/10.1002/cyto.a.22433
10. Li H, Shaham U, Stanton KP et al (2017) Gating mass cytometry data by deep learning. Bioinformatics 33(21):3423–3430. https://doi.org/10.1093/bioinformatics/btx448
11. Maecker HT, Rinfret A, D'Souza P et al (2005) Standardization of cytokine flow cytometry assays. BMC Immunol 6:13. https://doi.org/10.1186/1471-2172-6-13
12. McNeil LK, Price L, Britten CM et al (2013) A harmonized approach to intracellular cytokine staining gating: results from an international multiconsortia proficiency panel conducted by the cancer immunotherapy consortium (CIC/CRI). Cytometry A 83(8):728–738. https://doi.org/10.1002/cyto.a.22319
13. Verschoor CP, Lelic A, Bramson JL et al (2015) An introduction to automated Flow cytometry gating tools and their implementation. Front Immunol 6:380. https://doi.org/10.3389/fimmu.2015.00380
14. Rebhahn JA, Roumanes DR, Qi Y et al (2016) Competitive SWIFT cluster templates enhance detection of aging changes. Cytometry A 89(1):59–70. https://doi.org/10.1002/cyto.a.22740
15. Aghaeepour N, Finak G, Flow CAPC et al (2013) Critical assessment of automated flow cytometry data analysis techniques. Nat Methods 10(3):228–238. https://doi.org/10.1038/nmeth.2365
16. Malek M, Taghiyar MJ, Chong L et al (2015) flowDensity: reproducing manual gating of flow cytometry data by automated density-based cell population identification. Bioinformatics 31(4):606–607. https://doi.org/10.1093/bioinformatics/btu677
17. Finak G, Frelinger J, Jiang W et al (2014) OpenCyto: an open source infrastructure for scalable, robust, reproducible, and automated, end-to-end flow cytometry data analysis. PLoS Comput Biol 10(8):e1003806. https://doi.org/10.1371/journal.pcbi.1003806
18. Aghaeepour N, Chattopadhyay P, Chikina M et al (2016) A benchmark for evaluation of algorithms for identification of cellular correlates of clinical outcomes. Cytometry A 89(1):16–21. https://doi.org/10.1002/cyto.a.22732
19. Finak G, Langweiler M, Jaimes M et al (2016) Standardizing flow cytometry Immunophenotyping analysis from the human Immunophenotyping consortium. Sci Rep 6:20686. https://doi.org/10.1038/srep20686
20. Weber LM, Robinson MD (2016) Comparison of clustering methods for high-dimensional single-cell flow and mass cytometry data. Cytometry A 89(12):1084–1096. https://doi.org/10.1002/cyto.a.23030
21. Rogers WT, Holyst HA (2009) FlowFP: a bioconductor package for fingerprinting flow cytometric data. Adv Bioinforma 2009:193947. https://doi.org/10.1155/2009/193947
22. Aghaeepour N, O'Neill K, Jalali A (2014) flowType: phenotyping Flow cytometry assays. R package version 2.14.0
23. Van Gassen S, Vens C, Dhaene T et al (2016) FloReMi: flow density survival regression using minimal feature redundancy. Cytometry A 89(1):22–29. https://doi.org/10.1002/cyto.a.22734
24. Anchang B, Hart TD, Bendall SC et al (2016) Visualization and cellular hierarchy inference of single-cell data using SPADE. Nat Protoc 11(7):1264–1279. https://doi.org/10.1038/nprot.2016.066
25. van der Maaten LJP, Hinton GE (2008) Visualizing high-dimensional data using t-SNE. J Mach Learn Res 9:2579–2605
26. Bruggner RV, Bodenmiller B, Dill DL et al (2014) Automated identification of stratifying signatures in cellular subpopulations. Proc Natl Acad Sci U S A 111(26):E2770–E2777. https://doi.org/10.1073/pnas.1408792111
27. Aghaeepour N, Jalali A, O'Neill K et al (2012) RchyOptimyx: cellular hierarchy optimization for flow cytometry. Cytometry A 81(12):1022–1030. https://doi.org/10.1002/cyto.a.22209
28. O'Neill K, Jalali A, Aghaeepour N et al (2014) Enhanced flowType/RchyOptimyx: a BioConductor pipeline for discovery in high-

dimensional cytometry data. Bioinformatics 30(9):1329–1330. https://doi.org/10.1093/bioinformatics/btt770

29. Bendall SC, Simonds EF, Qiu P et al (2011) Single-cell mass cytometry of differential immune and drug responses across a human hematopoietic continuum. Science 332(6030):687–696. https://doi.org/10.1126/science.1198704
30. Qiu P, Simonds EF, Bendall SC et al (2011) Extracting a cellular hierarchy from high-dimensional cytometry data with SPADE. Nat Biotechnol 29(10):886–891. https://doi.org/10.1038/nbt.1991
31. Lee JA, Spidlen J, Boyce K et al (2008) MIFlowCyt: the minimum information about a flow cytometry experiment. Cytometry A 73(10):926–930. https://doi.org/10.1002/cyto.a.20623
32. Courtot M, Meskas J, Diehl AD et al (2015) flowCL: ontology-based cell population labelling in flow cytometry. Bioinformatics 31(8):1337–1339. https://doi.org/10.1093/bioinformatics/btu807
33. R Programming. https://www.coursera.org/learn/r-programming. Accessed 19 Dec 2017
34. Statistics and R. https://www.edx.org/course/statistics-r-harvardx-ph525-1x-0. Accessed 19 Dec 2017

Chapter 17

Analysis of Mass Cytometry Data

Christina B. Pedersen and Lars R. Olsen

Abstract

The CyTOF system produces single cell protein expression data similar to that from flow cytometry, but with an increased number of features measured. Traditionally, analysis of these data is carried out using manual gating, but with the increased dimensionality, manual gating becomes a suboptimal analysis strategy in some cases. To address this, a number of data analysis tools for tasks such as clustering, differential abundance analysis, and visualization have been developed and made freely available. We here introduce some of the more popular tools for CyTOF analysis and exemplify their utility in a common analysis workflow.

Key words Mass cytometry, Dimensionality reduction, Clustering, Differential expression

1 Introduction

The analysis of mass cytometry data is nontrivial, but is facilitated by a large number of publicly available analysis tools. The analysis tasks and methods required depend, of course, on the biological question being asked. Having pre-processed the data from a series of electronic pulses into an FCS file containing an expression matrix, the possibilities are plentiful using even basic bioinformatics tools. Additionally, a lot of effort has been invested in developing tools specific for the common questions relating to single cell analysis of hematopoietic cells and cancers. A number of publications catalog and review these tools [1], and it is thus not within the scope of this article to do so. Instead, we will discuss a few of the more common algorithms and tools in the context of the problems they were designed to answer.

The first step, which can be seamlessly carried out in the CyTOF data acquisition software, is to transform the estimated ion counts using an Arcsinh transformation. This transformation retains the linearity of the counts in the low end of the spectrum, while resembling a log transformation for higher values. The cutoff

Helen M. McGuire and Thomas M. Ashhurst (eds.), *Mass Cytometry: Methods and Protocols*, Methods in Molecular Biology, vol. 1989, https://doi.org/10.1007/978-1-4939-9454-0_17,

for low/high in the expression spectrum is determined by dividing counts by a cofactor (typically around 5 for CyTOF v1 and v2, and around 2 for Helios). The cofactor alters the visual appearance of the data by de-emphasizing the noise that appears in the very low end of the spectrum.

In addition to the transformation, the Helios software also does a slight randomization of the estimated ion counts, such that each nonzero value, x, is randomized to x-1 and x. This is mostly done for aesthetic purposes, as count data produces "picket fenced" histograms. The randomization smooths the histograms out, and ensures unique events, which is important for a number of the downstream analysis tools. It is, however, adding noise to already noisy data, so many opt to deselect this default setting.

In addition to Arcsinh (or other transformations), some users choose to perform other normalization such as Z-scoring on the data prior to the methods discussed below [2]. Since the signal from different markers can vary by at least a factor of 100, this has the advantage of putting the data on a more unified scale. However, this does have the effect of amplifying the noise associated with a low-intensity marker, and potentially removing some resolution of other markers that may have legitimate variations in signal intensity (e.g., true zero/low/medium/high). Similarly, the previously discussed transformations can have the effect of partially rescaling the data [3], so further rescaling is most likely not required.

After transformation and normalization, analyses can be started. Which analyses to conduct and which tools to use for the job depends on the question being asked. There are, however, analysis tasks and visualizations that are universally applicable to most, if not all, research questions, namely cell subset detection by clustering and visualization of the clusters using dimensionality reduction. A standard analysis workflow could consist of: (1) cell subset detection using clustering algorithms, (2) differential cell subset abundance and/or differential feature expression in the different subsets across different conditions using statistical testing, and (3) visualization of cell subsets using dimensionality reduction. Since a number of clustering algorithms rely on dimensionality reduction, we will start by going through the two most commonly used algorithms: principal component analysis (PCA) and the t-Distributed Stochastic Neighbor Embedding (t-SNE) and then move on to clustering, differential abundance testing, and additional common visualizations.

2 Dimensionality Reduction

The most popular algorithm for dimensionality reduction of biological data is PCA. Being a linear dimensionality reduction algorithm, there is a limit to how much variance can be captured in the three visualizable dimensions. For CyTOF data, a PCA will usually capture around 40–50% of the variance on the first two principal components (for transcriptomics data, 60–70% is not unusual). This is usually not quite enough to provide good visual separation of clusters (Fig. 1a). Since the distance between events is linear, data on a nonlinear manifold will be misrepresented—this is best exemplified with the "Swiss roll" data example [1]. Whether this is an issue with CyTOF data or not depends heavily on the source of data. For example, as cells in the hematopoietic system develop in a hierarchical structure, the transition between some cells will result in a continuum rather than distinct clusters, which may not be entirely captured by PCA.

The t-SNE dimensionality reduction algorithm is a more recent addition to the dimensionality reduction toolbox [4], but has quickly gained popularity for cytometry data. The distances

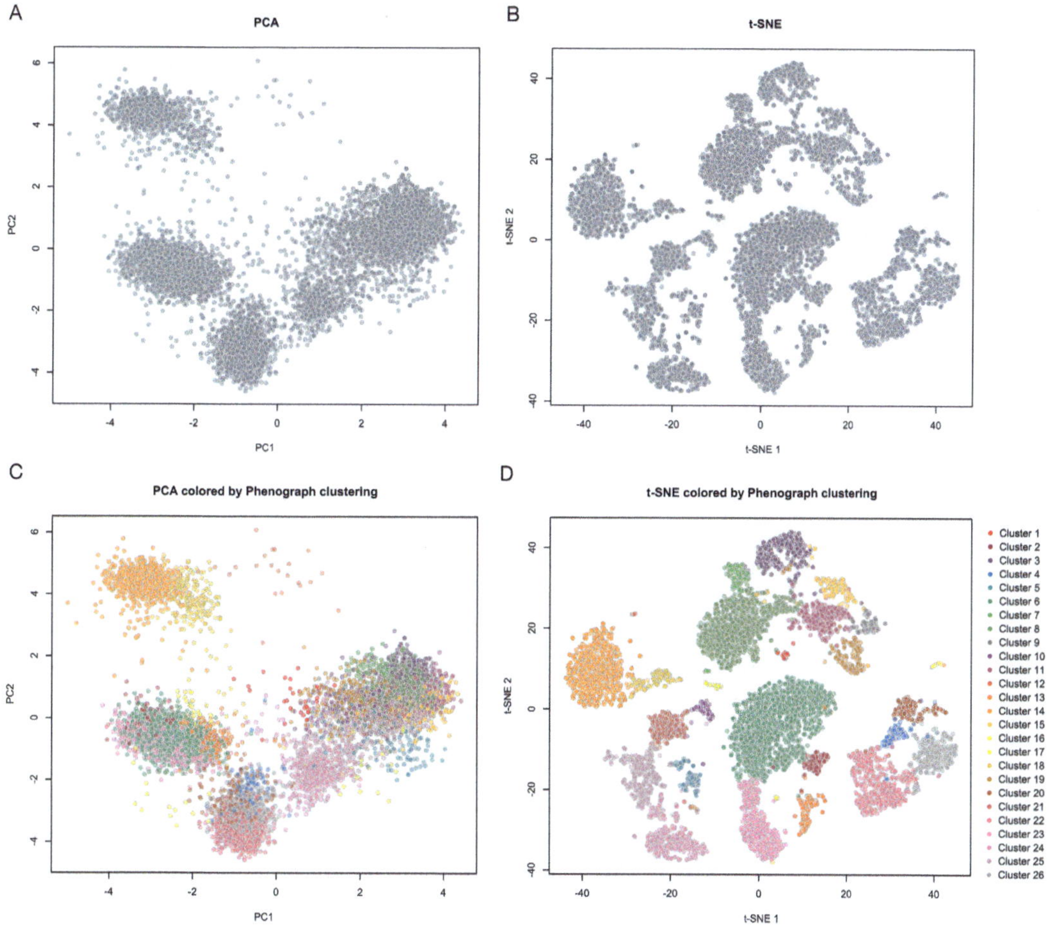

Fig. 1 (**a**) The two principal components of a PCA of mass cytometry data. (**b**) A t-SNE visualization of mass cytometry data reduced to two dimensions. (**c**) PCA with clusters determined using Phenograph. (**d**) t-SNE with clusters determined using Phenograph

between data points are represented by a t-distribution, meaning that true distances are reflected to a greater degree than in PCA when dimensions are reduced. However, since t-SNE does not linearly preserve distances (and by extension; densities) in the data, it is, for example, not possible to draw conclusions about the difference between two clusters based on their distance, or a cluster's diversity based on its density in t-SNE space. Since the algorithm is not deterministic (it is based on a random start with multiple "correct" solutions), users cannot expect to get the same result from two different runs of the algorithm (unless same seed and iteration number is selected) and t-SNE plots of two different samples are not comparable in terms of location of clusters or distance between data points. As a final quirk of the t-SNE algorithm, it does not allow for identical observations, which means that if identical cells exist in the data, expression values must be slightly scrambled or duplicates removed. The t-SNE algorithm usually captures the clusters more clearly than PCA and is an excellent visualization for mass cytometry data (Fig. 1b), but for the reasons mentioned above, deeper interpretation should be done with care. The widely used MATLAB graphical user interface (GUI) implementation of t-SNE is referred to as viSNE [5].

3 Cell Subset Detection (Clustering)

Unless one is working with cells in monoculture, deconvoluting a mixture of cells to the populations of interest is the first step in analysis of cytometry data. While for flow cytometry this is usually carried out by manual gating on biaxial plots, determining the cell subsets in mass cytometry data is often done using multidimensional clustering algorithms. This approach has considerable advantages over biaxial gating. Most notably, it is far less labor intensive. With the number of features being measured exceeding 40 in some instances, manual gating can be incredibly time consuming. More notably, clustering algorithms are not biased by user decision-making to the same degree as manual gating. Clustering algorithms are therefore much more likely to reveal populations that were not explicitly the subject of the investigation, and results are, at least mathematically, more robust across samples analyzed by different researchers.

A large number of clustering algorithms for mass cytometry have been published in recent years, and the accuracy of these algorithms has been the subject of much debate. The primary reason for this is the lack of a ground truth for evaluating the performance. Manually gated subpopulations are often the basis for comparison, and since these data are (a) biased by the researcher performing the gating, (b) prone to exclusion of cells, and (c) prone to assigning some cells to multiple different clusters, it is not the optimal benchmark. This lack of a clear performance measure is perhaps the biggest drawback of clustering algorithms, but there are also additional potential drawbacks to consider. Firstly, many clustering algorithms necessitate some degree of user decision-making, by tweaking parameters, the function of which may not be immediately obvious to nonexperts. Secondly, clustering algorithms are designed to "force" all cells into distinct clusters, which may not reflect reality. The reason for cell loss in manual gating is often the understanding that cell populations exist in relatively fluid continuums rather than well-defined discrete clusters, so researchers will "tighten" gates to subset only the most well-defined members of a given subpopulation for further analysis. Current clustering algorithms, on the other hand, are not designed to exclude transitional cell types, and therefore force these cells into the nearest cluster or assign them to their own cluster, which may or may not be a biologically meaningful thing to do.

As mentioned, many basic clustering algorithms have been applied to mass cytometry data, and a growing number of advanced algorithms have been defined explicitly for these data. It is beyond the scope of this chapter to review and compare them all, but we will focus on a select few algorithms in the next sections.

Since the hematopoietic continuum is a cellular hierarchy, the obvious first choice is simply hierarchical clustering. However, this method is considered by some to be suboptimal, usually ascribed to the need to manually select the number of clusters based on a distance cutoff in the resulting dendrogram. This forces the user to guess the number of populations in the sample: selecting too many will over-separate the data into meaningless small clusters; selecting too few will agglomerate smaller, but potentially interesting clusters into larger ones. Additionally, with a computational complexity of O(n) ("order-n," indicating a linear relationship between the size of the dataset and performance) and a typical yield of ~200,000 cells per sample, this can be a time consuming algorithm to run on CyTOF data. However, multiple tools still utilize hierarchical clustering in their algorithms. SPADE [6], for example, overcomes the computational challenges and the risk of missing rare populations by performing density-dependent downsampling before hierarchical clustering. Utilizing hierarchical clustering enables SPADE to capture the continuity of phenotypes in the hematopoietic system. Another simple algorithm that comes to mind is k-means clustering. This algorithm is, however, not optimal without modifications due to its tendency to favor clusters of equal size, which causes rare populations to be absorbed by larger neighbors.

Clustering, like any other learning algorithm, thus becomes a balance between sensitivity and specificity—a balance which is hard to strike considering that the distance between the true clusters differs greatly. A large number of algorithms have thus been developed with the challenges specific to mass cytometry data in mind. The most common and best benchmarked include FlowSOM [7] and Phenograph [8]. FlowSOM is based on the self-organizing map algorithm, which is an unsupervised, artificial neural network-based algorithm for reducing dimensionality in high-dimensional data. The algorithm is very fast and eliminates the need for downsampling. The major drawback for many users is the need to define the number of clusters in the output, which introduces some level of bias in the analysis. Phenograph is a graph-based approach, which, in contrast to FlowSOM, attempts to determine the number of clusters found in the data automatically. It works by forming a graph (network) by connecting similar cells to each other, and then identifies substructures in the graph resembling unique cell subsets. For a thorough review and benchmark, *see* [3]. Using Phenograph, the PCA and t-SNE visualizations of the data can be color coded by the clusters found by the algorithm (Fig. 1c, d).

4 Visualization of Protein Expression in Clusters

Following clustering of the cells of a mass cytometry experiment, a natural next step is to investigate which cell type(s) are placed in each of the clusters. A visual way to do this is to generate a heatmap showing median expression of the markers in each cluster. Using the median expression level, however, is associated with a loss of the information contained in the expression variance for each cluster—a metric that may in fact be very relevant to consider (*see* below). After generation of a heatmap it is possible to use expression of lineage markers to label clusters. In an example with a single healthy PBMC sample and clustering using Phenograph, the marker expression looks as shown in Fig. 2. From this plot it is very clear that the B-cell populations are found in clusters 12, 18, and 14 based on CD19/CD20 expression. Since cluster 18 is IgD^-CD27^+, it may be labeled as class-switched memory B-cells [9]. Cluster 14 is IgD^+CD27^- and should represent naive B-cells, whereas cluster 12 is IgD^+CD27^+ and likely represents non-switched memory B-cells. However, caution is strongly advised for this type of labeling, since it may be ambiguous. As an example, cluster 12 actually seems to be $CD3^+CD4^+$, which is not expected from B-cells. This could be due to experimental artifacts [10] or the cluster containing a mixed set of cells.

While using median marker expression levels as the main descriptive statistic of a cluster remains state of the art in processing of mass cytometry data, it is in fact rather problematic, because it disregards the variance of each cluster. If clustering is too coarse, or is not including a particular marker, there is a chance that two distinct subpopulations may be clustered together, and this can lead to bimodal distributions of the marker expression, which are not at all captured when using the median. A way to overcome mixed clusters is to include *all* the relevant markers, and rather generate

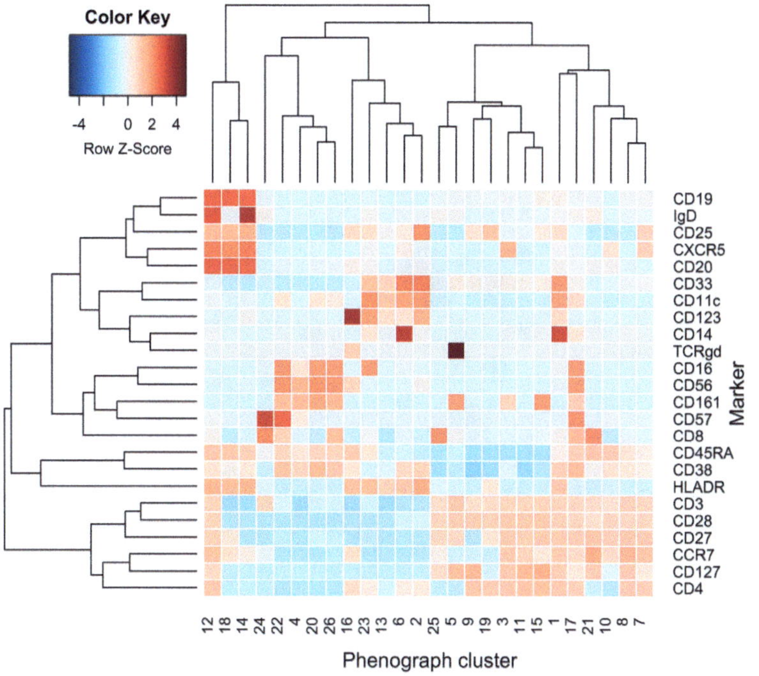

Fig. 2 Heatmap showing marker expression for clusters in a healthy PBMC sample

more than fewer clusters. Too few clusters can result in grouping of nonbiologically relevant subsets, while it is easy to post-process clustering results and merge some clusters with similar features. A heatmap may provide guidance for this.

An alternative to heatmaps is to use Radviz (https://cran.r-project.org/web/packages/Radviz/vignettes/single_cell_projections.html), in which each event may be plotted into 2D space by reducing the measured markers into two-dimensional anchors, which are ordered and placed on a circle. Each event is then placed in the circle based on its expression of the markers. To visualize clusters directly, Radviz includes the option to plot cluster centroids, and use the size of the cluster to define the size of the point (Fig. 3). This type of visualization is valuable for direct comparison of clusters to identify their differences, but may prove less useful in labeling the cell type placed in each cluster.

5 Analysis of Differential Abundance of Cells and Proteins

Many mass cytometry experiments are conducted with the purpose of studying the effect of a particular treatment or stimulation or to compare groups of samples or different time points. When carrying out the analysis of such experiments, the main goal is to identify changes between groups—that being abundance of a cluster

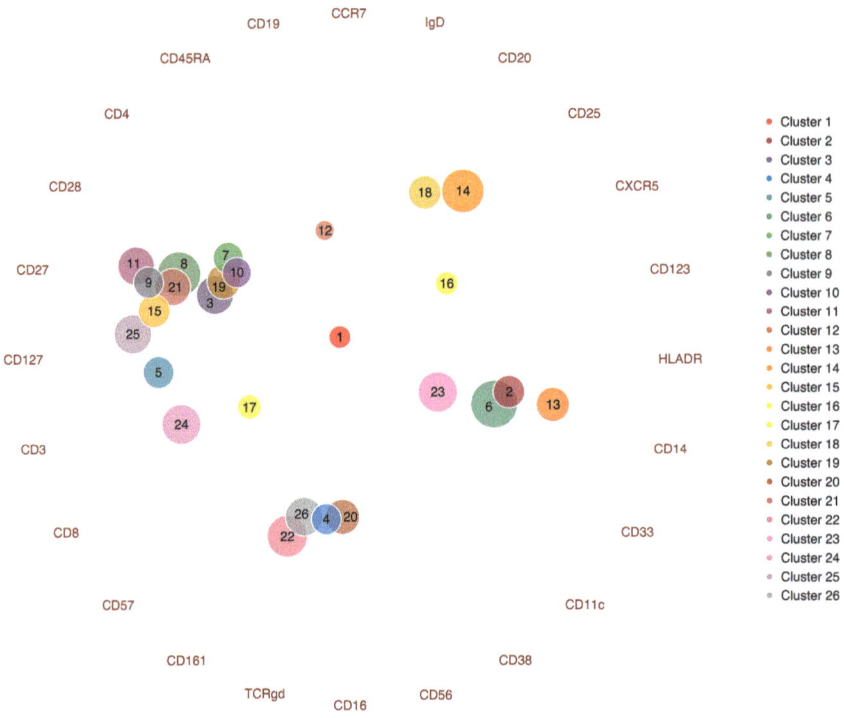

Fig. 3 Radviz bubble plot showing marker expression for clusters in a healthy PBMC sample

representing the same cellular subset, or the actual expression of protein markers in such a cluster. These two types of analysis correspond to common transcriptomics analyses known from both microarray and RNA-seq data, and consequently the same tools and methods may be applied. Citrus [11] combines hierarchical clustering and analysis of differential abundance and expression. Starting with pre-gated cells from samples of two or more groups, hierarchical clustering is performed using user-defined markers. Typically, those are the markers one would also use for manual population gating. The clustering requires a minimum cluster size, which should be set such that rare populations may be accounted for, but while knowing that having many small clusters will reduce the statistical power of the downstream analysis. Following clustering, it is possible to detect either differential abundance or expression using up to three statistical models: Significance Analysis of Microarrays (SAM), Nearest Shrunken Centroid (PAMR), and L1-Penalized Regression (glmnet). SAM was, as indicated by the name, originally developed for microarray data [12], and it is a correlative model that will identify features differing between groups, while the two latter are predictive models which will detect features that are predictive for each of the groups—these features can then later be used to classify new samples. For analysis of differential expression, the Citrus methods all work by comparing median expression levels for each cluster, which is associated with several problems, as discussed above. All of Citrus' three statistical models may also be used as stand-alone analysis tools for mass cytometry data in the case in which one is not interested in using the built-in Citrus clustering. This also allows for more direct control of the test, since the problem type may be customized to the given experiment, e.g., in the case of paired or time-course data. Furthermore, the Citrus output is mainly graphic and some users may find it more difficult to interpret than explicitly printed q-values and fold changes. An example of employing SAM separately from Citrus is shown in the Statistical Scaffold method [13]. As a side note to the problem of applying medians, the case of differential expression analysis is special, in that the models used assume that the distribution of the medians is Normal, which will not be the case when using raw expression values. Accordingly, data should be Arcsinh transformed before performing the statistical tests for differential expression. Additionally, it is worth noting that the markers used for clustering should not be included in the differential expression testing.

Differential abundance in clusters can be used to study coordination of cellular responses in which changes in two populations are correlated. This can also be applied to differentiation patterns. Surveying this is often done using generalized linear models as implemented in several legacy tools including edgeR and limma, since the distribution of counts follows a negative binomial distribution. Here, it is important to either apply normalization or downsampling to make counts comparable between samples. The cydar tool [14] was specifically developed for mass cytometry data to compare abundance between sample groups. This tool is independent from clustering of the data, but instead relies on assigning cells to hyperspheres and then testing for differential abundance within these and correcting for multiple testing using the edgeR package [15]. Each cell may be assigned to multiple hyperspheres, and as such the hyperspheres are hard to interpret biologically. However, as illustrated by Lun and colleagues [14], cydar is actually applicable to both time-course and paired data, but unlike Citrus it does not come with a GUI, which can make the tool inaccessible to some.

In general, it is important to have a reasonable amount of samples in each experimental group when performing statistical tests for differential abundance/expression. Citrus claims that eight samples per group is the minimum for robust results, and generally, more statistical power will be achieved using more samples. Also, beware that results may be spurious for very small clusters. Using paired reference and BCR-crosslinking PBMC data from Bodenmiller et al. [16], the above methods may be illustrated. First, data is clustered, and differential expression is surveyed for each cluster using SAM. In the $IgM^{+}CD20^{+}HLA\text{-}DR^{+}$ (B-cell) cluster, nine markers with differential expression were identified: pS6, pPlcg2, pp38, pLat, pAkt, pErk, pStat3, pStat5, and pNFκB. This may be illustrated using density plots, which have the added bonus of acting as a visual control for effects from the uncaptured variance (example in Fig. 4).

Differential abundance for the same clusters was quantified using edgeR, and after correction for multiple testing, 16/23 clusters displayed differential abundance between conditions. The largest numerical fold change is seen for cluster 2, whose size decreases with BCR-crosslinking (Fig. 5).

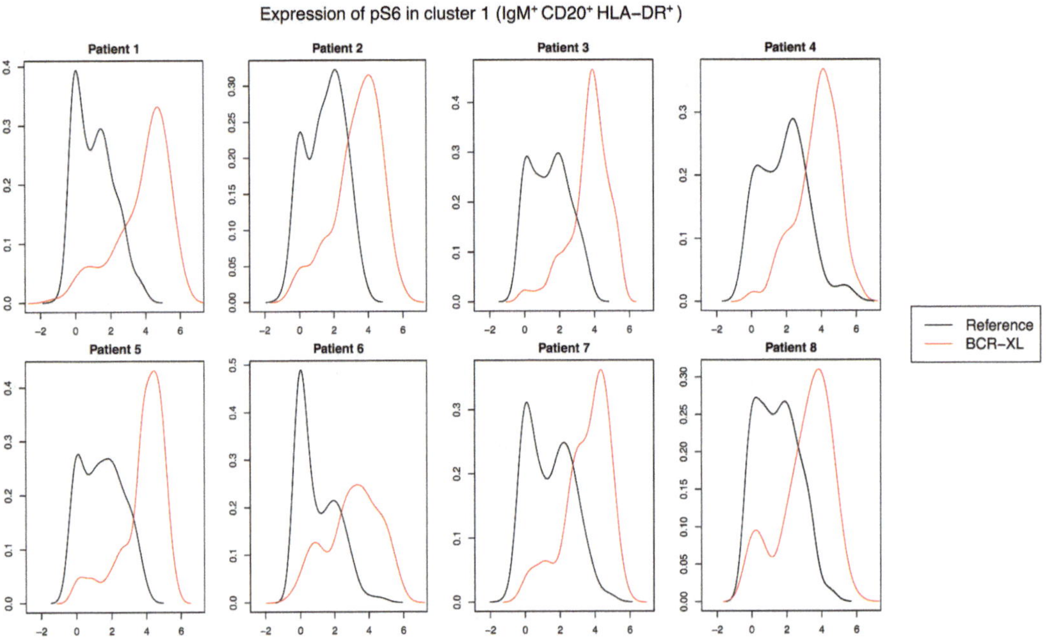

Fig. 4 Distribution of pS6-expression in one of the clusters. There is a clear upregulation in the BCR-crosslinked condition compared to the reference state in all eight samples

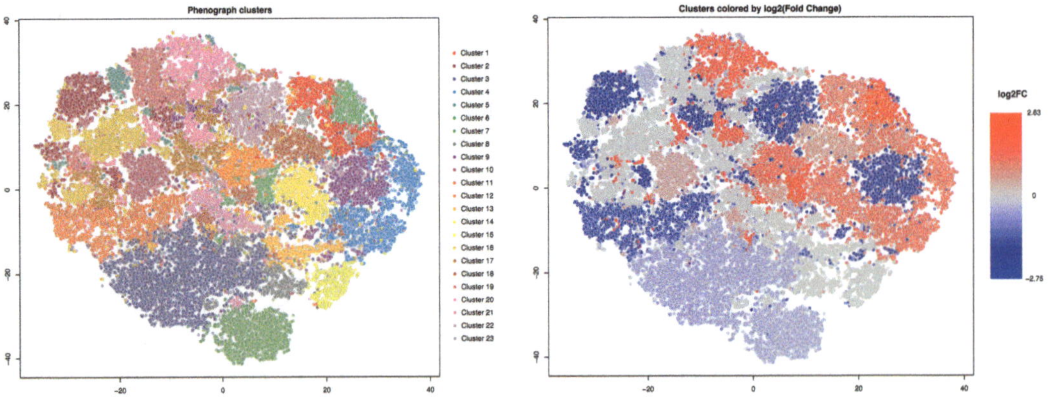

Fig. 5 Differential abundance analysis shown as t-SNE plots of the clusters. The right plot is colored by log2(Fold Change) between Reference and BCL-crosslinked conditions

6 Cellular Hierarchies

Because all immune cells represent a hierarchy of cells derived from the same common progenitor, the hematopoietic stem cell, it can be useful to study how novel populations fit in the hierarchy. Furthermore, several diseases are associated with alterations in cellular differentiation and hierarchical studies may help provide insight into such changes. SPADE [6] provides the option to visualize the results of a hierarchical clustering as a minimum-spanning tree, in which the most similar clusters are connected to each other via edges. Since it relies on density-dependent downsampling, rare populations should be maintained in the output. FlowSOM [7] generates similar plots, but has replaced the hierarchical clustering of SPADE with a self-organizing map—a form of neural network-based clustering that does not require any downsampling. Because of this, FlowSOM runs are faster and they come with the added bonus of star charts, which can visualize the marker expression patterns of each node. An example FlowSOM tree for a healthy PBMC sample is shown in Fig. 6. Group 2 appears to contain the B-cells ($CD19^{+}CD20^{+}$), and the IgD^{+} population is situated at the end of the orange branch separately from the IgD^{-} population.

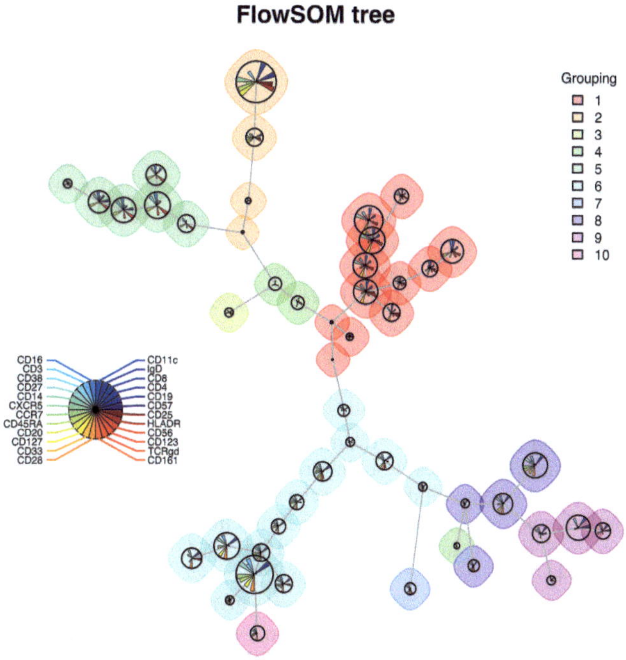

Fig. 6 FlowSOM tree generated using the FlowSOM R-wrapper. The marker expression for each cluster is shown as a star chart

For studies of developmental trajectories, the most applied tools are Wanderlust [17] and Wishbone [18]. While Wanderlust provides the ability to study the development of a cellular subset (e.g., B-cells) starting from a user-defined population, it has a major drawback in not allowing for branching trajectories. Wishbone allows for branching into two paths, but is otherwise a similar tool. Both algorithms rely on nearest-neighbor graphs and orders cells using the shortest path in the graph.

7 Conclusion

The majority of the tools discussed here can be applied directly to FCS files and run via graphical user interfaces by people with little or no programming experience. These efforts enable anyone to perform exploratory analyses of mass cytometry data from basic experimental setups. However, the user friendliness does come at the expense of flexibility and analysis of data from more complex experimental setups can benefit greatly from the collaboration with experienced bioinformaticians.

References

1. Chester C, Maecker HT (2015) Algorithmic tools for mining high-dimensional cytometry data. J Immunol 195(3):773–779. https://doi.org/10.4049/jimmunol.1500633
2. Shekhar K, Brodin P, Davis MM et al (2014) Automatic classification of cellular expression by nonlinear stochastic embedding (ACCENSE). Proc Natl Acad Sci U S A 111(1):202–207. https://doi.org/10.1073/pnas.1321405111
3. Weber LM, Robinson MD (2016) Comparison of clustering methods for high-dimensional single-cell flow and mass cytometry data. Cytometry A 89(12):1084–1096. https://doi.org/10.1002/cyto.a.23030
4. van der Maaten L, Hinton G (2008) Visualizing data using t-SNE. J Mach Learn Res 9:2579–2585
5. Amir el AD, Davis KL, Tadmor MD et al (2013) viSNE enables visualization of high dimensional single-cell data and reveals phenotypic heterogeneity of leukemia. Nat Biotechnol 31(6):545–552. https://doi.org/10.1038/nbt.2594
6. Qiu P, Simonds EF, Bendall SC et al (2011) Extracting a cellular hierarchy from high-dimensional cytometry data with SPADE. Nat Biotechnol 29(10):886–891. https://doi.org/10.1038/nbt.1991
7. Van Gassen S, Callebaut B, Van Helden MJ et al (2015) FlowSOM: using self-organizing maps for visualization and interpretation of cytometry data. Cytometry A 87(7):636–645. https://doi.org/10.1002/cyto.a.22625
8. Levine JH, Simonds EF, Bendall SC et al (2015) Data-driven phenotypic dissection of AML reveals progenitor-like cells that correlate with prognosis. Cell 162(1):184–197. https://doi.org/10.1016/j.cell.2015.05.047
9. Morbach H, Eichhorn EM, Liese JG et al (2010) Reference values for B cell subpopulations from infancy to adulthood. Clin Exp Immunol 162(2):271–279. https://doi.org/10.1111/j.1365-2249.2010.04206.x
10. Nagel A, Mobs C, Raifer H et al (2014) CD3-positive B cells: a storage-dependent phenomenon. PLoS One 9(10):e110138. https://doi.org/10.1371/journal.pone.0110138
11. Bruggner RV, Bodenmiller B, Dill DL et al (2014) Automated identification of stratifying signatures in cellular subpopulations. Proc Natl Acad Sci U S A 111(26):E2770–E2777. https://doi.org/10.1073/pnas.1408792111
12. Tusher VG, Tibshirani R, Chu G (2001) Significance analysis of microarrays applied to the ionizing radiation response. Proc Natl Acad Sci U S A 98(9):5116–5121. https://doi.org/10.1073/pnas.091062498

13. Spitzer MH, Carmi Y, Reticker-Flynn NE et al (2017) Systemic immunity is required for effective cancer immunotherapy. Cell 168(3):487–502.e415. https://doi.org/10.1016/j.cell.2016.12.022
14. Lun ATL, Richard AC, Marioni JC (2017) Testing for differential abundance in mass cytometry data. Nat Methods 14(7):707–709. https://doi.org/10.1038/nmeth.4295
15. Robinson MD, McCarthy DJ, Smyth GK (2010) edgeR: a Bioconductor package for differential expression analysis of digital gene expression data. Bioinformatics 26(1):139–140. https://doi.org/10.1093/bioinformatics/btp616
16. Bodenmiller B, Zunder ER, Finck R et al (2012) Multiplexed mass cytometry profiling of cellular states perturbed by small-molecule regulators. Nat Biotechnol 30(9):858–867. https://doi.org/10.1038/nbt.2317
17. Bendall SC, Davis KL, Amir el AD et al (2014) Single-cell trajectory detection uncovers progression and regulatory coordination in human B cell development. Cell 157(3):714–725. https://doi.org/10.1016/j.cell.2014.04.005
18. Setty M, Tadmor MD, Reich-Zeliger S et al (2016) Wishbone identifies bifurcating developmental trajectories from single-cell data. Nat Biotechnol 34(6):637–645. https://doi.org/10.1038/nbt.3569

Chapter 18

Analysis of High-Dimensional Phenotype Data Generated by Mass Cytometry or High-Dimensional Flow Cytometry

Branko Cirovic, Natalie Katzmarski, and Andreas Schlitzer

Abstract

Recent advances in single cell multi-omics methodologies significantly expand our understanding of cellular heterogeneity, particularly in the field of immunology. Today's state-of-the-art flow and mass cytometers can detect up to 50 parameters to comprehensively characterize the identity and function of individual cells within a heterogeneous population. As a consequence, the increasing number of parameters that can be detected simultaneously also introduces substantial complexity for the experimental setup and downstream data processing. However, this challenge in data analysis fostered the development of novel bioinformatic tools to fully exploit the high-dimensional data. These tools will eventually replace cumbersome serial, manual gating in the two-dimensional space driven by a priori knowledge, which still represents the gold standard in flow cytometric analysis, to meet the needs of the today's immunologist. To this end, we provide guidelines for a high-dimensional cytometry workflow including experimental setup, panel design, fluorescent spillover compensation, and data analysis.

Key words Mass cytometry, Flow cytometry, High-dimensional cytometry, Data analysis

1 Introduction

Recently the ability to investigate heterogeneous cell populations with single cell resolution on the epigenetic, transcriptomic, and phenotypic level has transformed the understanding of cellular development, function, and phenotypic complexity, especially in the field of immunology. However, analyzing such high-dimensional data with thousands of single cells and up to thousands of data points per single cell creates enormous data complexity which needs to be analyzed appropriately. Traditional analyzing strategies relying on a priori knowledge and investigator-driven analysis can be challenging to implement and will not utilize the information within such high-dimensional data to the full extent. Therefore, novel unbiased analysis techniques are needed to account for the complexity of single cell resolution data.

Helen M. McGuire and Thomas M. Ashhurst (eds.), *Mass Cytometry: Methods and Protocols*, Methods in Molecular Biology, vol. 1989, https://doi.org/10.1007/978-1-4939-9454-0_18,

Traditionally, cellular phenotyping strategies in immunology relied heavily on the use of flow cytometry with 10–18 phenotypic markers analyzed. Recently technological advances have pushed the number of analyzable parameters upward reaching an upper limit of 50, utilizing high-dimensional flow cytometers or mass cytometers. Here we outline a user-friendly methodology to analyze such high-dimensional phenotyping data either generated using mass cytometry or next generation flow cytometry.

2 Methodology

In this section, we present a basic, exemplary analysis workflow using the freely available R/Bioconductor package "cytofkit" [1]. This analysis tool comprises a highly customizable data processing pipeline, including sample multiplexing, data transformation, dimensionality reduction, marker visualization, cluster identification, as well as estimation of subset progression. Moreover, this package allows a flexible approach to the processing steps depending on the data analysts' background and preference, including the integrated graphic user interface (GUI), which is recommended for familiarization with the package functions and which will be the main focus here. The command line option, however, is the method of choice to facilitate analysis standardization and documentation. Finally, the primary data output of the cytofkit core function represents a starting point for flexible downstream analysis by using other R/Bioconductor packages, using the included Shiny application, or reimporting data into programs, such as FlowJo (https://www.flowjo.com), for classical evaluation of analysis results. Of note, this analysis pipeline applies both for mass and flow cytometric data.

The data we use for demonstration here contains mass cytometric data of PBMC-derived CD14-enriched human monocytes from three donors that have been cultured in the presence of M-CSF, GM-CSF or GM-CSF and IL-4 for 6 days including 36 markers of interest in addition to live/dead (cisplatin) and lineage markers for exclusion of contaminating cells. For additional information concerning the samples and the scientific context, we refer to the article by Sander and colleagues [2].

2.1 Preparing the Input Data for Analysis

If required, the primary data should be pre-filtered to focus on cell populations of interest. As our parameter set includes lineage markers, we first gated on live, CD45 positive, lineage negative (CD3, CD7, CD15, CD19, CD20) cells in FlowJo and exported this population as new .fcs file (*see* **Note 1**).

2.2 Installing and Launching Cytofkit and GUI in R

The cytofkit package can be installed using the following commands in R:

source("https://bioconductor.org/biocLite.R")

```
biocLite("cytofkit")
```

The commands have only to be used once for installation of the package. Mac OS users are advised to verify installation of required software dependencies (*see* **Note 2**). The latest development version with additional features is available from Github (https://github.com/JinmiaoChenLab/cytofkit).

To load the package run:

```
library("cytofkit")
```

The GUI is then initialized with the code:

```
cytofkit_GUI()
```

2.3 Setting the Parameters in the GUI

Once the GUI is initialized in R, a separate window appears that contains all relevant core parameters to be set for a complete analysis (Fig. 1). The first input (Fig. 1a) indicates location of the .fcs files followed by the selection of the files that will be included and merged during analysis (Fig. 1b). In this example we inserted 12 .fcs files—resulting from the four conditions for each of the three donors. The available parameters (markers) are automatically listed and can be selected by highlighting (Fig. 1c, *see* **Note 3**). Choosing a results directory (Fig. 1d) will define the location of the output files which are generated by default once analysis is complete (*see* Subheading 2.4). All results files will contain the character string as chosen in Fig. 1e as prefix. The cytofkit package offers different

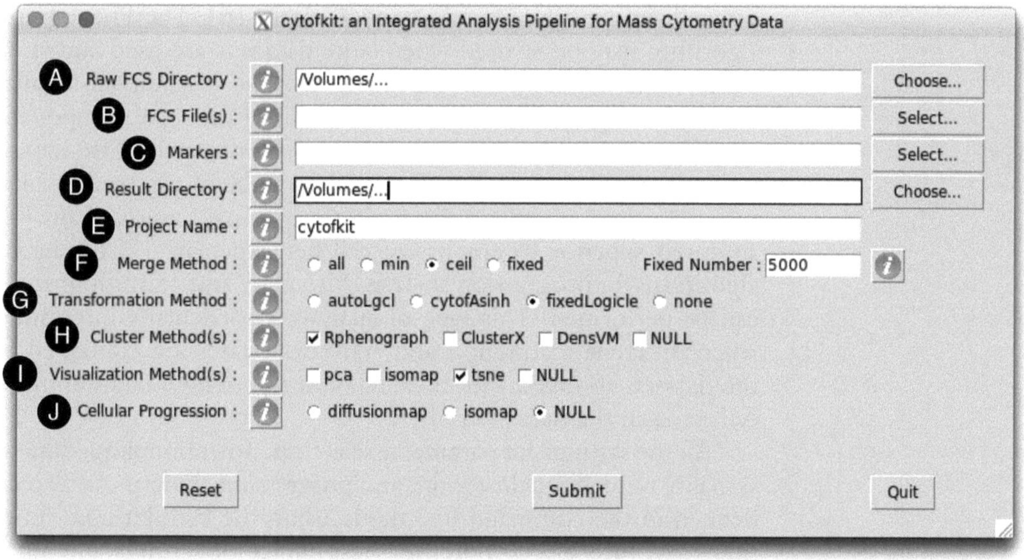

Fig. 1 GUI Interface of the cytofkit package to set parameters for analysis. (**a**) directory where the fcs files to be analyzed are located, (**b**) the files to be analyzed, (**c**) markers to be used for clustering, (**d**) directory for output files, (**e**) name of the analysis, (**f**) method for managing the number of cells to be merged from multiple files, (**g**) method of transformation, (**h**) methods for clustering, (**i**) methods for visualization, and (**j**) methods for analyze cellular progression

methods to merge the individual data files (Fig. 1f). Selecting "all" will simply include all data points (i.e., events measured) from the .fcs files, while "min" sets a minimal number of randomly sampled cells defined by the number of cells contained in the smallest sample file. The option "fixed" allows you to define a given cell number. Cells from each file will be resampled, if necessary, to reach the defined number of cells. For the present analysis, we selected "ceil" together with a fixed cell number of 1000 cells that will be randomly chosen from each file without replacement (*see* **Note 4**). Furthermore, several methods for data transformation are available (Fig. 1g). The general recommendation is to choose "autoLgcl" for flow cytometric data to account for the inherent high dynamic range of fluorescent signals and to improve the visualization of negative and low signal values. Similarly, a modified version of a hyperbolic transformation is classically used for mass cytometric data ("cytofAsinh") which corrects for negative signal artifacts generated during acquisition. These recommendations may be considered as first starting points as the most suitable mode of transformation seems to depend on the individual data set [3]. In addition, the third transformation setting "fixedLogicle" allows customization of transformation parameters. In latter case a new window will open where the parameters can be directly set. For clustering and subset identification three algorithms are available including the two density peak-based algorithms "DensVM" and "ClusterX" and the graph-based algorithm "Phenograph" (Fig. 1h) [1, 4, 5]. In a recent comparison, all three methods showed similar high precision and accuracy [6]. The Phenograph algorithm may be favored when large data sets are used and large cluster size variability is expected. Options for dimensionality reduction (Fig. 1i) include the classical linear principal component analysis (pca) method and nonlinear methods, such as isomap as well as t-distributed stochastic neighbor embedding (tsne), which recently gained popularity for multidimensional single cell analysis appeared superior to emphasize cell heterogeneity [7, 8]. Lastly algorithms for progression analysis ("isomap" and "diffusionmap") can be performed. This type of analysis is particularly interesting when different maturation and activation states are contained in the datasets to investigate their relation assuming a continuum of cell states in the data [9–11].

All the settings for parameter selection, downsampling, dimensionality reduction, clustering, and progression analysis can also be defined in the command-line mode using the cytofkit core function (*see* **Note 5**). Of note, for best choice of settings and algorithms the results should be carefully validated for each data set. One first approach is to reimport the data into FlowJo to manually gate well-defined cell subsets in a standard workflow and subsequently overlay and compare with the analysis results from the cytofkit algorithm (*see* below).

2.4 Evaluation of the Default Output Results

After submission of the parameters defined in the previous section, the requested analysis will be computed in relatively short time depending on the options chosen and automatically saved in the results directory (*see* **Note 6**). Default output files can be classified into four categories. The pdf files contain visualizations of the dimensionality reduction (here pca and tsne) and refer to the clustering results as shown by color coding and numbering (Fig. 2a). Different shapes of the data points relate to the origin of .fcs files. Additional heatmaps and hierarchical clustering indicate either the contribution of .fcs file samples or marker parameters to each individual cluster (Fig. 2b, c). The table files cover the merged and transformed expression data as well as cluster affiliation, contribution and coordinates of dimensionality reduction. Conveniently, a new set of .fcs files will be exported to define the set of cells that have been included in the analysis by sampling. Moreover, new parameters are introduced into the exported .fcs files covering dimensionality reduction coordinates and cluster affiliations which allows further inspection of the results in standard FlowJo gating workflow. Lastly, an RData file is generated, which enables further analysis with other R packages or the integrated Shiny application.

2.5 Options for Post-processing of Primary Results

As implied in the previous section, there are several options for the post-processing of the primary results according to personal preference. One possibility is to import and concatenate the output .fcs files in FlowJo. Now the parameter selection allows the selection of tsne coordinates for the x/y axes (*see* **Note 7**). Importing and overlaying the individual output .fcs files in the layout area of FlowJo helps to easily define the position of the individual samples in the tsne space. The new color map axis function in recent FlowJo

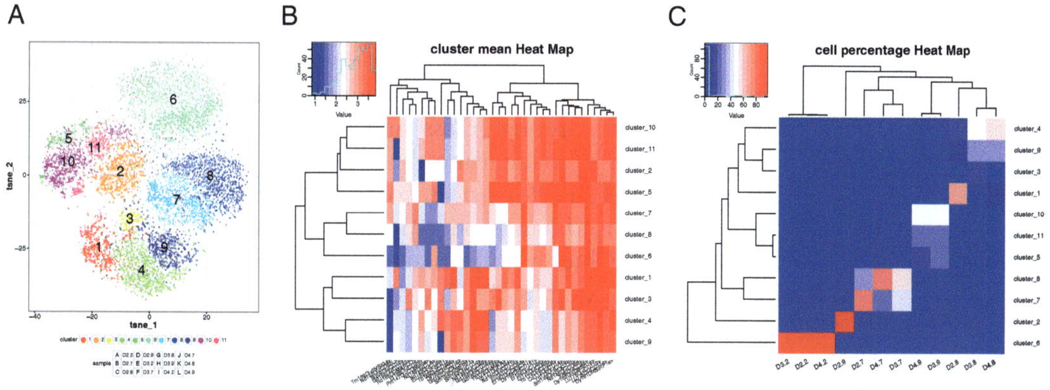

Fig. 2 Examples of default output figure (**a**) visualization of dimensionality reduction using tsne. In addition, results of the Phenograph cluster analysis are represented with different colors and numbering. Sample affiliation is indicated by different letters that constitute the data points. Here, the letters *A–L* represent 12 fcs input files. (**b**) Hierarchical clustering of markers and clusters or (**c**) fcs samples and clusters to indicate cluster similarity and how different markers (**b**) or samples (**c**) relate to cluster specification

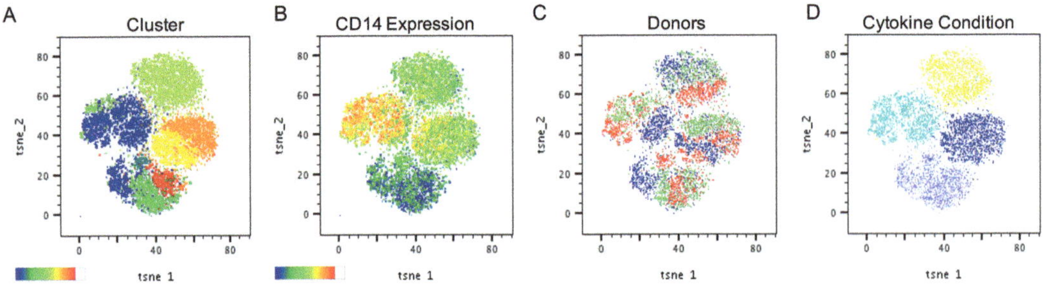

Fig. 3 Post-processing in FlowJo. Reimporting data into FlowJo allows further analysis with classical flow cytometric workflow strategies. Here, we use the color map axis function to mark Phenograph cluster affiliation (**a**) or CD14 marker expression (**b**). Overlaying individual files and variations of color coding clearly mark the negligible contribution of individual donors (**c**) in comparison to the strong effect of different cytokine conditions (**d**)

versions (≥10.4) additionally allows for screening of marker expression or cluster affiliation (Fig. 3). Classical gating and back-gating strategies can further help to elucidate the nature of the tsne space.

Alternatively, the generated RData file can be imported into the integrated shiny application for interactive data exploration (Fig. 4). Run the following command to open the cytofkit shiny application in a web browser.

cytofkitShinyAPP()

Most of the functions can be intuitively used and the effect will be instantly shown as this application works in a responsive mode. Therefore, a new user is advised to try all the options—the analysis can be reset using the control panel A at any point.

Lastly, to acquire the most control over the visualization of the results, the experienced R user can extract the output data from the "analysis_results" object contained in the RData file. Functions from the ggplot2 package are a good starting point to discover additional layers in the data (Fig. 5; *see* **Note 8**).

The basic workflow overview presented here should encourage the data analyst with any background to perform a basic analysis with multidimensional mass and flow cytometry data. A first approach can be made using the interactive tools, but every single step can also be performed in command line for standardized routine and reduced hands-on time data analysis. For details, we recommend the highly informative quick-start guides and package vignettes provided by the developers.

3 Considerations for the Setup and Analysis of High-Dimensional Flow Cytometry Data

Besides the development of mass cytometry, there have been several strides in fluorescence flow cytometry concerning the instruments as well as the reagents to the same extent. These advances

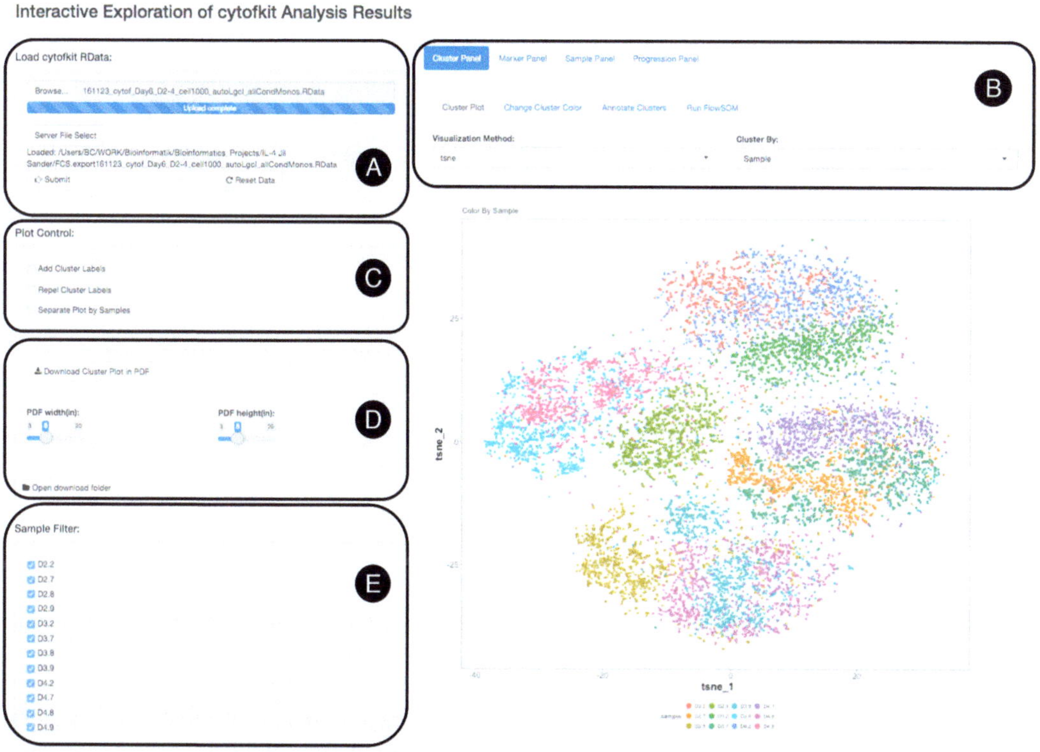

Fig. 4 The interactive cytofkit application consists of several control panels. (**a**) An interface to import the RData results file. In the top row of (**b**) the main category of the setting can be chosen focusing on clusters, markers, samples and progression inference. Below, the tabs further control visualization type and mode including the option to re-annotate cluster, markers and samples for clearer representation of the data. Accordingly, appearance of (**c**) will change to fit the selected tables (**d**) options to export figures and (**e**) options to toggle on and off individual samples

allow researchers to perform complex multicolor experiments with up to approx. 30 parameters leading to a more defined delineation of single cells and a clearer more unbiased definition of cell populations. Additionally, the parallel examination of several cell types and markers within one sample leads to more reproducible data, the need for less sample material, and greater correlation of data within samples. However, by increasing the complexity of this experimental approach, new challenges and obstacles arise. In the following sections, we will provide some guidelines to succeed with high-dimensional multicolor fluorescence flow cytometry, with the aim to generate high-quality data sets which are suitable to be analyzed using the analysis technologies described before. In order to generate a dataset suitable for high-dimensional analysis techniques, the creation of an optimized staining panel is essential. Although potentially time consuming, this is essential for proper data collection, interpretation, and analysis.

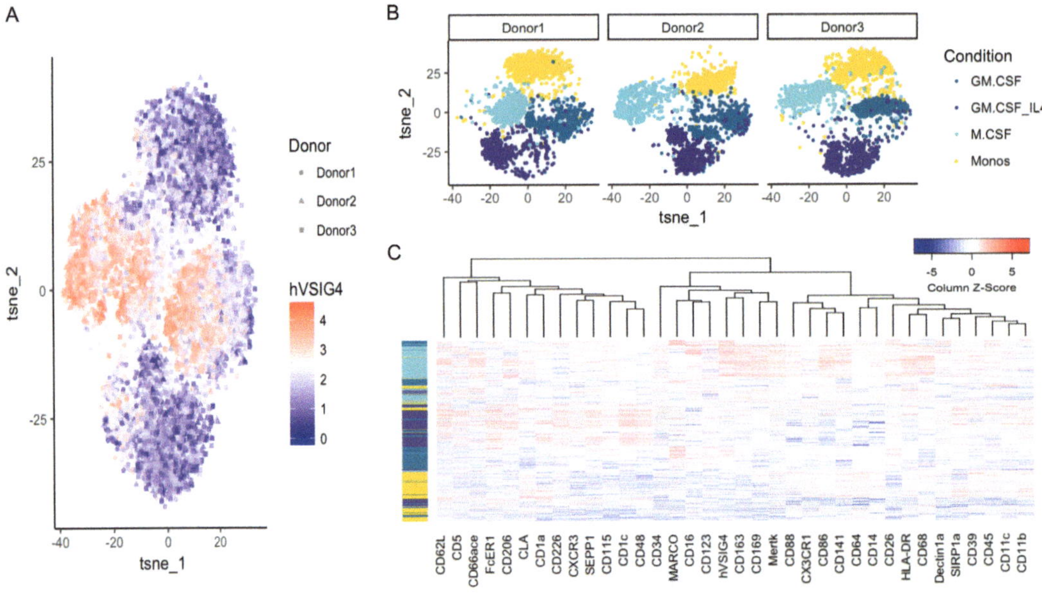

Fig. 5 Customized figures can be generated with basic data visualization commands. (**a**) tsne coordinates (analysis_results$dimReducedRes$tsne) as well as expression data (analysis_results$expressionData[,"hVSIG4"]) extracted for the gglot2 function to visualize the 12,000 cells in the reduced dimension and to overlay the expression data for the hVSIG4 protein. A variable corresponding to the three individual donors has been introduced to indicate donor affiliation by different shapes of the data points. (**b**) Here the color code corresponds to the cytokine condition. Furthermore, plots have been separated into the individual donor fractions. (**c**) Alternatively, the complete expression table is used for hierarchical clustering and heatmap visualization applying the heatmap.2 function from the gplots package. Row colors correspond to the four cytokine condition as in (**b**)

The basic principles for the design of a high-dimensional multicolor flow cytometry panel are:

- Be informed of available instruments.
- Use the instrument-specific spillover-spreading matrix [12].
- Match the brightness of fluorophores to the density of antigens (pair bright fluorophores with low-density antigens, and dim fluorophores with high-density antigens).
- Adjust selections to reduce spillover/spreading error.
- Use tandem dyes with care.
- Use appropriate controls.

3.1 Instrumentation

An understanding of the cytometry instrument to be used is a critical prerequisite for setting up multi-color flow cytometry panels. Critical aspects are the lasers, as well as the detectors and available filter sets build into the cytometer. For multi-color assays, it is recommended to establish a quality control system for the instrument that results in a consistent instrument setup. This quality control system aims to track and adjust basic instruments settings, sensitivity, and day-to-day variations in measurements. Essentially, a mixture of non-labeled and fluorescence-labeled synthetic beads corresponding to the build-in laser/detector system should be used to standardize and optimize the instrument setup. Normally, most suppliers provide a ready-to-use mixture of beads for an instrument quality control procedure, which is usually automated. However, due to the recent expansion of new fluorescent dyes, the ready-to-use mixtures may not adequately cover relevant parts of the visible spectrum. Because of this, performance and instrument settings may have to be monitored manually.

3.2 Fluorophores, Spillover, and Spreading Error

The first step to create a panel is to match the brightness of fluorophores to be included in a panel with the density of the antigens of interest. The simplest approach is to match the brightest fluorophores with the lowest expression antigens, and vice versa. The brightness of fluorophores is described by the stain index (SI), which is specific for an instrument with particular lasers and detectors [13]. The stain index captures the difference between the medians of the positive and negative peak divided by the peak of the background peak.

$$\mathrm{SI} = \frac{D}{W} = \frac{\mathrm{Median}_{\mathrm{Positive}} - \mathrm{Median}_{\mathrm{Negative}}}{2 \times \mathrm{rSD}_{\mathrm{Negative}}}$$

Additionally, the potential spillover of individual fluorophores can be considered in advance by using available databases (e.g., BioLegend Spectra Analyzer, BD Bioscience Spectrum Viewer, ThermoFischer FluorescenceViewer). These databases indicate the spectral overlap of selected fluorophores. Minimizing spillover is important because the greater the spillover of a fluorophore into a nontarget detector, the greater the spread of the positive population could be in that detector (although greater spillover doesn't necessarily result in high spreading error, spreading error can only occur where there is spillover). This will affect the resolution of dim signals as they will not be able to be resolved above background in the affected channels. Tandem conjugates should be used with care within a panel as they are more light-, temperature- and fixation-sensitive compared to most other available fluorophores. Furthermore, due to their sensitivity tandem conjugates show a higher variation between different experiments and production batches. As such, these require attention when generating single-color reagents for compensation, to ensure fluorophores in compensation controls are treated the same way as the real samples.

3.3 Controls

Single stains for each fluorophore-labeled antibody and an unstained control sample are mandatory to create a suitable compensation matrix. Especially for markers with low levels of expression, the accurate discrimination between positive and negative signal can be challenging. Fluorescence minus one (FMO) controls are an excellent way to ensure accurate discrimination between positive and negative stained cells. FMO controls are created by including all markers except the one of interest and provide actual indications to set the threshold between positive and negative signal.

3.4 Data Quality Control

The next step after successful panel design and acquisition of samples on a cytometer is to subject the data to a quality control pipeline. As a first step the scatter values should be plotted against time to filter out regions with abnormal behavior, such as clogs or air bubble measurement. This step can be done utilizing the flowClean [14] or flowQ [15] packages. Additionally, this quality check pipeline should include considerations of artifacts due to improper compensation, background noise, or experimental batch effects. These steps can be automatically performed in one pipeline, such as with the Bioconductor flowCore packages [16]. Afterward, samples are further processed by removing doublets, debris, and dead cells. This pre-processing step can be performed manually or automatically by using openCyto [17] or flowDensity [18] packages. If a dataset passes these quality control steps, high-dimensional analysis techniques can be used as outlined for the mass cytometry dataset above.

3.5 Data Analysis

As with mass cytometry, flow cytometry data can be analyzed by using visualization tools based on dimensionality reduction technique and cluster-based techniques. pca and tsne are dimensionality reduction visualization techniques which are widely used to analyze high-dimensional cytometry data. In contrast to dimensionality reduction techniques, clustering-based techniques first group cells into cell type specific clusters in the high-dimensional space and use subsequently algorithms to visualize these cell clusters in lower dimensional-space. One of the most prominent clustering-based methods to visualize flow cytometry data is SPADE (spanning tree progression of density normalized events) [19]. SPADE is based on a hierarchical clustering approach to group cells into cell type specific clusters before they are visualized in a minimal spanning tree (MST). An alternative is FlowSOM [20] which is similar to SPADE, but uses a self-organizing map instead of the hierarchical clustering. The results can be visualized using a tsne- or MST-based algorithm. FlowSOM is efficient in processing millions of cells while tsne-based methods and SPADE are only able to process fewer cells (a few tens of thousands of cells). While clustering-based methods reveal an average picture for each cluster, methods based on dimensionality reduction plot each single cell. The advantage of clustering-based methods is that it provides groups of cells which can be interpreted as cell types whereas techniques based on dimensionality reduction need a further clustering step to allocate each cell to a cell type. Methods based on tsne project data in a way that cells that are similar in high-dimensional space will tend to be close in two-dimensional space. However, cells close in the two-dimensional plot are not necessarily similar in the high-dimensional space. A similar notion applies to MSTs created by SPADE or FlowSOM. These spanning trees do not capture actual developmental hierarchies. This should be kept in mind when visualized mass or flow cytometry data is interpreted.

To discover biomarkers for certain treatments or diseases from flow or mass cytometry data, Citrus [21] or flowType-RchyOptimyx [22] algorithms can be applied. Both pipelines use clustering and a subsequent selection step to extract the most relevant populations. Citrus makes use of hierarchical clustering to extract information about cell populations which are then used in a regularized regression model while flowType-RchyOptimyx uses exhaustive splitting of negative and positive populations and uses the most informative populations to make predictions.

Another approach of high-dimensional analysis are techniques used for cell development modeling, such as the Wanderlust algorithm [23]. This technique attempts to follow gradients in the data to model cellular developmental processes, reconstructing the developmental pathway a cell would follow. Such an approach operates on the assumption that the dataset is a mixture of cells representing different developmental stages, with no other populations present. The reconstruction is based on a user-defined starting cell which represent either the most mature or immature cell within the dataset.

To overcome limitations set by manual gating, the use of automated gating tools can be a solution. During the past decade, several of these unbiased gating tools were developed, such as FLAME [24], flowMeans [25], or FLOCK [26]. They can be differentiated based on the clustering method they are using. While FLAME uses a model-based clustering skewed t-distribution on individual samples followed by a step to model mapping between the samples, flowMeans uses a K-means clustering algorithm which is representative-based. In FLAME and flowMeans the number of clusters is not automatically detected in contrast to FLOCK were autodetection is available. FLOCK uses clustering based on centroids derived from density estimation in a hypergrid. However, as flow data is inherently variable, these tools face a number of challenges. In particular, variability of cell subsets and rarity of cell types are often major problems for automated gating tools. In general, the use of high-dimensional analysis techniques for data created by fluorescence-based flow cytometry is possible. However, the creation of a reproducible data set that passes all quality controls is depended on many parameters, as discussed above. As the number of markers of interest to define the phenotype of a single cell increase, so too do the number of fluorophores that are used, requiring greater care to ensure that the measured fluorescent signal is not due to spillover, spreading error, or noise. Often flow cytometry data is suitable for manual gating and analyses, but the distinction between negative and positive signal is not clear enough, or to variable, to run a faithful computational analysis. To overcome this obstacle, carefully created multi-color flow cytometry panels should be complimented by standardization of sample preparation and a quality control pipeline to analyze the raw data.

4 Notes

1. Dispensable parameters, such as lineage markers or live/dead markers, can be dropped during exporting the data. A first quality check may also be performed before export. If necessary, the .fcs files can be cleaned up in order to remove artifacts by visual inspection in the graph window (with obvious outliers in the axes dimension). In addition, automatic and standardized quality controls can be performed using, for example, the flowAI library package [27].
2. If using a mac, verify installation of XQuartz (available at https://www.xquartz.org/) and Xcode (available on the Mac App Store).
3. Our analysis contained 36 parameters. Lineage markers were not included in the marker selection as lineage positive cell were already removed during cleanup and export and are not of interested for downstream analysis. The user should closely consider which markers (and or events) potentially provide the information for the analysis and which should rather be dismissed. In common practice, FSC and SSC parameters are omitted in this context (applies to flow cytometry data only).
4. For standard analysis, it is recommended to use the same number of cells from each sample file. The total number of cells that should be included depends on the expected cell heterogeneity and cell subtype frequency.
5. The arguments of the function *cytofkit()* will define all necessary parameters as set in the GUI. Run the command? *cytofkit* for further details.
6. Computation of the 12,000 cells taking into account the 36 parameters, omitting progression analysis and isomap visualization took around 15 min on a standard laptop computer.
7. For this step the axis transformation has to be reset to "linear" for tsne and pca coordinates using the transform ('T') button in the graph window of FlowJo.

8. The figures displayed in Fig. 5 are generated with the following lines of code:

```
(A)
     library(ggplot2)
ggplot(as.data.frame(analysis_results$dimReducedRes$tsne),aes(x=tsne_1,
y=tsne_2))+
geom_point(alpha=0.5,size=1.2,aes(shape=Donor,colour=analysis_resul
ts$expressionData[,"hVSIG4"]))+
theme_classic()+ scale_colour_gradient2(low = ("blue"),mid = "white",
midpoint = 2.5,high=("red"))+
labs(colour="hVSIG4")

(B)
library(ggplot2)
ggplot(as.data.frame(analysis_results$dimReducedRes$tsne),aes(x=tsne_1,
y=tsne_2))+
geom_point(size=1,aes(colour=Condition))+
theme_classic()+
scale_colour_manual(values=c(colors()[c(125,107,45,144)]))+
facet_grid(~c(Donor))

(C)
library(gplots)
heatmap.2(analysis_results$expressionData,
     scale = "column",
     col=bluered(30),
     trace = "none",
     density.info="none",
     dendrogram = "column",
     RowSideColors = allDonor_ID,
     key.title = "",
     labRow = NA)
```

References

1. Chen H, Lau MC, Wong MT et al (2016) Cytofkit: a bioconductor package for an integrated mass cytometry data analysis pipeline. PLoS Comput Biol 12(9):e1005112. https://doi.org/10.1371/journal.pcbi.1005112
2. Sander J, Schmidt SV, Cirovic B et al (2017) Cellular differentiation of human monocytes is regulated by time-dependent interleukin-4 signaling and the transcriptional regulator NCOR2. Immunity 47(6):1051–1066. e1012. https://doi.org/10.1016/j.immuni.2017.11.024
3. Finak G, Perez JM, Weng A et al (2010) Optimizing transformations for automated, high throughput analysis of flow cytometry data. BMC Bioinformatics 11:546. https://doi.org/10.1186/1471-2105-11-546
4. Becher B, Schlitzer A, Chen J et al (2014) High-dimensional analysis of the murine myeloid cell system. Nat Immunol 15(12):1181–1189. https://doi.org/10.1038/ni.3006
5. Levine JH, Simonds EF, Bendall SC et al (2015) Data-driven phenotypic dissection of AML reveals progenitor-like cells that correlate with prognosis. Cell 162(1):184–197. https://doi.org/10.1016/j.cell.2015.05.047
6. Weber LM, Robinson MD (2016) Comparison of clustering methods for high-dimensional

single-cell flow and mass cytometry data. Cytometry A 89(12):1084–1096. https://doi.org/10.1002/cyto.a.23030
7. Amir el AD, Davis KL, Tadmor MD et al (2013) viSNE enables visualization of high dimensional single-cell data and reveals phenotypic heterogeneity of leukemia. Nat Biotechnol 31(6):545–552. https://doi.org/10.1038/nbt.2594
8. van der Maaten L, Hinton G (2008) Visualizing data using t-SNE. J Mach Learn Res 9:2579–2605
9. Haghverdi L, Buettner F, Theis FJ (2015) Diffusion maps for high-dimensional single-cell analysis of differentiation data. Bioinformatics 31(18):2989–2998. https://doi.org/10.1093/bioinformatics/btv325
10. Tenenbaum JB, de Silva V, Langford JC (2000) A global geometric framework for nonlinear dimensionality reduction. Science 290(5500):2319–2323. https://doi.org/10.1126/science.290.5500.2319
11. Haghverdi L, Buttner M, Wolf FA et al (2016) Diffusion pseudotime robustly reconstructs lineage branching. Nat Methods 13:845. https://doi.org/10.1038/nmeth.3971
12. Nguyen R, Perfetto S, Mahnke YD et al (2013) Quantifying spillover spreading for comparing instrument performance and aiding in multicolor panel design. Cytometry A 83(3):306–315. https://doi.org/10.1002/cyto.a.22251
13. Parks D (2004) Presented at the XXII congress of the International Society for Analytical Cytology. Montpellier, France
14. Fletez-Brant K, Spidlen J, Brinkman RR et al (2016) flowClean: automated identification and removal of fluorescence anomalies in flow cytometry data. Cytometry A 89(5):461–471. https://doi.org/10.1002/cyto.a.22837
15. Le Meur N, Rossini A, Gasparetto M et al (2007) Data quality assessment of ungated flow cytometry data in high throughput experiments. Cytometry A 71(6):393–403. https://doi.org/10.1002/cyto.a.20396
16. Hahne F, LeMeur N, Brinkman RR et al (2009) flowCore: a Bioconductor package for high throughput flow cytometry. BMC Bioinformatics 10:106. https://doi.org/10.1186/1471-2105-10-106
17. Finak G, Frelinger J, Jiang W et al (2014) OpenCyto: an open source infrastructure for scalable, robust, reproducible, and automated, end-to-end flow cytometry data analysis. PLoS Comput Biol 10(8):e1003806. https://doi.org/10.1371/journal.pcbi.1003806
18. Malek M, Taghiyar MJ, Chong L et al (2015) flowDensity: reproducing manual gating of flow cytometry data by automated density-based cell population identification. Bioinformatics 31(4):606–607. https://doi.org/10.1093/bioinformatics/btu677
19. Qiu P, Simonds EF, Bendall SC et al (2011) Extracting a cellular hierarchy from high-dimensional cytometry data with SPADE. Nat Biotechnol 29(10):886–891. https://doi.org/10.1038/nbt.1991
20. Van Gassen S, Callebaut B, Van Helden MJ et al (2015) FlowSOM: using self-organizing maps for visualization and interpretation of cytometry data. Cytometry A 87(7):636–645. https://doi.org/10.1002/cyto.a.22625
21. Bruggner RV, Bodenmiller B, Dill DL et al (2014) Automated identification of stratifying signatures in cellular subpopulations. Proc Natl Acad Sci U S A 111(26):E2770–E2777. https://doi.org/10.1073/pnas.1408792111
22. O'Neill K, Jalali A, Aghaeepour N et al (2014) Enhanced flowType/RchyOptimyx: a BioConductor pipeline for discovery in high-dimensional cytometry data. Bioinformatics 30(9):1329–1330. https://doi.org/10.1093/bioinformatics/btt770
23. Bendall SC, Davis KL, Amir el AD et al (2014) Single-cell trajectory detection uncovers progression and regulatory coordination in human B cell development. Cell 157(3):714–725. https://doi.org/10.1016/j.cell.2014.04.005
24. Pyne S, Hu X, Wang K et al (2009) Automated high-dimensional flow cytometric data analysis. Proc Natl Acad Sci U S A 106(21):8519–8524. https://doi.org/10.1073/pnas.0903028106
25. Aghaeepour N, Nikolic R, Hoos HH et al (2011) Rapid cell population identification in flow cytometry data. Cytometry A 79(1):6–13. https://doi.org/10.1002/cyto.a.21007
26. Qian Y, Wei C, Eun-Hyung Lee F et al (2010) Elucidation of seventeen human peripheral blood B-cell subsets and quantification of the tetanus response using a density-based method for the automated identification of cell populations in multidimensional flow cytometry data. Cytometry B Clin Cytom 78(Suppl 1):S69–S82. https://doi.org/10.1002/cyto.b.20554
27. Monaco G, Chen H, Poidinger M et al (2016) flowAI: automatic and interactive anomaly discerning tools for flow cytometry data. Bioinformatics 32(16):2473–2480. https://doi.org/10.1093/bioinformatics/btw191

Chapter 19

Computational Analysis of High-Dimensional Mass Cytometry Data from Clinical Tissue Samples

Sam Norton and Roslyn Kemp

Abstract

The advent of mass cytometry has resulted in the generation of high-dimensional, single-cell expression data sets from clinical samples. These data sets cannot be effectively analyzed using traditional approaches. Instead, new approaches using dimensionality reduction and network analysis techniques have been implemented to assess these data. Here, detailed methods are described for analyzing immune cell expression from clinical samples using network analyses. Specifically, details are given for performing SCAFFoLD and CITRUS analyses. The methods described will use immune cell tumor infiltrate as an example.

Key words Mass cytometry, CYToF, SCAFFoLD, CITRUS, Immunology, Clinical, Tissue, Analysis

1 Introduction

Data gathered from mass cytometry experiments are too vast to be assessed by traditional analysis methods such as those used for flow cytometry data. Instead, population-based cluster analyses are more useful.

Network analyses are 2-D visual representations of the relationships between data points. When considering cells, this translates to a visualization of the relatedness between clusters of unique phenotypes [1]. Phenotypes that are closely related sit close together, whereas phenotypes that are unrelated are more distant. This results in an easily interpretable visual landscape of a wide and phenotypically diverse community of cells. Changes in these cellular landscapes can then be compared between different conditions and stimuli, or, for example, between disease states and treatments to assess overarching changes [1–3].

Many different algorithms have been used to produce network analyses. There are several broad concepts that are used in most of these algorithms, many of which are concepts drawn from Newtonian physics. The most important is that of repellent and attractive force between nodes. Each node represents a unique

Helen M. McGuire and Thomas M. Ashhurst (eds.), *Mass Cytometry: Methods and Protocols*, Methods in Molecular Biology, vol. 1989, https://doi.org/10.1007/978-1-4939-9454-0_19, © Springer Science+Business Media, LLC, part of Springer Nature 2019

cellular phenotype that may represent one or several hundred cells. Each of these nodes pushes away from all other nodes based on a defined constant rate; this is the repellent force (Fig. 1). However, each node is also connected to other nodes by edges. Edges have a defined "weight" based on the relatedness between a pair of nodes; an edge between two highly similar nodes has a high weight value. The attractive force between two nodes is proportional to the specific edge weight, although the strength of the weight factor is often modulated between algorithms.

There are currently three prominent network analysis packages designed for analyzing mass cytometry data. These are SPADE (spanning-tree progression analysis of density-normalized events), CITRUS (cluster identification, characterization, and regression), and SCAFFoLD (Single-Cell Analysis by Fixed Force- and Landmark-Directed maps). Spitzer et al. [3] discovered a switch from a high myeloid presence to a high lymphoid presence from morning to night in the lungs of C57BL/6 mice using the SCAFFoLD package.

1.1 SCAFFoLD Network Analysis

SCAFFoLD is a force-directed cluster analysis, with coding available at https://github.com/nolanlab/scaffold. This approach uses supervised gating, and subsequent algorithmic clustering, to produce a fixed map, or "scaffold," and data can be applied upon this scaffold. Fixed points are dictated by pre-gated (using cytometry analysis programmes such as FlowJo™) named populations of choice. Raw data from individual or concatenated tissue samples is applied to the scaffold. Populations are placed around the supervised

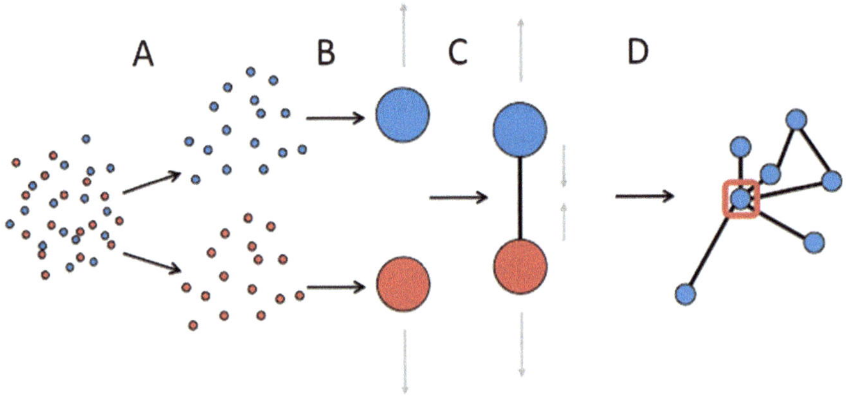

Fig. 1 Network analyses of single-cell data sets. Model of the basic process of converting raw single-cell data into a network analysis. (**a**) Different cell populations are sorted into clusters based on phenotype. (**b**) Clusters are redefined as nodes with values for mean expression of each parameter and values for abundance. Nodes repel each other at a constant rate. (**c**) Edges between related nodes are defined. Each edge has a weight corresponding to the relatedness of the two nodes. Edges pull nodes together based on their weight. (**d**) In some algorithms, edge weights for edges to nodes with a high degree (node in red box) are divided by the degree. The above steps are calculated for all nodes and edges creating a network analysis

nodes based on their similarity, with closer proximity between any two nodes indicating a close relationship in terms of cell marker expression. The size of each node is indicative of relative population frequency. This generates a fixed map which can be directly compared between tissue types based on the landmark (supervised) nodes. Community wide population shifts can then be observed within the whole immune compartment and within more specific populations. The unsupervised clusters can be further colored based on expression of any given marker, with red meaning high expression and gray low expression. Selected clusters can also be plotted, with related landmark nodes, in boxplots depicting expression of every marker included in the analysis. This allows easy comparison of all markers of interest between populations.

1.2 CITRUS Network Analysis

CITRUS is an algorithm that performs an unsupervised cluster analysis of all the data with a user set minimum percentage cluster size. For example, the minimum cluster size may be restricted to phenotypes that represent at least 5% of the assessed cells. This ensures that the assessed phenotypes are also biologically relevant in population frequency. Consistent differences are identified in either subset abundance or median marker expression within a subset, between sample types. It describes these differences as unique, sample-defining features and plots them on a box and whisker graph. Features are labeled with a four-digit identity tag that can be traced throughout the analysis. Each of these features is also described as a histogram overlay for each marker (red) against the average expression of that particular marker within the total population (blue). This algorithm describes the phenotype of cell subsets that are different between tissue types.

This chapter will provide a detailed method for analyzing mass cytometry data from clinical samples using a network analysis approach. While these analyses are carried out using R packages, no prior R expertise is required to perform them. In addition, methods will be described to prepare clinical samples for mass cytometry analysis.

2 Materials

2.1 Sample Preparation

1. Tissue samples.
2. 6-well tissue culture plates.
3. Sterile culture medium, such as RPMI-1640.
4. Collagenase.
5. 70 μm filters.
6. Sterile phosphate-buffered saline (PBS).
7. Ficoll-Hypaque.

8. Cell scraper.
9. 50 mL tubes.
10. Scalpel.
11. Freezing media: 100 μL of DMSO, 900 μL FCS.

2.2 Data Analysis

1. FlowJo software.
2. R and R studio software.
3. SCAFFoLD package.
4. CITRUS analysis tool, various platforms.

3 Methods

3.1 Transport and Storage of Samples [4]

1. Tissue samples acquired in surgery/clinic, identified and placed in labeled transport tube, e.g., Eppendorf tube.
2. Transport on ice, procedures should start as quickly as possible from surgery.
3. Prepare single-cell suspension before storage.

3.2 Enrichment of Cells from Clinical Tissue Samples

1. All procedures must take place in a Class II Biological Safety Cabinet, and all reagents should be sterile. Samples are usually unscreened and therefore represent a risk for transfer of communicable diseases (*see* **Note 1**).
2. Weigh each sample tube and record weight, date, and sample ID.
3. Put each sample into individual wells of a 6-well plate.
4. Add 2.5 mL sterile RPMI to each well (*see* **Note 2**).
5. Add 1 mg/mL collagenase to each well (*see* **Note 3**).
6. Weigh and record the weight of the empty tubes.
7. Incubate samples in 6-well plate for 45 min at 37 °C, 5% CO_2.
8. Cut up tissue specimens with a scalpel until they are small enough to pass through a Pasteur pipette.
9. Remove samples from wells and filter samples through a 70 μm filter into a 50 mL tube.
10. Scrape bottom of wells with a cell scraper, add 5 mL sterile PBS into bottom of well, then use that 5 mL sterile PBS to rinse filter (*see* **Note 4**).
11. Centrifuge cell suspension at 580 × *g* for 4 min (use these settings for all centrifugation unless otherwise stated) and discard supernatant.
12. Resuspend samples in 2.5 mL RPMI.
13. In the dark, add 5 mL Ficoll to a new 50 mL tube for each sample.

14. Using the gravity setting on the pipette, gently add 5 mL of a solution of 75% Ficoll: 25% RPMI.
15. Using the gravity setting on the pipette, gently add the 2.5 mL of RPMI containing the filtered sample cells.
16. Centrifuge the Ficoll gradient tubes at 800 × *g* for 20 min with low acceleration and no brake.
17. Extract the buffy coat using a Pasteur pipette, minimizing transfer of Ficoll layers (*see* **Note 5**).
18. Centrifuge the tube containing the buffy coat and discard supernatant.
19. Resuspend cells in 1 mL RPMI and count cells (*see* **Note 6**).
20. Centrifuge the cells and discard supernatant.
21. Resuspend cells in:
    (a) 5 mL RPMI for immediate use (for example, tissue culture or cytometry).
    OR
    (b) 1 ml freezing media for storage (*see* next section).

3.3 Freezing and Defrosting Cell Samples for Use

1. Place samples in cryotubes labeled appropriately (*see* **Note 7**).
2. Store cryotubes in appropriate container at −80 °C overnight (*see* **Note 8**).
3. After 24 h, transfer samples to liquid nitrogen storage.
4. For transport, completely frozen tubes can be kept on dry ice for up to 12 h [4] (this is based on how long dry ice lasts in a standard size ice box, not biological decay).
5. To defrost cells, remove individual cryotubes from liquid nitrogen and store on dry ice.
6. Defrost cells one cryotube at a time.
7. Add 1 mL room temperature RPMI to the cryotube and pipette the defrosting mixture into 50 mL RPMI.
8. Repeat **step** 7 until all cells are defrosted and resuspend in 50 mL.
9. Repeat **steps** 7 and **8** for all tubes.
10. Centrifuge cells and resuspend in RPMI for use.
11. Do not refreeze samples.
    Prepare samples as described previously in Chapter 9.
    Acquire mass cytometry data as described previously in Chapter 2.

3.4 Cleaning the Data Set [5]

1. This step can be completed in a program such as FlowJo™ (*see* **Note 9**).
2. Using the Ce140di (beads) and DNA1 channels, create a conservative (tight) gate on cells (beads negative, DNA positive).

3. Using DNA2 and cisplatin channels, create a conservative gate excluding dead cells (cisplatin negative, DNA positive).
4. If samples are barcoded (such as CD45 barcoding for immune cells), gate around the specific barcoded samples at this point (*see* Chapters 7 and 9).
5. Create a working directory (folder) at an easily accessible location (for example, "Desktop") in which to save all files and folders associated with the analysis.
6. Export individual live sample-specific cells to the working directory (*see* **Notes 10** and **11**).
7. Also export concatenated versions of all treatment/tissue types to the same folder (*see* **Note 12**).

3.5 Choosing the Appropriate Analysis [6, 7]

1. SCAFFoLD [3] (*see* **Note 13**).
2. CITRUS [8] (*see* **Note 14**).
3. Custom network analysis tools (*see* **Note 15**).
4. Non-network analyses (*see* **Note 16**).

3.6 SCAFFoLD Protocol

3.6.1 Running SCAFFoLD

1. Create a new FlowJo™ file with concatenated and single sample files.
2. Determine which major biological populations should be landmark nodes (*see* **Note 17**).
3. Gate around landmark populations (*see* **Notes 18–21**).
4. Organize the analysis directory.
5. Save all .fcs files of samples (concatenated or individual) to a new folder/directory (*see* **Note 22**).
6. Within that folder, create a new file named "gated."
7. Export each landmark population gate (just from one sample, the choice of sample will have no effect on outcome) to this gated folder. Each file must be named "gated_popname" (*see* **Note 23**). Accurate labeling of files and folders is essential throughout (*see* **Note 24**).
8. Install R and R studio if not already installed (*see* **Note 25**).
9. Within R studio, install the SCAFFoLD package. Details to do this can be found here: https://github.com/nolanlab/scaffold. All dependencies (other R packages) mentioned at the link will need to be installed.
10. Launch SCAFFoLD. Instructions for launching and running SCAFFoLD can be found here: https://github.com/nolanlab/scaffold. The below steps (within **step 11**) assume the method described in **step 9** (https://github.com/nolanlab/scaffold) is followed.

11. Use the SCAFFoLD interface to run the analysis. When selecting markers on the "Run Clustering" tab, select all markers of interest; do not select any CyTOF channels where no metals signals are present (*see* **Note 26**).
12. This first clustering step takes the most computing time (*see* **Note 27**).
13. As a result of this initial clustering step, the working directory with the files will now contain R data and cluster files. This step need only be completed once per sample set.
14. The next step is to create the SCAFFoLD (*see* **Note 28**). Again select all markers from the drop-down menu. Select gated from the next menu and leave all other options as default.

3.6.2 SCAFFoLD Quality Control

1. The SCAFFoLD plot is now able to be viewed.
2. It will initially load the default view for each sample. Here, the red dots are the landmarks and the blue are raw data points.
3. The first iteration is often hugely distorted and may look biologically improbable.
4. It is important to use common sense with these analyses (*see* **Note 29**). In this step subjective troubleshooting is required.
5. Major distortions of the network plot are usually due to problems with the definition of landmark nodes.
6. To fix this, alter the landmark gating approach in FlowJo™ (*see* **Note 30**).
7. Export these files to the gated file of the analysis directory as described previously in *Cleaning the data set* (Subheading **3.4**) (*see* **Note 31**).
8. At this point, relaunch SCAFFoLD and skip straight to the second, quicker, clustering step (Subheading 3.6.4).
9. After completing the above steps, reassess the plot. It may still appear biologically unlikely, in which case repeat **steps 3–8** again.
10. If the plot now looks correct, begin to assess the plot.

3.6.3 Analyzing SCAFFoLD Plots

1. SCAFFoLD plots contain a significant amount of data that can be challenging to fully comprehend. It is important to assess SCAFFoLD plots with a targeted question.
2. First, look at the concatenated files for each of the treatment/tissue types. Identify key characteristics between samples using the default view (*see* **Note 32**). Use the heat map marker expression coloring to assess expression in those major populations. The "plot clusters tool" is useful at this step—*See* protocol below (*see* **Note 33**).

3.6.4 Graphing Specific Clusters

1. Select the marker of interest to color-code the nodes.
2. Hold shift and click every node that has high or low expression of that marker (depending on the population/marker of interest).
3. Tick the box to combine clusters.
4. Select box and whisker from the menu and then plot clusters.
5. Repeat this with each sample.
6. This will show relative expression for all other markers as box and whisker graphs for both the populations selected (clusters) and any related landmark nodes.
7. This approach allows assessment of specific changes in marker expression within a given population, for example, $CD64^+$ macrophages (*see* **Note 34**).

3.7 CITRUS Protocol

3.7.1 Running CITRUS (*See* Notes 35 *and* 36)

1. CITRUS analysis requires individual sample files rather than concatenated files.
2. CITRUS can be run directly from R. This can often require some troubleshooting.
3. CITRUS can also be run from community servers such as Cytobank™. This approach is more reliable and assistance is more readily available. CITRUS has two main running options. *Abundance* which compares the relative abundance of major populations between sample types. *Expression* which compares changes in expression of any marker within given unsupervised populations between sample types.
4. Setup: Within the CITRUS user interface (UI) select all assessed markers form the drop-down menu to be included in the analysis. Select the SAM analysis for comparisons between sample types (*see* **Note 37**). CITRUS defaults to show populations of 5% or greater (*see* **Note 38**).

3.7.2 Interpretation: Abundance

1. CITRUS gives two variants of the cluster map in addition to the SAM results (*see* **Note 39**). The first is a network map colored by population abundance across all tissue samples. The second is repeated cluster maps colored by expression of a given marker on each map (one specific marker per repeated map).
2. SAM gives three main outputs, each with three different false detection rate (FDR) values (1, 5 and 10%): Clusters, feature plot, features.
3. First look at the feature plot. The feature plot shows the same network plot map structure as the basic CITRUS data plot. The plot will be colored based on significant differences in abundance of a particular cluster between sample types. If the plot is red then there is a difference, if the plot is blue, there is

no difference. The Features and Clusters files will only show clusters that are red on the features plot.

4. Next, look at the Clusters plots. These plots help to determine the phenotype of each given numbered cluster. The lilac histogram represents the average expression of that particular marker across all assessed cells. The red histogram gives the relative expression of that marker on that particular cluster. That cluster is then also labeled with a four-digit number that should be used as an identifier throughout the analysis. Use this data to define the clusters and understand what they biologically represent.
5. Finally assess the Features. These graphs show the relative abundance of each cluster as a box and whisker plot for each sample type. This shows the relative frequency of a given subset between sample types.
6. Overall. (1) it is possible to identify relevant populations across all patients, (2) it is possible to determine the phenotype of those populations, and (3) it is possible to determine whether that population is increased or decreased between sample types.

3.7.3 Interpretation: Median Expression

1. The results for this are displayed in the same manner as the abundance data and so a similar approach to interpretation is recommended.
2. However, here, there are several feature plots rather than one (*see* **Note 40**). Each of these plots shows the same network plot colored based on whether that particular marker is increased or decreased on a particular cluster between sample types.
3. The feature graphs are also different in this analysis type (*see* **Note 41**). Features now show median expression of a given marker for a given cluster for each sample type as box and whisker plots.

4 Notes

1. Local institutional health and safety and risk assessment procedures must be incorporated.
2. For larger samples, add 5 mL RPMI.
3. Other digestion reagents can be used, depending on tissue type; for example, DNAse, lipolase. For an analysis of these reagents, *see* Volovitz et al. [9].
4. Use cell scraper to gently push the fluid through the filter.
5. Buffy coat will appear as a fluffy white layer.

6. Cell count can be performed using Trypan blue exclusion or automated cell counter; viable cell count is important.
7. Each tube should contain 1 mL with ~1×10^7 cells. Label should include sample ID, date, cell number, name/initials of person who prepared sample.
8. Examples include Mr Frosty™ (Thermo Fisher) and CoolCell™ LX (Global Science).
9. FlowJo™ is used as an example cytometry analysis software through these methods. Specific instructions regarding FlowJo™ may not directly translate to other analysis software.
10. These data can be confusing to handle as a large number of files and folders are generated. Clear separation between analyses and filing is essential. Accurate labeling of files and folders is the most important aspect.
11. Keep all iterations in separate folders within an overarching principal directory.
12. Useful for some analyses (such as SCAFFoLD) but not others (CITRUS requires individual sample files). The option to concatenate files is found under the export menu in FlowJo™.
13. SCAFFoLD is best for assessing major changes in the cellular landscape between tissue/treatment types and for identifying novel populations that may express noncanonical marker combinations. SCAFFoLD is best used as a hypothesis generating tool.
14. CITRUS is most useful for major populations (>5% of total cells, for example, regulatory T cells). CITRUS is more suited to hypothesis answering than SCAFFoLD, as it highlights specific major population shifts between sample types. CITRUS highlights phenotypes that are significantly more prevalent in one tissue/sample than another tissue/sample.
15. Packages such as Igraph can be used in R to produce custom analysis tools. This requires more R coding knowledge, and is not discussed in detail here.
16. Dimensional reduction techniques (tSNE, PCA or traditional gating) may be applicable here, but are not discussed in detail. Refer to Chapters 12, 16–18.
17. Keep to broad lineages, as too much detail will obscure the variance in the raw data. It is better to use major lineages, for example, T cells or macrophages. If looking at just one lineage then "zoom" in further, for example, memory, naïve, effector T cells.
18. It is important to use both positive and negative gates to define each population.

19. Use a hierarchy of gates from the original populations rather than separate gates. For example, $CD3^+$ and $CD3^-$ gates, gate T cells from $CD3^+$ and everything else from $CD3^-$. This hierarchy helps to ensure that marker expression is associated with the correct phenotype.
20. For myeloid cells, CD11c or CD11b should not be used to define populations, as there is considerable overlap in myeloid populations making them hard to define [10].
21. Ensure that there is >1 event in each landmark gate in every sample. If this criterion is not met, SCAFFoLD will show a generic error when run and the analysis will fail; the error is not explained.
22. These data can be exported with only the appropriate channels to streamline later steps. To achieve this streamlining, select only the channels from the advanced options of the export menu associated with actual markers.
23. This process can be made quicker by changing the prefix (to gated) for the exported file under advanced options in the FlowJo™ export menu.
24. This streamlines and simplifies the analysis process. All names/labels are case sensitive.
25. For large sample sets (for example, above $n = 20$) it is recommended that a relatively powerful computer is used to reduce computing time. It can take over an hour to calculate on a middle of the line laptop (2.6 GHz Intel Core processor, 8 GB 1600 MHz DDR3 RAM).
26. It will also show every other channel that was recorded unless these were deselected when exporting the file.
27. To move these data (by copy and pasting to other directories), these files must remain with the original .fcs files.
28. This step takes much less computing time than the previous clustering step.
29. If the analysis shows that, for example, 90% of the immune infiltrate in a tumor is a single type of rare immune cell, it is probably not correct.
30. This may mean increasing or decreasing the number of positive or negative defining markers for each population.
31. Ensure that old gates are replaced—It is recommended that old gates are kept in a separate folder as a backup.
32. Be sure to take note of changes in abundance around a node as well as heterogeneity (spread) around a node when assessing the data for patterns. An example could be an increase in T cell heterogeneity (spread) and relative abundance (greater number and size of nodes around the T cell landmark).

33. An example of its use would be to compare expression of all assessed markers on, for example, $CD64^+$ cells between sample types.
34. Any of the approaches used on the concatenated files can also be used on the individual sample files. However, these files will have a lower cell number than the concatenated files potentially making analysis more spurious. Individual files can also be used to validate any major population shifts determined from the concatenated data sets. If a major shifted population exists across the majority of the samples, then it is probably real. However, some populations may exist only in one patient. This could have produced an artifact in the concatenated file and may only be representative of that patient, not the whole tissue/treatment in general.
35. An *n* of at least eight samples is needed for this approach to be accurate.
36. It is possible to have multiple sample groups rather than just two. It is important to know the question before deciding how many sample groups to include in one analysis. Including all sample groups could generate the most data, however, it may be less informative than simply comparing between two key sample types.
37. The other two options are predictive and therefore less relevant for analyzing fundamental biological data.
38. If no results are found by the SAM analysis, lowering this to 2% can be helpful. Lower values are not recommended without a considerably larger number of samples. A greater number of samples ensures that smaller populations are meaningful across all samples and not just indicative of outlier samples.
39. These plots give an idea of the data and what each clustered node represents but are not as informative as the SAM analysis as they do not show comparisons between samples.
40. Essentially, the plots show changes in expression of a single marker on a given population. This can make interpretation challenging as a change in a single marker could dictate a completely different cell type, making biologically logical assumptions from the data more difficult.
41. For biological data this approach can be less intuitive and often a less useful way of interpreting data directly than the abundance algorithm. It does however highlight the heterogeneity of biological populations.

General Notes

42. Mass cytometry data and network analyses are a way of looking at the big picture without sacrificing detail.

   (a) This is particularly relevant for immunology and tumor biology.

43. Always have a clear independent question when assessing the data.
   (b) There is too much to look at all at once, so a targeted approach is required.
   (c) However, always be aware of the other levels of complexity that may be missed using this approach.
44. If it does not make sense biologically, do not blindly trust the analysis.
   (d) Educated subjective analysis of the output is critical for interpreting these analyses.
   (e) However, do not dismiss results just because they do not conform with canonical principles.
45. There must be a strong fundamental understanding of the data set and every marker assessed within it.
   (f) Failure to have this can result in serious misinterpretation.

References

1. Ma X, Gao L (2012) Biological network analysis: insights into structure and functions. Brief Funct Genomics 11(6):434–442. https://doi.org/10.1093/bfgp/els045
2. Spitzer MH, Carmi Y, Reticker-Flynn NE et al (2017) Systemic immunity is required for effective cancer immunotherapy. Cell 168(3):487–502 e415. https://doi.org/10.1016/j.cell.2016.12.022
3. Spitzer MH, Gherardini PF, Fragiadakis GK et al (2015) IMMUNOLOGY. An interactive reference framework for modeling a dynamic immune system. Science 349(6244):1259425. https://doi.org/10.1126/science.1259425
4. Grizzle WE, Bell WC, Sexton KC (2010) Issues in collecting, processing and storing human tissues and associated information to support biomedical research. Cancer Biomark 9(1–6):531–549. https://doi.org/10.3233/CBM-2011-0183
5. Bendall SC, Simonds EF, Qiu P et al (2011) Single-cell mass cytometry of differential immune and drug responses across a human hematopoietic continuum. Science 332(6030):687–696. https://doi.org/10.1126/science.1198704
6. Irish JM, Doxie DB (2014) High-dimensional single-cell cancer biology. Curr Top Microbiol Immunol 377:1–21. https://doi.org/10.1007/82_2014_367
7. Leelatian N, Doxie DB, Greenplate AR et al (2017) Single cell analysis of human tissues and solid tumors with mass cytometry. Cytometry B Clin Cytom. https://doi.org/10.1002/cyto.b.21542
8. Bruggner RV, Bodenmiller B, Dill DL et al (2014) Automated identification of stratifying signatures in cellular subpopulations. Proc Natl Acad Sci U S A 111(26):E2770–E2777. https://doi.org/10.1073/pnas.1408792111
9. Volovitz I, Shapira N, Ezer H et al (2016) A non-aggressive, highly efficient, enzymatic method for dissociation of human brain-tumors and brain-tissues to viable single-cells. BMC Neurosci 17(1):30. https://doi.org/10.1186/s12868-016-0262-y
10. Misharin AV, Morales-Nebreda L, Mutlu GM et al (2013) Flow cytometric analysis of macrophages and dendritic cell subsets in the mouse lung. Am J Respir Cell Mol Biol 49(4):503–510. https://doi.org/10.1165/rcmb.2013-0086MA

Chapter 20

Supervised Machine Learning with CITRUS for Single Cell Biomarker Discovery

Hannah G. Polikowsky and Katherine A. Drake

Abstract

CITRUS is a supervised machine learning algorithm designed to analyze single cell data, identify cell populations, and identify changes in the frequencies or functional marker expression patterns of those populations that are significantly associated with an outcome. The algorithm is a black box that includes steps to cluster cell populations, characterize these populations, and identify the significant characteristics. This chapter describes how to optimize the use of CITRUS by combining it with upstream and downstream data analysis and visualization tools.

Key words CITRUS, Biomarker discovery, Supervised machine learning, viSNE

1 Introduction

Analyzing single cell data with traditional, biaxial approaches can lead investigators to focus only on cell populations that they are already aware of and might lead them to miss important discoveries in other populations [1, 2]. This is particularly important in studies that are designed to identify novel biomarkers for a specific clinical endpoint like disease status or response to therapy.

CITRUS [3] is a supervised machine learning algorithm for single cell biomarker discovery. It has been used to identify single cell biomarkers in many contexts. For example, CITRUS has been used to identify: biomarkers for response and resistance to CD19 CAR-T cell therapy in chronic lymphocytic leukemia [4]; age-related epithelial subpopulations in breast tissue [5]; peripheral blood biomarkers for response to anti-CTLA-4 and anti-PD-1 checkpoint blockade [6]; and differences in immune populations between placebo-responders and controls in a mouse model [7].

The CITRUS algorithm includes steps to cluster cell populations, characterize these cell populations, and identify the characteristics of the cell populations that are statistically significant biomarkers that distinguish outcome groups. In this way, CITRUS

Helen M. McGuire and Thomas M. Ashhurst (eds.), *Mass Cytometry: Methods and Protocols*, Methods in Molecular Biology, vol. 1989,
https://doi.org/10.1007/978-1-4939-9454-0_20,

is a black box that identifies statistically significant single cell biomarkers for any discrete outcomes of interest. Since the CITRUS algorithm seamlessly integrates these steps, it is able to search efficiently for the boundaries of a cell population that is significantly different between the outcome groups. The clustering step identifies nested clusters such that each "parent" cluster contains all of the cells within each of its "child" clusters. Then, the significance testing is performed for each of these nested levels of cell populations: clusters that are too narrow or too broad in their boundary of a population will not be identified as significant. Since CITRUS performs so many statistical tests, it uses permutations and the False Discovery Rate method [8] to report only single cell biomarkers that are significant after correction for multiple hypothesis testing. The single cell biomarkers it identifies can be either population abundances or population-specific median expression of functional markers. In addition to correlative analyses that identify all of the significant single cell biomarkers, CITRUS can build predictive models that identify a minimum set of biomarkers necessary to discriminate effectively between the groups, and estimate the accuracy with which these biomarkers will be able to do this.

This chapter describes our protocol for combining the CITRUS algorithm with upstream and downstream data analysis tools to take full advantage of CITRUS's unbiased biomarker discovery. With any machine learning method, data QC and cleanup is essential to generating meaningful results. In addition, effective exploratory data analysis is essential to understanding the biology and heterogeneity that drives those results (or lack of results). Thus, our protocol includes considerations for experimental setup, data tidying and quality control, visualization and exploratory data analysis, clustering and biomarker discovery, and visualization for communication of results (Fig. 1). As a part of this pipeline, we utilize viSNE [9, 10], biaxial plots, and heatmaps for the visualization and exploration of significant clusters. Together, these tools allow unbiased identification of significant single cell biomarkers and effective communication of these results.

2 Materials

CITRUS was developed in and may be run using the R programming language [3]. This chapter discusses CITRUS as implemented in Cytobank, a cloud-based informatics platform available at www.cytobank.org.

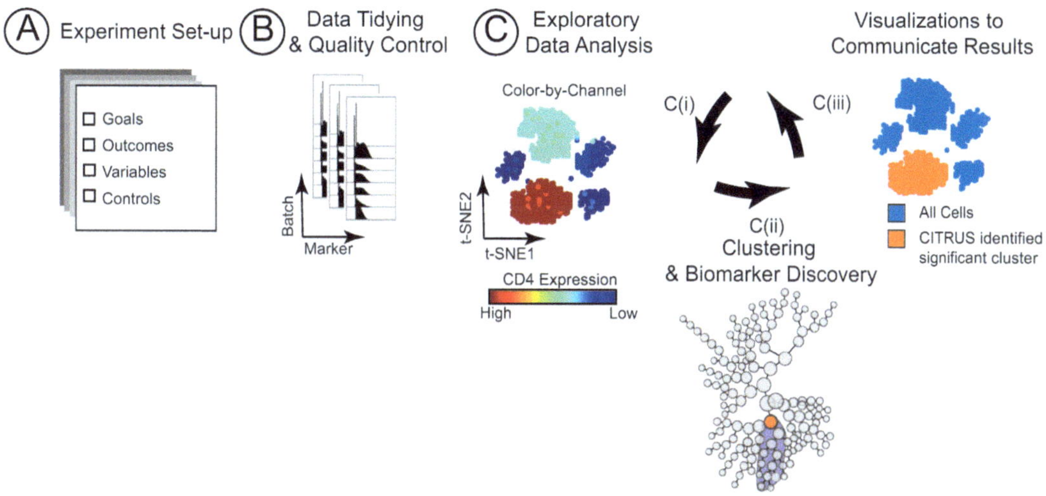

Fig. 1 Single cell biomarker discovery pipeline with CITRUS. Key steps outlined for biomarker discovery analysis pipeline. (**a**) Experimental setup (**b**) data tidying and quality control and (**c**) exploratory data analysis, clustering and biomarker discovery, and visualizations to communicate results. Steps (**a**) and (**b**) are required before initiating any of the components in step (**c**). Each part of step (**c**) (*i–iii*) may be initiated following experimental setup and data tidying/quality control

3 Methods

3.1 Considerations for Analysis Setup

Thinking through certain aspects of the analysis setup before beginning analysis will reduce time spent in later steps of the analysis.

1. During data acquisition, ensure that channel short names and long names are consistent across all of the files that will be analyzed together. As much as possible, include information about outcome groups, stratifying variables, and technical factors in file names.
2. Define the outcomes and stratifying dimensions you will compare. The outcomes should be discrete (i.e., have defined categories with groups of samples) and the samples in each group should be independent (i.e., not consist of multiple observations on the same individual, even under different conditions) (*see* **Note 1**). In addition to outcomes, there may be *stratifying variables*, or other dimensions in the data that you wish to compare across. These stratifying variables should also be discrete, but samples split across these stratifying variables do not have to be independent (*see* **Note 2**). An example of typical outcome, stratifying, and technical variables you might track is shown in Table 1.
3. Determine the number of samples you will have for each group of outcome and stratifying dimensions (*see* **Note 3**). CITRUS will be run for your outcome variable within each stratifying

Table 1
Example of Stratifying, Outcome, and Technical Variables in an Experiment

Stratifying or outcome variable	# in each category
Baseline (pre) or post-therapy	Pre: 40 files Post: 40 files
Therapy response	*Responders (R): 28 individuals (2 files each)* *Non-responders (NR): 12 individuals (2 files each)*
Patient ID	40 unique patient IDs (2 files each)
Staining batch (1–4), labeled by date	1 = 20 files 2 = 20 files 3 = 20 files 4 = 20 files

In this example study, patient peripheral blood mononuclear cells (PBMCs) have been acquired in batches for samples pre- and post-therapy. The study outcome variable to be used for CITRUS (highlighted in italics) is therapy response. Timepoint (pre- or post-therapy) is a stratifying variable in this study; one CITRUS run will be performed on the pre-therapy samples, and a separate CITRUS run will be performed on the post-therapy samples. Staining batch will be used to look for batch effects. Timepoint, therapy response, patient ID, and staining batch are annotated by Cytobank sample tags. The number of files expected in each category is outlined in the second column and used to check the files present in Cytobank before starting the analysis

group. For each of these comparisons, you need a minimum of three samples per group to run CITRUS, and we recommend that you have at least eight samples per outcome group to obtain reliable results.

4. Determine whether the samples were run in multiple batches and whether control samples are available. If the samples were run in multiple batches, determine whether the outcome and stratifying variables were distributed equally across the batches and whether a technical control sample was run with each batch (*see* **Note 4**). Identify any factors based on experimental setup or lab observations while processing samples and acquiring data that were tracked; these can be used later for sensitivity analyses.
5. Determine whether the scientific question requires a correlative or predictive analysis, or both. Correlative analyses will identify all of the significant biomarkers that can be detected in the data, while predictive analyses will identify a minimum set of markers needed to best predict an outcome (*see* **Note 5**). The choice of analysis will dictate which models to select in CITRUS (Fig. 2).

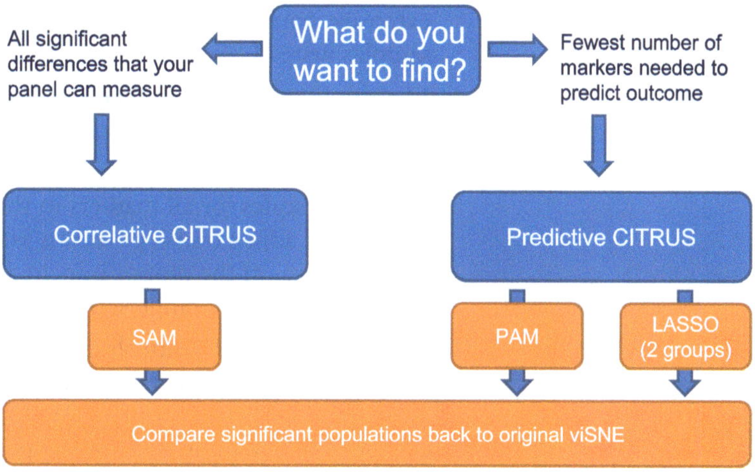

Fig. 2 Flowchart to determine correlative vs. predictive CITRUS and CITRUS models to use. The CITRUS models to select are determined by the overall analysis goals. Use the above flowchart to determine which models will address the scientific question. Regardless of the model selection, significant populations identified by CITRUS can be visualized on viSNE maps

6. Determine which markers in the panel will be used for phenotyping and which will be used as functional readouts (*see* **Note 6**).
7. Complete standard preprocessing steps for the mass cytometry data, which may include normalization for time and debarcoding if barcoding was used.

3.2 Data Tidying and Quality Control

The goal of the data tidying and quality control (QC) steps is to identify any problems in the data that may result in the exclusion of samples, markers, or batches of samples from analysis. For issues that are identified related to batches of samples, normalization may be performed in an attempt to control for batch effects.

1. Upload data to Cytobank and annotate the data according to outcome groups, stratifying variables, and technical factors using Cytobank's Sample Tags (*see* **Note 7**).
2. Check that the number of files in each tagged group matches what you expect according to your experimental design (Table 1) (*see* **Note 8**).
3. Check that the panels are annotated as expected based on the assay(s) that were run. Each CITRUS analysis should be performed on a set of samples run on a single panel (*see* **Note 9**). Fix any discrepancies in channel names that may have happened accidentally during acquisition. Identify any channels that will need to be excluded because they differed across samples in the experiment. If there are a few samples that have

extensive panel differences, you may choose to exclude these samples instead of excluding discrepant channels.

4. Identify files that are outliers for total event count and check whether there were issues acquiring data for that sample. These samples will likely need to be excluded from the analysis.
5. Check the scales for all markers in the panel across all samples. Perform or adjust transformations as required (*see* **Note 10**). If your data are off the plot or bunched on the axis you may need to adjust your scale range settings (*see* **Note 11**).
6. Examine technical controls for evidence of batch effects or problems with marker staining using histograms or other basic plot types (*see* example in Fig. 3a, **Note 12**). If you find markers for which staining does not appear to have worked (either within a particular batch or across all batches), these markers should be excluded from the CITRUS analysis.
7. If you think that observed differences across technical control replicates are due to batch effects, run viSNE on the technical control samples from each batch to identify whether observed differences between batches will impact downstream analysis results (*see* example in Fig. 3b, **Note 13**).
8. If you identify batch effects that will impact your downstream analysis results, one approach is to normalize the data [11, 12] (*see* **Note 14**). After normalization, reexamine the scales, rerun viSNE on the technical controls, and look again for evidence of batch effects to assess whether the normalization was successful (*see* example in Fig. 3b). Alternatively, batches for which large batch effects are identified may be excluded, so that only batches without any apparent differences across batches are included.

3.3 Pre-gating

Pre-gating to intact singlets or similar might be required in order to perform some of the investigations described above. For the rest of the CITRUS analysis, further pre-gating might be desired. This decision will depend on the cell populations the panel is designed to measure, the number of samples being comparing, and the scientific questions being answered. If a large number of samples are being compared, where differences in rare cell populations are being investigated, further pre-gating may be required to include enough cells per sample to detect these rare cell populations reliably.

1. At a minimum, pre-gate to the events the panel and scientific question are focused on. For example, gate to select CD45+ cells if lymphocytes are the focus, or gate to remove cells stained with antibodies in a dump channel.

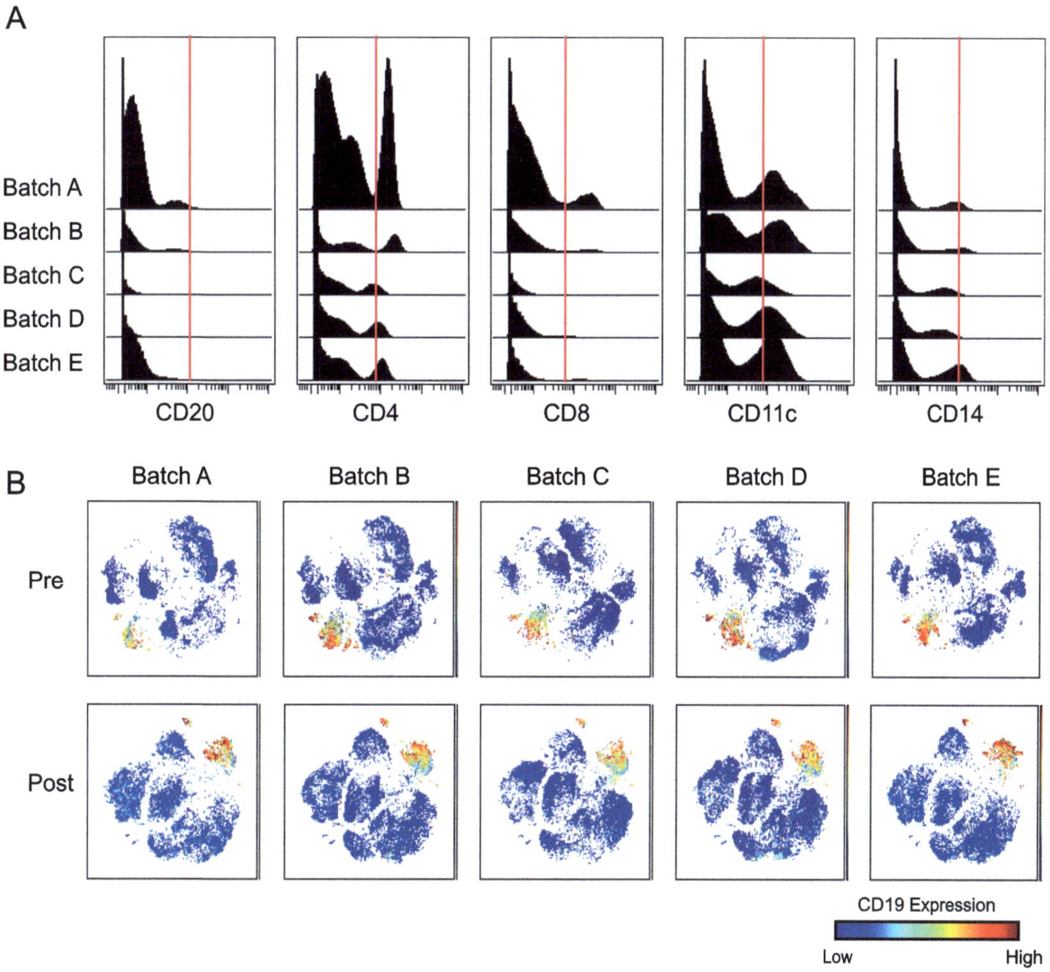

Fig. 3 Technical control evaluation. Peripheral blood from one healthy donor was collected and cryopreserved to use as a technical control in this study. A different antibody cocktail was prepared for each of the five indicated staining batches and evaluated for differences across batch using the technical control. (**a**) Pre-normalized data for technical control samples shown. Histograms displaying a few measured markers are examined across batch. Intensity of indicated marker and the negative and positive spread are compared across batch. The red line indicates the middle of the plot. Data reveal staining differences. (**b**) Pre and post-normalization viSNE maps are compared by coloring across channel. Here, only CD19 expression is displayed. Technical control samples per batch were assessed using viSNE as an exploratory data analysis tool. Row 1 show data pre-normalization and row 2 show data post z-score normalization by batch. Data post-normalization highlight more uniform intensity of CD19 expression and reveal more uniform viSNE "island" shapes. The effect of normalization can be quantified using a clustering algorithm to define cell populations and then measuring the correlation of the cell populations across the batches pre- and post-normalization

2. Determine the maximum number of events you can include in a single viSNE map. This is limited in Cytobank by your subscription options [13]. Care should be taken in interpreting viSNE plots of over 100,000 cells, as data crowding will start to skew population distributions.

3. Determine the number of samples to include in the viSNE map. At a minimum, this is the number of samples that will be included in a single CITRUS run for a set outcomes within a given stratifying variable group (*see* **Note 15**). Exclude samples based on the conclusions from data tidying and QC.
4. Determine the number of events per file in the initial pre-gated population. For the exercise below, use the file with the minimum number of events in the pre-gated population. If one or more files have a very low number of events in this population, plan to exclude them and use the next lowest number.
5. Determine the expected frequency of the rarest cell population of interest as a fraction of your initial pre-gated population.
6. Use the flowchart in Fig. 4 to determine whether you need to pre-gate further (*see* **Note 16**) and whether you will use multiple pre-gated populations for your analysis (*see* **Note 17**).
7. For the final pre-gated population(s), check whether any samples have pre-gated event counts lower than the number of events you determined was necessary per file. These samples should be excluded from the analysis.

3.4 Visualization and Exploratory Data Analysis

1. Run one viSNE for each pre-gated population and set of samples that are necessary for the data and scientific questions above. Exclude any samples selected for exclusion during file QC, data QC, and pre-gating. Set the total number of events based on pre-gating event count calculations, and select equal sampling. Select the channels that were identified as phenotyping markers in Subheading 3.1 as clustering channels, but exclude any channels used during pre-gating (for a discussion on exceptions *see* **Note 18**). If including a large number of total events, increase the number of iterations and perplexity (*see* **Note 19**).
2. Check the quality of the resulting viSNE map to ensure that no further tuning of any of the algorithm settings is required. In Cytobank, set up a working illustration that shows the viSNE map for every file on the rows colored by every channel on the columns. If this map is not high enough quality (*see* **Note 20**), adjust the advanced settings and rerun viSNE.
3. Create a concatenated viSNE map across all samples, and use the working illustration dot plots to color it by the expression of each of your clustering marker channels (*see* **Note 21**, and an example of separate viSNE maps in Fig. 5 compared to a concatenated viSNE map in Fig. 6c).
4. Create views of the viSNE map to help with exploration of the data. viSNE analysis, as described here, can also be used after

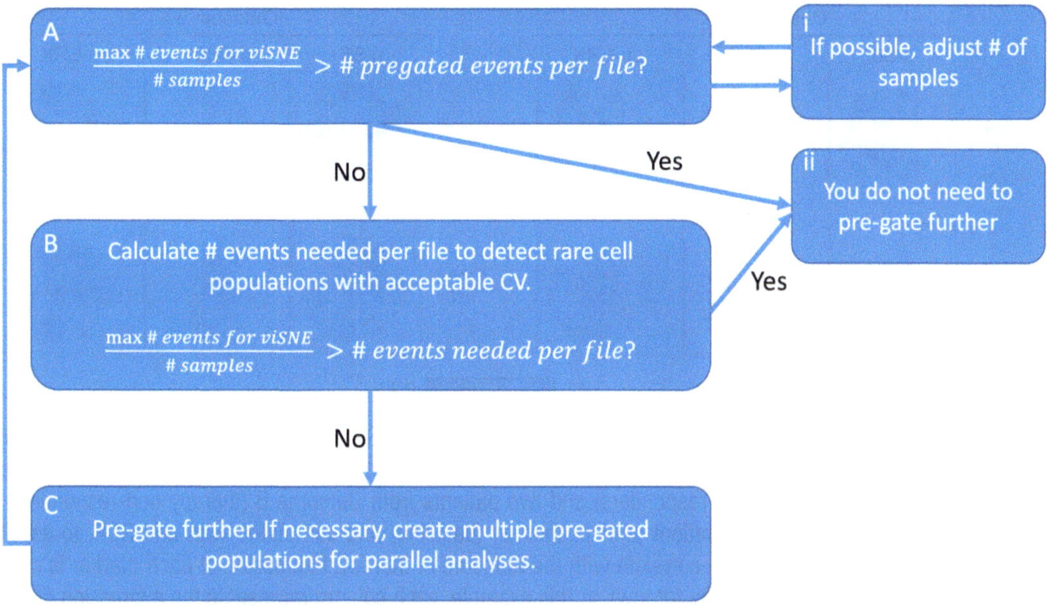

Fig. 4 Determine required pre-gating using event count. For example, we use the following logic to determine how many events are needed to be able to detect Tregs with a desired 5% CV starting with all CD3+ T cells for 50 samples. Let's say the file with the lowest number of CD3+ T cells has 50,000 events in this population, and the maximum number of events we can include on viSNE is 1.3 M. (**a**) 1.3 M max events/ 50 samples = 26,000 events, which is <50,000. (**b**) We estimate that Tregs have a frequency of about 2% of CD3+ T cells. Thus, the # of events needed per file to detect Tregs is 20,000. 26,000 events is greater than the 20,000 events needed per file. (*ii*) No additional pre-gating is required in this example. If our samples did not meet these conditions, we could (*i*) reduce the number of samples being analyzed, or (**c**) perform further pre-gating to isolate a smaller and more specific group of cells containing the Treg cells

performing CITRUS clustering (described below) to help explain the CITRUS results (*see* **Note 22**). Samples may be kept individual or concatenated in groups according to the outcome, stratifying, and technical variables you annotated with Cytobank's Sample Tags. To visualize differences in individual samples or groups of samples, use dot plots colored by marker expression to visualize heterogeneity. When using viSNE after CITRUS clustering (as described below) dot plots colored by marker expression can be used to visualize differences between and within cell clusters identified by CITRUS, and dot plots colored by density can be used to visualize differences in abundance of cell clusters identified by CITRUS.

5. If using some markers as functional readouts that weren't included in the parameters used to generate the viSNE map (such as phosphorylated antigens or cytokines) these can still

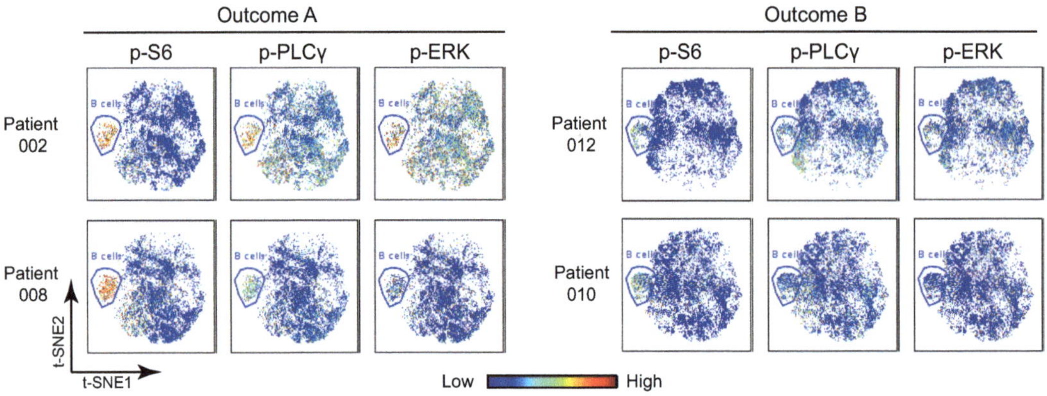

Fig. 5 Using viSNE for Exploratory Data Analysis. Synthetic example viSNE grid, colored by marker expression, shown for patients with B cell malignancies. Two individuals from each group of interest are shown—two patients from Outcome A (therapy responders) and two patients from Outcome B (therapy non-responders). PBMCs were isolated from each patient prior to therapy (baseline) and stimulated with anti-BCR, to see if baseline stimulation responses will correlate with the outcome of therapy. Expression of each marker is indicated by scale bar (red for high, blue for low). In this example, each column represents the expression of the indicated marker (p-S6, p-PLCγ, p-ERK) for intact live cells from peripheral blood mononuclear cells (PBMCs). The B cell "island" of the viSNE map is circled in blue and reveals the inter- and intra-heterogeneity in expression of measured markers between groups and across patients. This exploratory data analysis highlights how patients who responded to therapy (from outcome group A) were able to respond to BCR stimulation prior to therapy; whereas, patients from outcome group B were not. The strongest response from patients in outcome group A was for p-S6 expression

be visualized on the viSNE map by creating a colored dot plot (*see* example Fig. 5 and **Note 23**).

3.5 Clustering and Correlative Biomarker Discovery Analysis

1. Run one CITRUS analysis to compare a set of outcome groups (such as therapy responders vs. non-responders) within one stratifying variable (such as pre-therapy, or post-therapy) that was defined above. For each CITRUS, start from the viSNE experiment that contains the appropriate samples (*see* **Note 24**) and select the pre-gated population. For each CITRUS run, choose to either look at differences in the *abundances* of cell populations between your groups (as a percentage of your pre-gated population) or the *medians* of functional marker expression in cell populations between your groups (*see* **Note 25**).

Fig. 6 (continued) three cohorts. The parent clusters (which are "highest" up the tree), 1 and 2, are highlighted in orange and blue, respectively. (**b**) Boxplot displays show that the both clusters are most abundant in Group A and least abundant in Group C. (**c**) viSNE plots for each sample were concatenated and the expression of the displayed marker for the combined assessed patient population is displayed (e.g., CD11c, CD14). The significant clusters identified by CITRUS are shown on the viSNE coordinates in the bottom row to the right. Comparison of these plots helps identify the phenotype of these significant clusters. (**d**) Heatmap and (**e**) histograms show the expression of the indicated marker on the significant clusters (1 and 2) vs. all other non-T cells

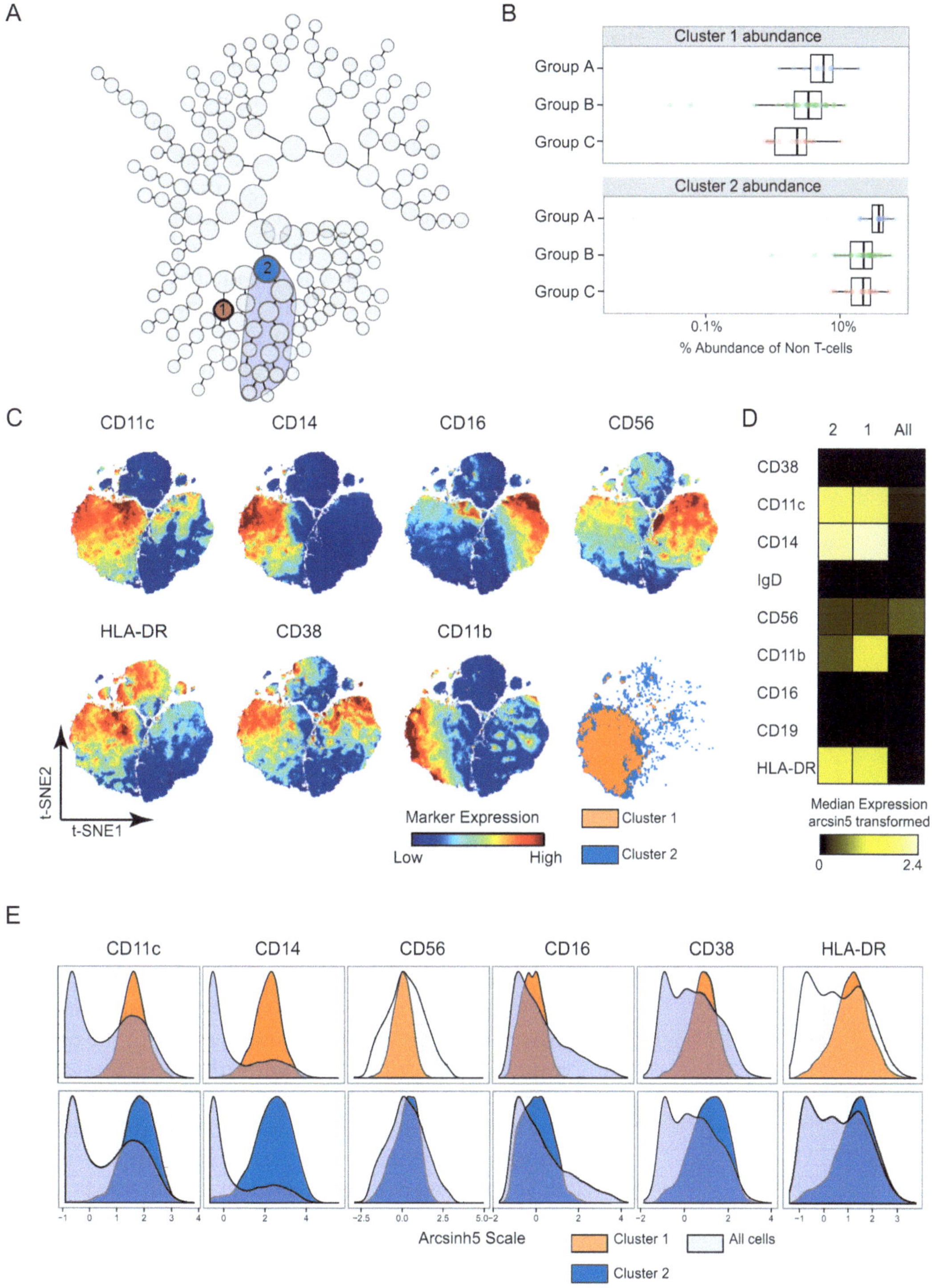

Fig. 6 Biomarker discovery results visualization. CITRUS using SAM with a 5% FDR was run on non-T cell subsets (i.e., CD3 negative pre-gated cells in PBMC samples) to look for any differences in abundances of non-T cells among three different disease types—Group A, Group B, Group C. (**a**) Display of the CITRUS defined clusters. Highlighted clusters show cell subsets that had a significantly different abundance among the

Regardless of which choice you make, select the clustering channels that you chose to use for phenotyping markers in Subheading 3.1. Assign the "file groupings" based on the outcome groups you are planning to compare (*see* **Note 26**). For correlative analysis, choose the SAM model [14] and in the "Additional Settings" box, make sure that the number of events sampled per file is large enough that all of the events from your viSNE will be included in the CITRUS. Set the minimum cluster size (%) to the approximate percentage of the lowest abundance population you are interested in detecting. Set the cross-validation folds to 1 and leave the FDR set at the default (*see* **Note 27**). Select "Normalize Scales."

2. Select a false discovery rate (FDR) level at which you will assess the SAM results (*see* **Note 28**).
3. Determine whether you have any significant results. If there are no significant biomarkers found, no model output will be returned (*see* **Note 29**).
4. If the model returns significant results, select the clusters to focus on. Use the "featurePlots" output from CITRUS for the chosen FDR and identify the significant nodes highlighted in red on the tree. When there are multiple significant clusters that are immediate neighbors of each other, CITRUS will draw a bubble around these neighbors. Choose the highest cluster on the tree within each bubble to focus on (*see* example in Fig. 6a and **Note 30**).
5. Review how these clusters differ between the outcome groups on the "features" boxplots output from CITRUS for the chosen FDR (*see* example in Fig. 6b and **Note 31**).
6. From the completed CITRUS page, export the clusters you selected to focus on to a new experiment along with the original files.
7. Map these clusters back to viSNE to understand their phenotypes (*see* example in Fig. 6c and **Note 32**). Concatenate the events separately for each selected cluster and all of the original files. Create a viSNE dot plot for all of the concatenated files with the significant cluster files overlaid on the original file (*see* **Note 33**). Compare this viSNE map showing the significant clusters to the concatenated viSNE expression maps that were created in Subheading 3.4 to help explain the phenotype of the significant CITRUS clusters.
8. Examine the selected clusters on the "markerPlots" (*see* **Note 34**) and the "clusters" histograms (*see* example in Fig. 6e and **Note 35**) output by CITRUS. Check that this expression matches what you see on the viSNE map in the region where the significant clusters are overlaid.

9. Create a heatmap in the working illustration (*see* example in Fig. 6d) of the exported clusters experiment with the clustering channels as the rows and the concatenated files for each significant cluster as the columns. Check that this expression matches what is seen on the viSNE map in the region where the significant clusters are overlaid (*see* **Note 36**).
10. Back-gate the significant CITRUS clusters onto biaxial plots using color overlay dot plots (*see* **Note 37**).

3.6 Clustering and Predictive Biomarker Model Development

1. Run one CITRUS for each predictive model you would like to build (*see* **Note 38**). Set the parameters for this CITRUS as for the correlative analysis (*see* Subheading 3.5), except: (1) choose the PAM model [15], and additionally choose the lasso model [16] if only two groups are being compared (*see* **Note 39**); and (2) in the "additional settings" box, set the cross-validation folds to at least 5 and no more than 10 (more will take longer) and set the FDR to the significance level at which you would like to evaluate your results. Take note of the number of samples in each CITRUS group (*see* **Note 40**).
2. Determine whether the biomarkers included in the predictive model are able to discriminate the groups using the model error rate plot (*see* example Fig. 7 and **Note 41**). Identify the green dot representing the CV min constrained model (*see* **Note 42**) and calculate the cross-validated estimation of the accuracy of your model to discriminate your outcome groups as one minus the model cross-validation error rate (y-axis) at this point. Determine whether the accuracy of your model is "good" given the context of your analysis (*see* **Note 43**).
3. Determine whether the predictive model results are also significant (when using PAM) by comparing the predictive model constrained by the minimum cross-validated error rate (the CV min model, green circle) to the model constrained by the false discovery rate (the FDR model, yellow triangle) on the model error rate plot (*see* **Note 44**). Check that the FDR constrained model contains more features than the CV min constrained model (i.e., the yellow triangle should be further to the right on the X-axis); this indicates that all of the features in the predictive model are statistically significant.
4. As described for the correlative analysis results in Subheading 3.5, select the clusters to focus on, review how these clusters differ between the outcome groups, and map the clusters back to the viSNE map to understand their phenotypes. Compare the results of the predictive and correlative analyses and of all of the predictive models that you build (*see* **Note 45**).

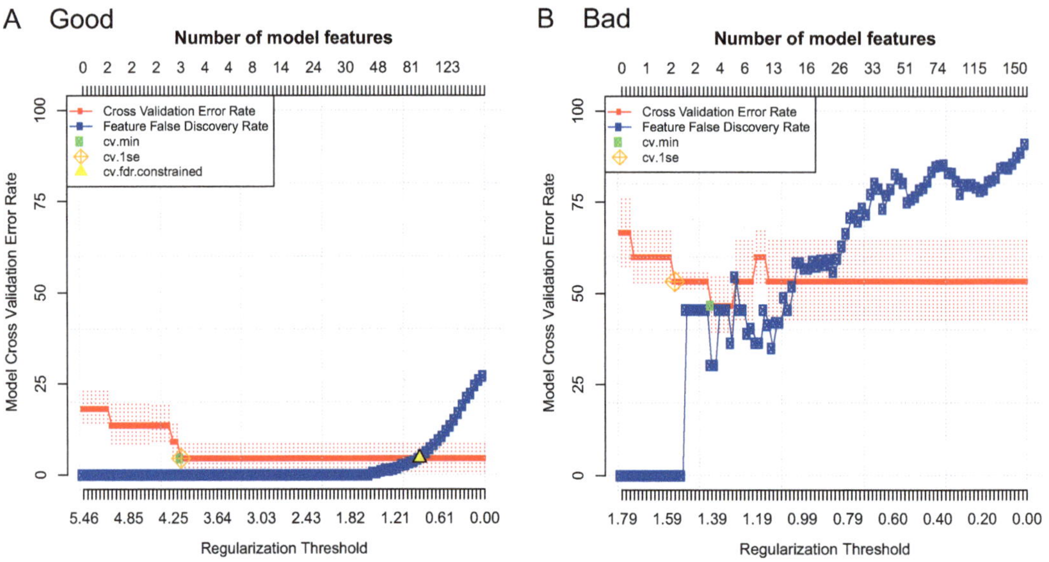

Fig. 7 "Good" vs. "bad" example model error rate plots. Two examples of model error rate plots output from predictive CITRUS analyses using PAM and an FDR of 5% with equally balanced groups. In each plot, the cross-validation error rate (red line) and false discovery rate (blue line) are shown as the number of features in the model is increased (top *x*-axis) according to the regularization threshold used to build the model (bottom *x*-axis). (**a**) A "good" model error rate plot. An FDR constrained model (yellow triangle) was identified with more features than the CV constrained models (cv.min, cv.1se). The CV min model had a CV error rate of around 5% using three features (cluster abundances in this example), which was considered excellent for this context. (**b**) A "bad" model error rate plot. The CV error rate never drops below 50%, meaning the model fails to accurately predict the groups 50% of the time. No FDR constrained model was identified (no yellow triangle), meaning that the features included in the CV constrained models are not statistically significant

3.7 Further Analyses to Validate the CITRUS Results

There are several reasons why additional CITRUS analyses might be run, in order to understand how your results are affected by certain factors in your data. Generally, sensitivity analyses involve excluding certain samples or markers from the CITRUS analysis and checking that you identify the same trends that you see when you use all of the data to ensure the effects that you find are consistent across your entire dataset and are not being driven by outlying samples or markers (*see* **Note 46**). In addition to the sensitivity analyses, once you have identified biomarkers of interest, you should repeat your CITRUS analysis to guard against false positives due to stochasticity (*see* **Note 47**).

1. Rerun CITRUS excluding any samples or markers for which there are concerns about data quality. Such data quality concerns may have been identified during the data tidying and QC phase.
2. Rerun CITRUS excluding any batches of samples for which there are concerns about batch effects, or, if several samples per group within each batch are available, rerun CITRUS stratifying analyses by batch.

3. Rerun CITRUS excluding, combining or splitting samples from certain outcome groups if there are more than two (*see* **Note 48**).
4. If a predictive model is being built with unbalanced groups, rerun CITRUS many times, each time arbitrarily selecting subsets of the samples so that the groups used to build the model and assess accuracy are equal in size.
5. Rerun the entire viSNE + CITRUS analysis pipeline described above to reproduce any results of interest. Do not set the same seed for viSNE (as rerunning with the same seed will result in an identical analysis). For correlative CITRUS analyses, significant clusters with the same phenotype in each repeat should be identifiable (*see* **Note 49**). For predictive CITRUS analyses, the cross-validated model error rate seen on the model error rate plot should be similar for each repeat (*see* **Note 50**).
6. Experimental replication of findings in additional samples is also an important step in any project.
7. Use exploratory data analysis on the viSNE map as described in Subheading 3.4 to assist in understanding why a particular CITRUS cluster is significant or not in its clinical, biological, and experimental context.

4 Notes

1. Examples of typical outcomes include different disease categories (e.g., healthy donors vs. disease or two different disease groups), different response categories to a therapy, or different disease severity categories. You can compare two or more outcomes within a single CITRUS run, but there may be additional considerations for sensitivity analyses and interpretation of results when you have more than two groups (discussed below). While other implementations of CITRUS can manage continuous outcomes, discrete outcomes are necessary for analysis in Cytobank.
2. Examples of typical stratifying variables include time points (e.g., samples from the same patients across time in a clinical study, or the same biological sample aliquoted and measured at various time points after stim) and experimental conditions (e.g., the same samples measured under different culture or stim conditions).
3. Generally, the number of samples you have in each group will be driven by the study design and (in human studies) recruitment and any loss to follow-up. When the goal is to build a predictive model with CITRUS (as opposed to running a correlative analysis), interpretation becomes more straightforward

if there is an equal (or roughly equal) number of samples in each group.

4. If there will be multiple batches of samples, randomizing samples across batches according to outcome groups, stratifying variables and including a technical control sample that is run in every batch can help to assess and control batch effects. Ideally, the technical control sample will be the same sample aliquoted so that one aliquot is run with each batch (e.g., a pool of healthy donor peripheral blood mononuclear cells (PBMCs), if the cell populations being measuring are present in healthy PBMCs).
5. Even if the project goal is to build a predictive model, we often find it useful to also run a correlative analysis to identify all of the significant biomarkers in the dataset. Since predictive analyses return a *minimum* set of markers that can best predict the differences between the groups, there are often additional markers that are significantly different between the groups.
6. The markers to use for phenotyping and functional readouts can be decided based on their traditional use, but we often find it useful to use at least some of the non-traditional phenotyping markers in a panel as clustering markers for viSNE and CITRUS. Since these methods can define populations even when there is not a clear split between positives and negatives, including markers that would traditionally be functional readouts for phenotyping can help identify more nuanced cell types based on co-expression of functional markers and determine how these cell types vary between sample groups.
7. If you have carefully named your files to include information about all of these variables, this will expedite the process of creating sample tags since Cytobank will automatically put files into sample tag groups as you create them if the group name is in the file name. Otherwise, you can use a spreadsheet to import the sample tags from a CSV file; this matches the file name to the sample tags.
8. The experiment summary page in Cytobank can be used to quickly review the number of files in the experiment, sample tags associated with each file, and the total event count per file.
9. If multiple panels were run on the sample set, a separate CITRUS analysis can be performed for each panel, and the results combined for a final report or publication.
10. The arcsinh transformation helps accommodate the negative values that may be exported from flow cytometry instruments, while displaying the data in a log-like manner (for more information, *see* http://blog.cytobank.org/2012/03/30/making-beautiful-plots-data-display-basics). The arcsinh scale argument (or cofactor) should be set so that the distribution of

each marker is as expected based on the known biology of that marker and all collected data are visible. The higher the cofactor, the greater the compression of low-value data to zero. The default axis transformation for mass cytometry data in Cytobank is arcsinh with a cofactor of 5. For flow cytometry data, cofactor values up to 2000 are commonly used. As an example, peripheral blood mononuclear cells (PBMCs) should exhibit negative, intermediate, and high expression of CD4 on non T cells, monocytes, and CD4+ helper T cells, respectively. If CD4 expression on monocytes was apparent before axis transformation, but absent after, then the cofactor may be set too high.

11. The range of the scales should be set broadly enough that you can see most of the data on the plot. In Cytobank, if the range is too narrow, the data will appear stacked next to the limits of the plot.

12. Create a working illustration displaying all of the technical control files on the rows and all of the markers on the columns. Either histograms or dot plots with a widely expressed marker like CD45 on the y-axis work well for these plots. Looking down the rows for each marker, identify whether there are any technical controls that appear to have either poor staining (i.e., little to no expression) for a marker or have batch effects (i.e., an overall higher or lower distribution of expression of the marker compared to the other batches).

13. Run viSNE on intact singlets (or similar) with equal sampling from all of the technical control samples using the clustering markers you will use for your main CITRUS analysis. Create a working illustration displaying this viSNE map with each of the technical control samples on the rows; look at it using density dot plots, contour plots colored by density, and colored by the expression of each clustering marker. If you find variation in distribution of the markers or in the viSNE map across technical controls, this indicates batch effects that are large enough to influence analysis results.

14. There are several methods available for normalizing data for batch effects and a thorough review of these is outside the scope of this chapter. Any method that returns normalized data for every cell across each channel can be applied as part of the CITRUS pipeline. One class of methods will be applied one-by-one to each batch, with the goal being that the batches may be combined for analysis after normalization. One method of this type that has worked well is z-score normalization (or "mean scaling") [11]. To use this method, concatenate the events from all of the samples in a given batch. Then, normalize each value in a given channel by subtracting the mean of

that channel and dividing by the standard deviation of that channel. Finally, split the data back into separate files for continued analysis. Normalization using the 5th and 95th quantiles of the data for each channel may also work well and may be less sensitive to outlier cells. A second class of methods will be applied sample by sample, with the goal being to identify and align peaks for each marker [12]. We have seen good results with these methods, which must be run in R at the moment as far as we are aware (unpublished data), but will not be discussed in detail here. A third class of normalization methods takes into account cell populations when normalizing the data; these are typically not as applicable to the CITRUS analysis pipeline because they usually return cell populations rather than normalized per-cell data [17, 18].

15. If possible (depending on the calculations made for the number of events required per sample), it can be beneficial to include the samples for CITRUS runs in multiple stratifying variable groups in the same viSNE map.
16. To determine the number of events needed per file to detect rare populations with an acceptable percent CV, we recommend following the advice of [19]. For example, to detect a population that is 1% of the pre-gated population with a 5% CV, you need to include 40,000 events per file in the viSNE and CITRUS, whereas to detect a population that is 2% of the pre-gated population with the same CV, you only need to include 20,000 events per file.
17. If you find that you cannot include enough cells per file in the analysis to detect the rare populations you are interested in, but you have acquired enough cells per file to do detect these populations, you can start parallel CITRUS analysis pipelines for different pre-gated populations. For example, you might pre-gate to T cells and B cells separately and start CITRUS analysis pipelines for each of these subsets separately.
18. It may be helpful to include a pre-gating marker (e.g., CD45 or CD3) as a clustering channel depending on your particular data and biology. For example, T cells exposed to a chronic inflammatory environment may downregulate expression of the ζ-chain of CD3 [20]. Therefore, if collected data include T cells from unique environments, varying expression of CD3 may indicate an important biological phenomenon that would be helpful to capture by clustering on that marker.
19. It is best to set iterations and perplexity according to some rough guidelines rather than to simply maximize them because they can dramatically increase run time. A good starting point for iterations is roughly 1000 iterations for every 100k events. Perplexity is more difficult to set guidelines for because the

required perplexity is influenced greatly by the specific data set. Increasing perplexity can also have a large impact on algorithm run time. We recommend starting with a perplexity of 30 if the number of events in the viSNE is less than 1 M, 50 for 1 M to 1.5 M events, and 70 for more than 1.5 M events [13].

20. In a high quality viSNE map, cells with similar expression of clustering markers will form either separate viSNE islands (if they are very different from other cell types) or separate regions within viSNE islands. A poor quality viSNE map will have overlapping and poorly formed islands that don't separate the expression of a single marker into distinct locations on the map. Cells expressing a given marker may appear in a string-like or spindly pattern. If the viSNE map is poorly converged, rerun viSNE with additional iterations. viSNE maps can also be converged with the islands not very well separated. In this case, rerun viSNE with higher perplexity. Note that if starting with a distinct pre-gated population (e.g., CD4+ T cells) it is expected that the viSNE map will only have one island; expression of the clustering markers should define regions of the island rather than separate islands.
21. This is often done with the files separate as part of assessing viSNE map quality, but a better publication visual can be made on the concatenated file across all groups. This will be useful for visually explaining the phenotype of significant clusters.
22. We recommend making this exploratory data analysis with the viSNE map an iterative process that is used to explore the data before and after running CITRUS.
23. To keep track of the cell populations being examined, it may be helpful to manually gate the islands of the viSNE map and then display these on the working illustration. Do this by selecting "show gates" and using the gate label "gate name."
24. One CITRUS analysis is required for each separate stratifying variable that is included in a single viSNE. For example, if included healthy and disease samples from two stimulation conditions in the same viSNE, start one CITRUS comparing healthy and disease samples for each of the stimulation conditions. Alternatively, depending on your event count calculations done above, these stimulation conditions may be in separate viSNEs.
25. If both are desired, these can be done in separate CITRUS runs. If using medians, be sure to select equal sampling and specify the functional readout markers that were identified in Subheading 3.1 as the functional markers to characterize.
26. Since equal sampling was chosen for viSNE, all of the files listed should already have the same number of events.

27. Neither cross-validation fold nor FDR influence the SAM model.
28. CITRUS results are statistically significant below a FDR threshold which you select. FDR is similar to a p-value that's adjusted for multiple hypothesis testing, but it is less conservative (i.e., it will identify more significant biomarkers than, for example, a Bonferroni correction) and therefore useful for biomarker discovery. The SAM model outputs results at FDR thresholds of 1, 5, and 10% for every run. One of these can be selected to report as "these results are statistically significant below a false discovery rate threshold of xx%," where xx is the threshold you select.
29. There are several reasons why no model output will be returned by CITRUS. First, there may not be any significant difference between the outcome groups that can be measured by the panel that was used. Second, CITRUS might not find any significant biomarkers if there is an insufficient number of samples per group. Like any statistical test, whether or not a significant difference was detected depends on the magnitude of the difference in the biomarker between groups, the variation in the biomarker within each group, and the number of samples measured. We recommend running CITRUS with at least eight samples per group. Third, CITRUS might not find any significant biomarkers if an insufficient number of cells were included from a population that would otherwise exhibit useful biomarkers. Being able to include sufficient cells depends on collecting enough events in the experiment and pre-gating enough that enough events can be included to detect them. Finally, CITRUS might not find any significant biomarkers if the analysis combines more than two outcome groups that interact with each other. If including outcome groups that represent a single outcome (such as disease vs. healthy) and something that might be a stratifying variable (such as young vs. old) in the same CITRUS, it might be difficult for CITRUS to identify significant differences between the outcome groups if they are also impacted by the stratifying variable. If this is possibly the case, split the samples into multiple CITRUS runs and use the viSNE map to compare results and see if the same cell populations are significant at each level of the stratifying variable.
30. Since the clusters are nested, this cluster includes all of the cells within the lower clusters on the tree in this bubble.
31. These boxplots will show which outcome group has greater abundance or medians (whichever was used for CITRUS) for these clusters, and how spread out these values are within each group. Identifying trends in these differences across more than

two outcome groups or across different significant biomarkers can be helpful to explain the biological importance of the results and build confidence in them.

32. Since CITRUS was started from within the viSNE experiment in Cytobank, the tSNE channels that give the viSNE map coordinates are contained within the CITRUS results files, even though they haven't been used for clustering.
33. The dot size and/or plot size can be increased or decreased to make the visualization larger or smaller.
34. The "markerPlots" show the same CITRUS trees that were examined to find the significant nodes to focus on colored by marker expression.
35. The "clusters" histograms output by CITRUS for significant clusters show the expression of each clustering marker in a given significant cluster as compared to cells in all other clusters that were used in the CITRUS analysis ("background").
36. This expression should match what is seen on the viSNE map in the region where the significant clusters are overlaid, but provides a summary of the median expression of each marker in each cluster.
37. Back-gating the CITRUS clusters onto biaxial plots will help with understanding how the significant clusters might have been identified or overlooked in a biaxial gating hierarchy.
38. Typically, build one predictive model for each outcome group comparison within each stratifying variable defined above (just like for the correlative analysis). However, if a correlative analysis was run prior to building any predictive models, and that analysis did not find any significant results, a predictive model should not be built. Furthermore, based on the results of the correlative analysis, it might be suitable to build a predictive model for the outcome groups at a single level of the stratifying variables using either abundances or medians. For example, a predictive model might be built using median signaling marker expression at baseline to predict an outcome measured at a 2 month time point if significant marker expression in the corresponding correlative analysis was seen and if the scientific goal is to find baseline predictors of the 2-month outcome.
39. When building a predictive model, it is permissible to run multiple model-building algorithms (PAM and lasso) and select the one that gives the best prediction (lowest cross-validated model error rate) of the outcome. If the two models are equal, it is permissible to choose the simplest model with the fewest biomarkers.
40. When assigning "file groupings" for predictive models, it is best to have balanced groups with the same number of samples

per group. CITRUS uses overall accuracy predicting all groups to select the markers to use in the final model. Thus, if one group has a much larger number of samples than another group, the group with the larger number of samples will have more influence on this overall accuracy calculation.

41. In general, as the number of features shown across the top of the plot increases, the cross-validation error rate (red line) should decrease up to a point. If it does not, you most likely have not measured biomarkers with any ability to discriminate your outcome groups.

42. The CV min constrained model is the model with the minimum number of features required to achieve the best cross-validated estimation of model accuracy possible with the dataset. This model is what you might carry forward to use in a withheld validation/testing dataset.

43. In a clinical context where the best previously known predictor of response has a 50% accuracy (or 50% model error rate), a 70% accuracy (or 30% model error rate) would be an outstanding improvement, but in other contexts this accuracy might be considered poor. If the outcome groups are not balanced, it may be difficult to determine whether the accuracy is good or not, since the accuracy is determined across all samples and will therefore be influenced more heavily by groups with more samples. One option to avoid this concern is to arbitrarily select the same number of samples from each group to run the analysis, and then to perform some repeated analyses with different included samples to test the effect of arbitrary sample selection.

44. If there is no FDR constrained model shown on the model error rate plot, then none of the biomarkers in your predictive model have statistically significant differences between your outcome groups.

45. If you are building more than one predictive model, for example, across different levels of a stratifying variable, you might identify different biomarkers in each of the predictive models even if the same markers were present in the correlative analysis. This is expected since CITRUS will select a set of the fewest markers necessary to best predict the outcome, and there may be more than one equivalent set.

46. In these sensitivity analyses, the effects that you see may not be significant (particularly with fewer samples), but the trends should be the same.

47. Since CITRUS includes some stochastic elements, it's important to ensure that any significant biomarkers you identify are not false positives resulting from this stochasticity. If no downsampling of events was performed during the creation of the

viSNE map or CITRUS analysis, there is no need to repeat the pipeline.

48. For example, parallel CITRUS analyses that break a response variable into four categories (e.g., progressive disease, stable disease, partial response, complete response) or two categories (e.g., nonresponse or response) can be considered.
49. Since the events are down-sampled to create equal sampling for the viSNE map, the clusters may not consist of identical events, but the expression pattern of markers in these clusters should be very similar. To evaluate this, it may be helpful to use all of the tools outlined above for confirming the phenotype of CITRUS clusters. In addition, it may be helpful to compare numerical summaries (e.g., median) of surface marker expression that go along with heatmaps and the numerical summaries of cluster abundance or marker expression differences between groups in the CITRUS results to make sure that these are consistent.
50. For predictive models built with CITRUS, the included biomarkers are NOT necessarily expected to be identical between repeated CITRUS runs because there may be more than one set of equally predictive markers.

References

1. Kvistborg P, Gouttefangeas C, Aghaeepour N et al (2015) Thinking outside the gate: single-cell assessments in multiple dimensions. Immunity 42(4):591–592. https://doi.org/10.1016/j.immuni.2015.04.006
2. Newell EW, Cheng Y (2016) Mass cytometry: blessed with the curse of dimensionality. Nat Immunol 17(8):890–895. https://doi.org/10.1038/ni.3485
3. Bruggner RV, Bodenmiller B, Dill DL et al (2014) Automated identification of stratifying signatures in cellular subpopulations. Proc Natl Acad Sci U S A 111(26):E2770–E2777. https://doi.org/10.1073/pnas.1408792111
4. Fraietta JA, Lacey SF, Orlando EJ et al (2018) Determinants of response and resistance to CD19 chimeric antigen receptor (CAR) T cell therapy of chronic lymphocytic leukemia. Nat Med 24(5):563–571. https://doi.org/10.1038/s41591-018-0010-1
5. Pelissier Vatter FA, Schapiro D, Chang H et al (2018) High-dimensional phenotyping identifies age-emergent cells in human mammary epithelia. Cell Rep 23(4):1205–1219. https://doi.org/10.1016/j.celrep.2018.03.114
6. Subrahmanyam PB, Dong Z, Gusenleitner D et al (2018) Distinct predictive biomarker candidates for response to anti-CTLA-4 and anti-PD-1 immunotherapy in melanoma patients. J Immunother Cancer 6(1):18. https://doi.org/10.1186/s40425-018-0328-8
7. Ben-Shaanan TL, Azulay-Debby H, Dubovik T et al (2016) Activation of the reward system boosts innate and adaptive immunity. Nat Med 22(8):940–944. https://doi.org/10.1038/nm.4133
8. Benjamini Y, Hochberg Y (1995) Controlling the false discovery rate: a practical and powerful approach to multiple testing. J R Stat Soc 57(1):289–300
9. van der Maaten LJP, Hinton GE (2008) Visualizing high-dimensional data using t-SNE. J Mach Learn Res 9:2579–2605
10. Amir el AD, Davis KL, Tadmor MD et al (2013) viSNE enables visualization of high dimensional single-cell data and reveals phenotypic heterogeneity of leukemia. Nat Biotechnol 31(6):545–552. https://doi.org/10.1038/nbt.2594
11. Knapp D, Kannan N, Pellacani D et al (2017) Mass cytometric analysis reveals viable activated caspase-3(+) luminal progenitors in the normal adult human mammary gland. Cell Rep 21(4):1116–1126. https://doi.org/10.1016/j.celrep.2017.09.096

12. Hahne F, Khodabakhshi AH, Bashashati A et al (2010) Per-channel basis normalization methods for flow cytometry data. Cytometry A 77(2):121–131. https://doi.org/10.1002/cyto.a.20823
13. Cytobank (2018) How to configure and run a viSNE analysis. https://support.cytobank.org/hc/en-us/articles/206439707-How-to-Configure-and-Run-a-viSNE-Analysis. Accessed 27 July 2018
14. Tusher VG, Tibshirani R, Chu G (2001) Significance analysis of microarrays applied to the ionizing radiation response. Proc Natl Acad Sci U S A 98(9):5116–5121. https://doi.org/10.1073/pnas.091062498
15. Tibshirani R, Hastie T, Narasimhan B et al (2002) Diagnosis of multiple cancer types by shrunken centroids of gene expression. Proc Natl Acad Sci U S A 99(10):6567–6572. https://doi.org/10.1073/pnas.082099299
16. Tibshirani R (1996) Regression shrinkage and selection via the lasso. J R Stat Soc Ser B 58:267–288
17. Finak G, Jiang W, Krouse K et al (2014) High-throughput flow cytometry data normalization for clinical trials. Cytometry A 85(3):277–286. https://doi.org/10.1002/cyto.a.22433
18. Van Gassen S, Gaudiliere B, Dhaene T, et al (2017) A cross-sample cell-type specific normalization algorithm for clinical mass cytometry datasets. Paper presented at the 32nd congress of the International Society for Advancement of cytometry, Boston, MA
19. Hoy T (2006) Rare-event detection. In: Wulff S (ed) Guide to flow cytometry. Dako, Carpinteria, CA, pp 55–58
20. Baniyash M (2004) TCR zeta-chain downregulation: curtailing an excessive inflammatory immune response. Nat Rev Immunol 4(9):675–687. https://doi.org/10.1038/nri1434

Correction to: Quantitative Measurement of Cell-Nanoparticle Interactions Using Mass Cytometry

Andrew J. Mitchell, Angela Ivask, and Yi Ju

Correction to:
Chapter 15 in: Helen M. McGuire and Thomas M. Ashhurst (eds.), *Mass Cytometry: Methods and Protocols*, Methods in Molecular Biology, vol. 1989, https://doi.org/10.1007/978-1-4939-9454-0_15

This chapter was inadvertently published with the acknowledgement section leaving out the following sentence: "This work received funding from South Australian Government PRIF program Project "International Cluster on Nanosafety" of Nicolas H. Voelcker and Enzo Lombi." This correction has been updated in the chapter.

The updated online version of this chapter can be found at
https://doi.org/10.1007/978-1-4939-9454-0_15

Helen M. McGuire and Thomas M. Ashhurst (eds.), *Mass Cytometry: Methods and Protocols*, Methods in Molecular Biology, vol. 1989,
https://doi.org/10.1007/978-1-4939-9454-0_21,

INDEX

A

Acquisition 7, 8, 10, 11, 13–31, 43, 44, 48, 56, 61, 73, 87, 91, 97, 106, 120, 121, 125, 126, 132, 134, 144, 156, 175, 179–181, 190, 203, 230, 231, 233, 237–239, 251, 253, 267, 284, 290, 311, 313
Adherent 220, 221, 229, 230, 234–236
Algorithm 22, 102, 161, 184, 190, 214, 219, 220, 237, 245, 246, 249, 251, 253–262, 267–272, 278, 284, 290, 291, 295–297, 306, 309, 310, 315, 316, 327, 329
Aliquot 9, 18, 21, 27, 65, 91, 98–100, 104, 126–128, 165, 166, 169, 186, 195, 201, 210, 224, 229, 323, 324
Analysis .... v, 4, 13, 35, 47, 55–79, 83, 93, 111, 125–134, 152, 193–214, 218, 228, 245–263, 267–278, 281–293, 295–307, 310
Analyze 7, 13, 14, 44, 56, 60, 62, 73, 74, 148, 197, 198, 203, 205, 212, 228, 234, 246, 247, 251, 256, 270, 283, 289, 290, 309, 311
Antibody
 conjugating 47, 52, 53, 143
 tagging 47–54, 56, 224
Antigen 7, 10, 36, 48, 55, 57, 61, 63, 69, 84, 89–91, 95, 106, 117, 119, 147, 148, 194, 197, 199, 201, 202, 205, 208, 211, 212, 253, 288, 289, 317
Antigen-specific T cell analysis
 multiplex tetramer 150
Apoptotic 197, 204, 207, 210, 211
Approach .... v, 13, 22, 27, 36, 48, 57, 60, 68, 74, 83, 94, 95, 119, 121, 126, 148, 153, 154, 160–162, 164, 168, 177, 182–183, 185, 186, 189, 190, 194, 207, 209, 210, 213, 218, 227, 228, 240, 245, 249, 253–257, 270, 272, 282, 284, 286, 287, 289–291, 296, 297, 301–303, 306, 307, 309, 314
Argon gas 8, 9
Automated cell processing 111–122
Automatic
 clustering 27, 161, 249, 256, 272, 291
 gating 22, 27, 117, 120, 121, 245–247, 249, 251, 253–257, 260–262, 291

B

Barcoding
 CD45-labelled 21, 24, 94–98, 126, 130, 132, 133, 300
 Pd-labelled 132
Batch 7, 53, 66, 89, 118, 126, 165, 196, 246, 253, 289, 290, 312–315, 322, 324, 325
Bead 4, 6, 7, 10, 15–19, 26–29, 39, 44, 73, 84, 87, 103, 114, 117, 128, 132, 141, 142, 156, 167, 180, 181, 203, 220, 224, 246, 289, 299
Best practices
 SRL 3
β2-microglobulin (B2M) 95–99, 102–104
Bioconductor 245–253, 257, 262, 263, 282, 290
Bioinformatics 246–250, 255, 262, 267, 278
Biomarker discovery 93, 250, 257, 309–331
Bismuth 49, 52, 66, 69, 78
Block 9, 43, 84, 85, 87, 90, 91, 98–102, 106, 114, 127–130, 133, 134, 174, 187
Blood 9, 40, 41, 44, 60, 70–71, 74, 84, 89–91, 95–97, 114, 115, 126, 133, 139, 141, 144, 152, 166, 170, 172, 173, 187, 196, 199, 200, 218–220, 224, 234, 238, 309, 312, 315, 318, 324
Bone marrow (BM) 60, 96–97, 161, 162, 165, 169–171, 175, 181–184, 187, 191, 194, 199, 201, 218, 219, 224
Bovine serum albumin (BSA) 27, 49, 57, 65, 77, 84, 99, 102, 112, 119, 128, 140, 152, 154, 155, 196, 213, 219, 222, 223, 236
Bromodeoxyuridine (BrdU) 160, 163–171, 177, 178, 185, 186, 188–190, 193, 209
Buffer
 Ficoll 220, 299
 permeabilization 43, 94, 114, 145, 155, 164, 185, 187, 188
 saponin 86, 94, 114, 164, 185, 187–190
 Smart Tube 84, 89, 91, 195, 196

C

Calibration 15, 16, 27, 39, 44, 84, 87, 114, 117, 128, 132, 141, 142, 167, 203, 220, 224
Cancer 56, 93, 147, 148, 205, 267
Capillary 5, 9, 10, 238
Cell cycle analysis
 BrdU 160, 164, 167, 171, 193, 209
 cyclin 161, 176, 180, 194, 196, 197, 203–205, 212
 G0 .... 197, 204, 205, 208–210, 212
 G1 .... 197, 204, 205, 208–210, 212
 G2 .... 180, 197, 204, 205, 208, 211, 212

Helen M. McGuire and Thomas M. Ashhurst (eds.), *Mass Cytometry: Methods and Protocols*, Methods in Molecular Biology, vol. 1989, https://doi.org/10.1007/978-1-4939-9454-0, © Springer Science+Business Media, LLC, part of Springer Nature 2019

Cell cycle analysis (*cont.*)
IdU 160, 164, 167, 182–183, 194–196, 198–201, 203–205, 208–210, 212, 213
Ki67 ....160, 162, 164, 182–183, 185, 203, 204, 210, 211
M-phase 180, 197, 205, 208, 211
phosphorylated histone H3 (pHH3) 161, 162, 164, 180, 194, 196, 204, 208, 211
pRb 161, 162, 164, 182–183, 185, 194, 196, 203–207, 210–212
Cell-id™ 38, 41, 43, 44, 84, 87, 114, 141, 219
Cell recovery 18, 61, 70, 106, 126, 132, 187
Cell staining media (CSM) 196, 199, 201, 202, 213
Cell-surface staining 43, 94, 102, 201, 222
Cell types/subsets
dendritic 217, 221
erythrocyte 175, 181–184
hematopoietic 261
leukocyte 95–97, 175, 181–184
macrophage 218
MDSCs 218
monocyte 184, 210
myeloid phagocytes 217
myeloid regulatory cells 217
progenitors 160, 181
Cellular 13, 55, 95, 162, 193, 217, 227, 259, 271, 281, 295
Centrifugation 11, 50, 54, 67, 68, 115–117, 119, 131, 132, 144, 201, 202, 212, 213, 230, 236, 298
Channel 14, 36, 47, 57, 95, 115, 126, 143, 148, 182, 194, 230, 253, 289, 300, 311
Chelate 57
Chemokine 85, 86, 91, 94
Cisplatin 26, 38, 39, 41, 44, 84, 85, 90, 94, 95, 127, 128, 131, 152, 154, 156, 166, 174, 180, 181, 198, 219, 229, 230, 236, 282, 300
Citrus
predictive 259, 260, 274, 310, 312, 313, 321–323, 330, 331
stratify 249, 311, 312, 316–318, 322, 326–328, 330
Clean 5–7, 9, 10, 28, 63, 100, 126, 142, 180, 181, 185, 230, 231, 233, 238, 252, 292, 299–301, 310
Clinical 111, 139, 248–250, 260, 261, 295–307, 309, 323, 330
Cloud 14, 21–24, 27, 28, 30, 97, 100, 180, 231, 237, 239, 310
Cluster 27, 161, 196, 254, 268, 282, 295, 309
Cocktail 43, 44, 47, 69, 72, 73, 78, 85–87, 89, 90, 116, 127, 129, 131, 134, 154, 155, 199, 201, 202, 222–224, 315
Coefficient of variance (CV) 16, 17, 29, 117, 121, 122, 246, 317, 321, 322, 326, 330
Command 282, 284, 286, 288
Computational 104, 161, 182, 219, 247, 251–253, 258, 262, 271, 295–307
Concatenate 19–21, 88, 285, 296, 300–302, 304, 306, 316–321, 325, 327
Configuration 36–38, 126
Conjugation ....v, 4, 14, 21, 24, 38, 43, 47–49, 51–79, 83, 84, 89, 94–99, 101–106, 112, 128, 130, 140, 143, 160, 164–167, 174, 187, 205, 209, 219, 224, 228, 239, 289
Control 5–7, 10, 13–31, 35, 36, 40, 42, 49, 65, 69–74, 76, 78, 84, 86, 88–90, 103, 104, 112, 114, 118, 126–130, 134, 143, 144, 208, 211, 220, 246, 247, 252, 259, 275, 276, 286–292, 301, 309–315, 324, 325
Core
facility 282
panel design 37–38, 290
Curve 18, 22–25, 52, 76, 167, 238
Cytofkit 167, 182, 220, 282–284, 286, 287, 292
Cytometer 3–5, 9, 14, 18, 21, 27, 39, 61–63, 66, 69, 73, 74, 87, 102, 106, 115, 142, 144, 157, 160, 167, 175, 180, 190, 195, 203, 209, 212, 213, 220, 224, 228–230, 234, 235, 237, 239, 246, 252, 282, 288, 290
Cytometry
flow 4–7, 10, 13, 14, 26, 31, 47, 66, 93, 97, 102, 125, 126, 148, 150, 156, 167, 193, 194, 209, 212, 213, 227, 231–233, 237, 245–263, 270, 281–293, 295, 324
fluorescent 4, 193, 194, 209, 212, 213, 284, 289, 292
mass .... v, 3–11, 13–31, 35–44, 47–79, 83–107, 111–122, 125–134, 139–157, 193–214, 217–224, 227–240, 246, 267–278, 281–293, 295–307, 313, 325
Cytometry by time-of-flight (CyTOF) 3, 14, 15, 18, 19, 28, 40, 43, 44, 47, 56, 61–64, 66, 70–71, 74, 115, 125, 126, 128, 132, 134, 156, 167, 180, 181, 188, 189, 203, 218, 220, 224, 228, 237, 239, 256, 267, 269, 271, 301

D

Data
acquisition 8, 13–31, 61, 93, 97, 125, 180, 230, 231, 251, 253, 267, 284, 290, 311
collection 73, 125, 132, 237, 238, 287, 315, 325, 326
node 296, 297, 305, 306
PCA 156, 268–270, 272, 284, 285, 290
pipeline 182, 245, 246, 249, 257, 261, 282, 290–292, 310, 311, 326
processing 13–15, 26, 44, 249, 251, 282
quality control 7, 13–31, 49, 290–292, 310, 311, 313, 314, 316, 322
R 301
unsupervised analysis 60, 253, 254, 272, 297

Data analysis packages/software
  flowAI ... 251, 292
  flowDensity ... 258, 290
  FlowJo ... 14, 39, 44, 99, 104, 128, 132, 156, 167, 182–183, 191, 232–234, 237, 245, 282, 285, 298, 304
  FlowSOM ... 182–183
  flowType ... 261
  SCAFFoLD ... 296, 298
  SPADE ... 258, 296
  tSNE ... 182–183, 259, 285, 288
  viSNE ... 270, 314, 316, 323, 327
Data processing
  curve ... 18
  normalization ... 44
  residual ... 26
  size ... 271
  threshold ... 14
Deconvolution ... 22, 96–98, 102, 103, 106, 126, 133, 246
Dendritic ... 217, 221
Density ... 30, 38, 115, 144, 161, 218, 235, 240, 247–250, 254, 258, 270, 271, 275, 277, 284, 288–291, 296, 317, 325
Deoxyribonuclease (DNase) ... 152, 160, 162, 164–167, 177, 178, 185, 186, 189, 190, 194, 303
Detection ... v, 7, 16, 28, 29, 47, 56, 60, 63, 83, 94, 134, 162, 164, 165, 168, 175, 177, 194, 197, 199, 228, 229, 235, 238, 239, 251, 252, 256, 268, 270–272, 302
Detector ... 14, 16, 63, 112, 143, 213, 230, 231, 234, 235, 238, 239, 246, 288, 289
Dimensionality reduction ... 27, 161, 162, 182–183, 247, 260, 263, 268, 270, 282, 284, 285, 290
Discovery ... 93, 250, 254, 257, 260, 309–331
Disease ... 47, 60, 95, 103, 160, 193, 260, 277, 291, 295, 298, 309, 318–319, 323, 327, 328, 331
DNA ... 16, 17, 21, 26, 29, 30, 65, 73, 74, 96–97, 103, 105, 117, 127, 128, 132, 141, 142, 144, 152, 154–156, 160, 165, 166, 175, 176, 179–181, 189, 193, 194, 197, 204, 205, 209, 212, 229, 239, 300

E

Efficiency ... 16, 57, 70, 231, 232, 235, 238, 239, 246, 252, 253
Enrich ... 7, 43, 49, 52, 61, 62, 94, 100, 184, 218, 219, 282, 298, 299
Erythrocyte ... 172, 173, 175, 180–184, 187
Event length ... 14, 21, 22, 24–27, 29, 30, 96–97, 105, 115, 180, 181, 232–234, 237, 239
Expression ... 16, 47, 56, 95, 126, 147, 160, 205, 217, 253, 267, 285, 297, 310
Ex vivo
  stimulation ... 168

F

Factor ... 19, 38, 56, 63, 84, 86, 87, 112, 114, 140–146, 148, 151, 156, 160, 162–165, 173, 177, 185, 188–190, 217, 219, 268, 296, 311–313, 322
Ficoll ... 38, 41, 144, 218, 220, 297, 298
Filter ... 14, 44, 63, 84, 97, 128, 140, 152, 162, 195, 230, 251, 288, 297
Fixation ... 8, 18, 43, 70, 86, 94, 112, 126, 140, 160, 197, 223, 230, 289
flowAI ... 247, 251, 252, 292
Flow cytometry (FCM) ... 4, 13, 47, 66, 93, 125, 148, 193, 227, 245–263, 270, 281–293, 295, 324
flowDensity ... 249, 254–258, 290
FlowJo ... 14, 39, 99, 104, 128, 132, 156, 167, 181–183, 191, 232–234, 237, 245, 247, 282, 284–286, 292, 296, 298–301, 304, 305
FlowSOM ... 161, 162, 182–184, 190, 248, 256, 272, 277, 290, 291
flowType ... 248, 250, 257, 261, 291
Fluidigm ... 3, 14, 38, 48, 57, 84, 94, 112, 126, 141, 152, 166, 196, 219, 229
Fluorophore ... 48, 56, 83, 160, 164–167, 174, 187–189, 288, 289, 292
FoxP3 ... 38, 39, 43, 84, 87, 91, 140–146, 177, 185, 187, 189, 190
FoxP3/transcription ... 38, 84, 87, 140–146, 177, 185, 187, 189
Freeze ... 100, 101, 153, 165, 166, 186, 213, 220–222, 229, 299
Fusion ... 21–24, 27, 30, 180, 237, 239

G

G0 ... 197, 204, 205, 207–212
G1 ... 197, 204, 205, 207–212
G2 ... 180, 197, 198, 204, 205, 208, 210, 212
Gate ... 16, 22, 25, 26, 36, 42, 73, 96–97, 103, 176, 180, 181, 203–208, 210, 212, 233, 246, 247, 254–257, 271, 284, 300, 304, 305, 314, 316, 326, 327
Gaussian ... 21–23, 30, 117
Graphic user interface (GUI) ... 270, 275, 282–284, 292

H

Heatmaps ... 143, 272, 273, 285, 288, 310, 318–319, 321, 331
Helios ... 3, 5, 9, 14, 18, 19, 21, 27, 39, 40, 44, 47, 102, 156, 167, 180, 181, 228, 229, 231, 233–235, 237–239, 268
Hematopoiesis ... 56, 194
Heparin ... 40, 84, 85, 91, 199
Hierarchical ... 248, 249, 254, 269, 271, 274, 277, 285, 288, 290, 291
High-dimensional cytometry ... 160, 161, 256, 290
Human ... 28, 38, 56, 84, 95, 111, 128, 139, 148, 160, 194, 217, 239, 282, 323

I

Immune monitoring ... 28
Immunoglobulin (Ig) ... 55–79
Inflammatory ... 139, 326
Inhibitor ... 56, 127, 129, 130, 134
Injections ... 160, 168–170, 186, 195, 200, 232, 235
Intercalators ... 26, 27, 30, 38, 39, 43, 44, 65, 73, 74, 84, 87, 102, 114, 117, 127, 128, 132, 141, 142, 144, 152, 155, 156, 166, 175, 179, 180, 196, 202, 203, 219, 223, 229, 237, 238
Intracellular ... 7, 10, 39, 43, 55, 69, 73, 86, 87, 91, 94, 112, 114, 116, 117, 119, 127, 128, 131, 132, 134, 139–143, 152, 160, 162–165, 175, 177–179, 185, 188–190, 194, 197, 201, 211, 223, 224, 236, 250, 260
In vitro
  stimulation ... 218
In vivo
  stimulation ... 163, 193
Iododeoxyuridine (IdU) ... 160, 162–171, 177, 178, 181–183, 185, 186, 188–190, 194–196, 198–201, 203–210, 212, 213
Iridium (Ir) ... 73, 74, 79, 102, 128, 141, 144, 166, 179, 196, 202, 203, 219, 223, 229, 230, 237–239
Isothiocyanobenzyl-ethylenediaminetetraacetic acid (ITCB-EDTA) ... 94, 95, 98–101, 104, 105
Isotopes ... 4, 9, 16, 20, 21, 24, 28, 38, 43, 44, 55–79, 94, 95, 98–101, 104, 219, 224, 228–235, 237–240
ITCB-EDTA-Pd ... 98, 101, 105, 106

K

Ki-67 ... 113, 196, 197, 203–208, 210, 211

L

Lab automation ... 111
Lanthanides ... 48–50, 52, 57, 65, 83, 94, 95, 234, 237
Length
  event ... 14, 21, 22, 24–27, 29, 30, 96–97, 105, 115, 180, 181, 232, 233, 237, 239
Leukocytes ... 4, 95–97, 99, 172, 173, 175, 180–184, 187, 191, 218
Lineages ... 47, 160, 181, 184, 185, 210, 217, 272, 282, 292, 304
Liquid handling robot ... 118, 120, 122

M

Macrophages ... 217–222, 224, 234, 236, 302, 304
Maleimide ... 48, 52, 53, 57, 114, 234
MaxPar® ... 38, 41, 43, 48–50, 52, 57
Methanol ... 65, 70–72, 74, 84, 87, 91, 116, 118, 119, 126, 140, 142, 144–146, 162, 164, 165, 177, 178, 188–190, 195–197, 199, 201, 202, 210, 213, 219, 223
Monocytes ... 44, 74, 96–97, 103, 165, 169, 184, 187, 210, 217, 218, 220, 221, 224, 282, 325
Monomers ... 149, 150, 153, 154
M-phase ... 180, 197, 204, 205, 208, 210–212
Multiplex
  tetramer ... 147–157
Myeloid ... 36, 39, 70–71, 74, 96–97, 143, 144, 187, 194, 217–224, 296, 305
  regulatory cells ... 217–224
Myeloid-derived suppressor cells (MDSC) ... 218–221, 224

N

Nanoparticles ... 227–240
Nebulizers ... 5, 9, 10, 14, 21, 238
Nitrate
  bismuth ... 49, 52
Nodes ... 201, 210, 258, 277, 295–297, 300–302, 305, 306, 320, 329
Normalization
  data processing ... 44

O

Optimizes ... 13, 16, 57, 63, 106, 112, 119, 120, 152, 153, 190, 247, 249, 250, 262, 287, 289
Oxidation ... 16, 17, 28, 29, 35, 53, 57, 63, 210

P

Palladium (Pd) ... 25, 52, 94, 95, 98–102, 104–106, 126, 132
Panel design ... v, 7, 41, 63, 290
Paraformaldehyde (PFA) ... 10, 11, 38, 43, 44, 65, 70, 84, 87, 91, 114–116, 126, 128, 140, 142, 144, 145, 152, 155, 156, 160, 164, 175, 179, 185, 187, 188, 195–203, 219, 223
PBS + BSA ... 219, 223
PBS-BSA ... 152, 154, 155
PBS/BSA/azide ... 102
Peptide-MHC ... 148, 153, 154
Peptides ... 53, 148–151, 153
Performance ... 6, 7, 16–18, 22, 27–29, 58–59, 119, 165, 253–257, 270, 271, 289
Peripheral blood mononuclear cells (PBMCs) ... 39, 44, 70–71, 74, 84, 114, 115, 121, 126, 133, 141, 143–145, 152, 234, 312, 318, 324, 325
Permeabilization ... 26, 43, 70–72, 74, 94, 112, 114, 118, 119, 126, 145, 197–199, 201, 202, 213, 223
Phagocytosis ... 217–224
Phenograph ... 269, 272, 284–286
Phenotypes ... 55, 56, 60, 89, 95, 148, 154, 156, 159, 181, 184, 187, 217–219, 224, 257, 259, 261, 271, 281–293, 295–297, 303–305, 318–321, 323, 327, 331
Phosphorylated signaling markers ... 139–146
Phospho-specific ... 140, 142–145
Pipeline
  data ... 182, 184, 245, 246, 249, 290, 292, 310, 311, 326

Plasma
blood ... 220, 239
mass cytometry instrument ... 61
Plots ... 17, 21, 22, 24–26, 28, 70–71, 73, 74, 76, 88, 133, 156, 161, 191, 203–206, 213, 233, 234, 239, 240, 250, 270, 272–277, 288, 290, 297, 301–303, 306, 310, 314–323, 325, 329, 330
Polarization ... 218–222
Polymer
metal-chelating ... 48, 52
Predictive
CITRUS ... 259, 274, 310, 312, 313, 321–323, 330, 331
Principal component analysis (PCA) ... 156, 249, 268–270, 272, 284, 285, 304
Progenitors ... 61, 160, 162, 168, 174–176, 181, 184, 185, 206, 207, 277
Proliferation ... 47, 185, 193, 194, 200, 211, 212
Pulses ... 14, 15, 22–23, 141, 267
Pushes ... 14, 15, 21–23, 231, 237, 239, 282, 296, 303

Q

Quality control (QC)
data ... 13–31, 290–292, 310, 311, 313, 314

R

R
data analysis ... 117, 246–250
RchyOptimyx ... 260, 261, 291
Reduction ... 27, 48, 53, 57, 67, 83, 87, 115, 161, 162, 182–183, 189, 247, 260, 263, 268, 270, 282, 284, 285, 290, 304
Residual
data processing ... 26
Resolutions ... 9, 13, 17, 22, 25, 28, 36, 61, 76, 120, 121, 160, 194, 197–199, 203, 210, 212, 218, 268, 281, 289
Retinoblastoma protein (pRb)
cell cycle analysis ... 161, 194, 196
Robotics ... 114, 118–120, 122

S

Sample types
blood ... 9, 96–97, 115, 133, 139, 144, 218
bone marrow ... 95–97, 171, 174, 199, 218
cancer ... 93
clinical ... 111, 139, 295–307
dissociated tissue ... 141
MDSC ... 218
mouse samples ... 186, 200
PBMCs ... 84, 115, 144, 152, 312, 324
tumor ... 30, 152, 154
Saponin ... 86, 94, 114, 116, 118, 119, 126, 160, 164, 185, 187–190, 197
Scaffold-data analysis ... 298, 300
Sensitivity ... 6, 13, 16, 17, 28, 29, 52, 63, 112, 143, 161, 238, 246, 254, 271, 288, 289, 312, 322, 323, 330
Shared resource laboratory (SRL) ... 3, 5–8, 89
Signaling pathways-phosphorylation ... 139
Single-cells ... v, 13–15, 21, 22, 24, 28, 55–79, 83, 94, 96–97, 102, 103, 111, 125, 133, 139, 152, 153, 160, 161, 193, 199, 202, 219, 228, 229, 232, 237, 246, 258, 260, 267, 281, 284, 287, 290, 292, 296, 298, 309–331
Singlets ... 16, 17, 21–23, 26, 27, 29, 30, 176, 180–182, 314, 325
Size
data processing ... 271
Smart Tube ... 84, 89, 91, 195, 196, 199, 200, 213
Software ... 4, 14–16, 18–22, 24, 26, 28, 39, 44, 66, 73, 97, 99, 102, 115, 126, 128, 132, 156, 167, 180–183, 191, 231–235, 238, 239, 246–250, 255, 262, 263, 267, 283, 298, 304
Spanning-tree progression analysis of density-normalized events (SPADE) ... 156, 209, 213, 218, 219, 248, 258, 260, 271, 277, 290, 291, 296
Spectrometry ... 13, 56, 61, 227–229
S-phase ... 30, 160, 180, 186, 194, 198, 203–210, 212
Spillover ... 7, 28, 89, 194, 211, 288, 289, 292
Stabilization ... 49, 51, 54, 65, 69, 101, 246
Staining
cisplatin ... 236
DNA ... 30, 209
intracellular ... 39, 86, 87, 91, 117, 128, 131, 132, 140–142, 152, 163, 164, 177, 179, 185, 188, 189, 223
intranuclear ... 10, 147–157
Statistical ... 162, 172, 184, 246, 247, 249, 250, 257, 259, 268, 274, 275, 310, 328
Stimulation ... 70, 78, 84, 85, 90, 125, 127–130, 133, 134, 143, 218–220, 222, 250, 273, 318, 327
Stratify
CITRUS ... 311, 312, 316–318, 322, 326–328, 330
Streptavidins ... 148, 150, 151, 153, 154
Supervised machine learning ... 309–331

T

t-distributed stochastic neighbor embedding (t-SNE) ... 156, 219, 258, 259, 268–270, 272, 276
Tetramer
peptides ... 150
peptides-MHC ... 148, 153, 154
Thaw ... 84, 96–97, 112, 114, 115, 126, 128, 129, 141, 154, 165, 169, 186, 200, 229
Threshold
data processing ... 14

Titration
  antibody 7, 69, 74, 88, 167
Trajectory 47, 278
Transcription 38, 56, 86, 140–143, 148, 151, 161, 162, 164, 177, 185, 188–190
Transcription factors
  FoxP3 .... 38, 84, 86, 87, 140–146, 177, 185, 188–190
  Ki67 .... 161, 162, 164, 177, 185
Transformations 182, 190, 191, 247, 248, 251, 253, 267, 268, 282–284, 292, 314, 324, 325
Tumors 30, 55, 60, 95, 147, 148, 152, 154, 194, 218

U

Ultrapure water 6, 15, 140–142, 162, 166, 175, 179, 195, 229, 235
Unsupervised data analysis 253, 297

V

Visualization of t-distributed stochastic neighbor embedding (viSNE) 209, 210, 213, 218, 219, 270, 310, 314–321, 323–329, 331
Visualizations 20, 35–44, 161, 218, 250, 257–260, 268–270, 272, 273, 282–288, 290, 292, 310, 311, 316–319, 329

MIX
Papier aus verantwortungsvollen Quellen
Paper from responsible sources
FSC® C105338

If you have any concerns about our products,
you can contact us on
ProductSafety@springernature.com

In case Publisher is established outside the EU,
the EU authorized representative is:
Springer Nature Customer Service Center GmbH
Europaplatz 3, 69115 Heidelberg, Germany

Printed by Libri Plureos GmbH
in Hamburg, Germany